现代分析检测技术丛书

Advances in Flow Analysis
Volume 2 Advances in Detection Methods in Flow Analysis

流动分析技术
第二卷 应用与进展

［波］马雷克·特罗扬诺夫伊茨 (Marek Trojanowicz)　主编

刘 楠　张 威　何声宝　主译

中国轻工业出版社

图书在版编目（CIP）数据

流动分析技术. 第二卷, 应用与进展／（波）马雷克·特罗扬诺夫伊茨（Marek Trojanowicz）主编；刘楠，张威，何声宝主译. — 北京：中国轻工业出版社，2023.5
　　ISBN 978-7-5184-4365-9

Ⅰ.①流… Ⅱ.①马… ②刘… ③张… ④何… Ⅲ.①分析化学—分析方法—研究 Ⅳ.①O652

中国国家版本馆 CIP 数据核字（2023）第 060025 号

版权声明：

Advances in Flow Analysis：Volume 2：Advances in Detection Methods in Flow Analysis / Marek Trojanowicz.

Copyright © 2008 by John Wiley & Sons, Ltd. All Rights Reserved.

Authorised translation from the English language edition published by John Wiley & Sons Limited. Responsibility for the accuracy of the translation rests solely with China Light Industry Press Ltd. and is not the responsibility of John Wiley & Sons Limited. No part of this book may be reproduced in any form without the written permission of the original copyright holder, John Wiley & Sons Limited.

中文简体版经 John Wiley & Sons 公司授权中国轻工业出版社独家全球发行。未经书面同意，不得以任何形式任意重制转载。

本书封面贴有 WILEY 防伪标签，无标签者不得销售。

责任编辑：张　靓
文字编辑：王　婕　　责任终审：许春英　　封面设计：锋尚设计
版式设计：砚祥志远　　责任校对：朱燕春　　责任监印：张　可

出版发行：中国轻工业出版社（北京东长安街6号，邮编：100740）
印　　刷：三河市万龙印装有限公司
经　　销：各地新华书店
版　　次：2023年5月第1版第1次印刷
开　　本：787×1092　1/16　印张：18.5
字　　数：415千字
书　　号：ISBN 978-7-5184-4365-9　定价：98.00元

邮购电话：010-65241695
发行电话：010-85119835　传真：85113293
网　　址：http://www.chlip.com.cn
Email：club@chlip.com.cn
如发现图书残缺请与我社邮购联系调换

220742K1X101ZYW

本书翻译人员

主　译　　刘　楠（国家烟草质量监督检验中心）

　　　　　　张　威（国家烟草质量监督检验中心）

　　　　　　何声宝（国家烟草质量监督检验中心）

副主译　　王英元（国家烟草质量监督检验中心）

　　　　　　王　燃（河南农业大学）

　　　　　　冯晓民（国家烟草质量监督检验中心）

　　　　　　罗安娜（国家烟草质量监督检验中心）

　　　　　　徐如彦（江苏中烟工业有限责任公司）

　　　　　　张玉璞（吉林省烟草公司）

　　　　　　吴寿明（贵州中烟工业有限责任公司）

　　　　　　闫洪洋（中国烟草总公司职工进修学院）

　　　　　　安泓汛（四川中烟工业有限责任公司）

　　　　　　王红霞（河南中烟工业有限责任公司）

参　译　　王晓春（江苏中烟工业有限责任公司）

　　　　　　王　菲（河南省烟草公司）

　　　　　　韶济民（四川省烟草公司）

　　　　　　朱贝贝（四川中烟工业有限责任公司）

　　　　　　龚珍林（江苏中烟工业有限责任公司）

　　　　　　王春琼（云南省烟草质量监督检测站）

　　　　　　翟天瑗（四川中烟工业有限责任公司）

　　　　　　周　浩（河南中烟工业有限责任公司）

　　　　　　周　东（四川中烟工业有限责任公司）

尚　峰（河南卷烟工业烟草薄片有限公司）

王　毅（贵州省烟草公司黔南州公司）

张志灵（福建省烟草公司）

谷晓懂（四川中烟工业有限责任公司）

范月月（国家烟草质量监督检验中心）

张晓慧（国家烟草质量监督检验中心）

张　莉（江苏中烟工业有限责任公司）

钱建财（江苏中烟工业有限责任公司）

李玉娥（吉林烟草工业有限责任公司）

靳冬梅（四川省烟草公司）

冯国胜（河南省烟草公司驻马店市公司）

曹　煜（四川中烟工业有限责任公司）

江宇轩（四川宽窄优品商贸有限责任公司）

译者序

传统化学实验操作是通过滴管、移液管或药勺等器材,手动移取试验品混合到烧杯或锥形瓶等容器中,然后使其进行反应。此过程是在物理平衡下进行的完全反应。且仪器的调整、维护和使用需要操作者具有较高的专业技能才能保证检测具有较高的准确性,实验操作烦琐,费时、费力。因此,后来有化学家提出了一种能在非平衡状态下,将上述实验操作综合到一个实验装置中进行的方法——流动分析法。它的出现打破了人们的传统观念,使在非平衡状态下的定量分析成为可能。流动分析在混合过程与反应过程中的高度重现性,使它具有分析速度快、精度高、设备和操作简单、节省试剂与试样以及适应性广等优点。

流动分析技术发展迅速,它已被应用于很多分析领域:水质检测、土壤样品分析、农业和环境监测、发酵过程监测、药物研究、禁药检测、血液分析、食品分析、分光光度分析、火焰光度分析、质谱分析、原子光谱分析、荧光分析、生物化学分析等,以译者所在的烟草行业为例,与流动分析相关的国家和行业标准就有数十个之多。

华沙大学 Marek Trojanowicz 教授是国际知名的流动分析技术研究专家,其主编的 *Advances in Flow Analysis* 一书详细介绍了流动分析技术的理论、研究应用现状和发展趋势,是众多流动分析研究者不可多得的技术工具书,在国际流动分析学界有着巨大的影响力。流动分析技术在我国化学分析领域的应用越来越广泛,为便于广大流动分析检测技术人员学习流动分析技术理论,了解流动分析技术的研究应用现状,掌握流动分析技术的发展趋势,获得及时解决检测工作中"疑难杂症"的能力,译者进行了本书的翻译,本书分为两卷。第一卷是方法与设备,第二卷是应用与进展。

由于时间仓促以及水平有限,译文中难免存在疏漏或不当之处,恳请读者批评指正。

译者

本书编写人员

J. R. Albert-García
University of Valencia
Department of Analytical Chemistry
Dr. Moliner, 50
46100 Burjassot
València
Spain

Lúcio Angnes
Universidade de São Paulo
Instituto de Química
Av. Prof. Lineu Prestes 748
05508-900 São Paulo
Brazil

Alberto N. Araújo
Universidade do Porto
REQUIMTE
Departamento de Química Física
Faculdade de Farmácia
Rua Aníbal Cunha 164
4050 Porto
Portugal

Christopher M. A. Brett
Universidade de Coimbra
Departamento de Química
3004-535 Coimbra
Portugal

José L. Burguera
Los Andes University
Faculty of Sciences
Department of Chemistry
P. O. Box 542
Mérida 5101-A
Venezuela

Marcela Burguera
Los Andes University
Faculty of Sciences
Department of Chemistry
P. O. Box 542
Mérida 5101-A
Venezuela

J. Martínez Calatayud
University of Valencia
Department of Analytical Chemistry
Dr. Moliner, 50
46100 Burjassot
València
Spain

S. Cárdenas
Universidad de Córdoba
Department of Analytical Chemistry
Campus de Rabanales
14071 Córdoba
Spain

Andrea Cavicchioli
Universidade de São Paulo
Escola de Artes, Ciências e Humanidades
Rua Arlindo Béttio
1000 Ermelino Matarazzo
03828-000 São Paulo
Brazil

Víctor Cerdà
University of the Balearic Islands
Automation and Environment
Analytical Chemistry Group
Department of Chemistry
07122 Palma de Mallorca
Spain

Stuart J. Chalk
University of North Florida
Department of Chemistry and Physics
1 UNF Drive

Jacksonville
FL 32224
USA

Purnendu K. Dasgupta
University of Texas at Arlington
Department of Chemistry and
Biochemistry
Arlington
TX 76019-0065
USA

R. J. E. Derks
Vrije Universiteit Amsterdam
Division of Analytical Chemistry &
Applied Spectroscopy
De Boelelaan 1083
1081 HV Amsterdam
The Netherlands

José Manuel Estela
University of the Balearic Islands
Automation and Environment
Analytical Chemistry Group
Department of Chemistry
07122 Palma de Mallorca
Spain

Mário A. Feres Jr.
Universidade de São Paulo
Centro de Energia Nuclear
na Agricultura
Avenida Centenario, 303
Caixa Postal 96
13400-970 Piracicaba SP
Brazil

Juan Francisco García-Reyes
University of Jaén
Department of Physical and Analytical
Chemistry
Paraje Las Lagunillas S/N
23008 Jaén

Spain

Maria Fernanda Giné
University of São Paulo
Centro de Energia Nuclear
na Agricultura CENA
Avenida Centenario, 303
Caixa Postal 96
13400-970 Piracicaba SP
Brazil

Ivano G. R. Gutz
Universidade de São Paulo
Instituto de Química
Av. Prof. Lineu Prestes 748
05508-900 São Paulo
Brazil

Elo Harald Hansen
Technical University of Denmark
Department of Chemistry
Kemitorvet
Building 207
2800 Kgs. Lyngby
Denmark

Kees Hollaar
Skalar Analytical B. V.
Tinstraat 12
4823 AA Breda
The Netherlands

Fernando A. Iñón
University of Buenos Aires
Facultad de Ciencias Exactas y Naturales Pabellón 2
1428 Buenos Aires
Argentina

Hubertus Irth
Vrije Universiteit Amsterdam
Department of Analytical Chemistry &
Applied Spectroscopy
De Boelelaan 1083
1081 HV Amsterdam

The Netherlands

Takehiko Kitamori
The University of Tokyo
Department of Applied Chemistry
Graduate School of Engineering
7-3-1 Hongo
Bunkyo-ku
Tokyo 113-8656
Japan

Robert Koncki
University of Warsaw
Department of Chemistry
Pasteura 1
02-093 Warsaw
Poland

Jeroen Kool
Vrije Universiteit Amsterdam
Department of Analytical Chemistry &
Applied Spectroscopy
De Boelelaan 1083
1081 HV Amsterdam
The Netherlands

Pawel Kościelniak
Jagiellonian University
Faculty of Chemistry
R. Ingardena Str. 3
Kraków, 30-060
Poland

Petr Kubáň
Mendel University
Department of Chemistry
and Biochemistry
Zemědělská
Brno 61300
Czech Republic

José L. F. C. Lima
Universidade do Porto
REQUIMTE

Departamento de Química-Física
Faculdade de Farmácia
Rua Aníbal Cunha 164
4050 Porto
Portugal

Henk Lingeman
Vrije Universiteit Amsterdam
Department of Analytical Chemistry &
Applied Spectroscopy
De Boelelaan 1083
1081 HV Amsterdam
The Netherlands

Shaorong Liu
Texas Tech University
Department of Chemistry
and Biochemistry
Lubbock
TX 79409-1061
USA

R. Lucena
Universidad de Córdoba
Department of Analytical Chemistry
Campus de Rabanales
14071 Córdoba
Spain

Manuel Miró
University of the Balearic Islands
Department of Chemistry
Faculty of Sciences
Carretera de Valldemossa, km 7.5
07122-Palma de Mallorca
Spain

Antonio Molina-Díaz
University of Jaén
Department of Physical and Analytical
Chemistry
Paraje Las Lagunillas S/N
23008 Jaén

Spain

M. Conceição B. S. M. Montenegro
Universidade do Porto
REQUIMTE
Departamento de Química Física
Faculdade de Farmácia
Rua Aníbal Cunha 164
4050 Porto
Portugal

Shoji Motomizu
Okayama University
Department of Chemistry
Tsushimanaka
Okayama 700-8530
Japan

Bram Neele
Skalar Analytical B. V.
Tinstraat 12
4823 AA Breda
The Netherlands

Beata Rozum
University of Warsaw
Department of Chemistry
Pasteura 1
02-093 Warsaw
Poland

João L. M. Santos
Universidade do Porto
REQUIMTE
Departamento de Química-Física
Faculdade de Farmácia
Rua Aníbal Cunha 164
4050 Porto
Portugal

Javier Saurina
University of Barcelona
Department of Analytical Chemistry
Martí i Franquès 1-11

08028 Barcelona
Spain

B. M. Simonet
Universidad de Córdoba
Department of Analytical Chemistry
Campus de Rabanales
14071 Córdoba
Spain

Marek Trojanowicz
University of Warsaw
Department of Chemistry
Pasteura 1
02-093 Warsaw
Poland
and Institute of Nuclear Chemistry
and Technology
Dorodna 16
03-195 Warsaw
Poland

Manabu Tokeshi
Nagoya University
Department of Applied Chemistry
Graduate School of Engineering
Furo-cho
Chikusa-ku
Nagoya 464-8603
Japan

Mabel B. Tudino
University of Buenos Aires
Facultad de Ciencias Exactas y Naturales
Pabellon 2
1428 Buenos Aires
Argentinia

Lukasz Tymecki
University of Warsaw
Department of Chemistry
Pasteura 1
02-093 Warsaw

Poland

Miguel Valcárcel
Universidad de Córdoba
Department of Analytical Chemistry
Campus de Rabanales
14071 Córdoba
Spain

N. P. E. Vermeulen
Vrije Universiteit Amsterdam
Division of Molecular Toxicology
De Boelelaan 1083
1081 HV Amsterdam
The Netherlands

Elias A. G. Zagatto
Universidade de São Paulo
Centro de Energia Nuclear
na Agricultura
Avenida Centenario, 303
Caixa Postal 96
13400-970 Piracicaba SP
Brazil

前　言

　　化学分析是现代生活各个领域中不可或缺的一项技术。随着科学技术的进步，以及需求的增加，作为一门学科的分析化学及以实际应用为目的的化学分析方法和技术均取得了可观的进步。由于所需检测的样品数量越来越多，终端用户需求设计出可直接使用而无需专业实验室服务的分析仪器和方法，分析测定数据的质量要求也在提高，都导致了人们对分析测定的需求不断提升。根据应用领域的不同，这种需求包括缩短分析时间、减少分析所需的样品量、多组分测定更低的检测限或更好的选择性以及更好的精确度和/或准确度。

　　分析方法的进步以不同方式发生，是多种因素共同作用的结果。其影响分析方法的因素与影响分析测定结果的参数一样多。自然科学、材料科学、电子学和信息学的进步，材料和设备工程的进步，以及它们在分析程序中的应用，都会对分析方法的发展产生影响。人类创造和探索的欲望是无限的，而这正是人们进行科学研究的驱动力。因此，科学或技术发展的任何阶段都不是永恒不变的，这当然也包括分析化学以及化学分析方法和技术的发展。

　　起初，在流动模式下进行分析测定看似通过省略采样步骤对传统的非流动程序进行了简化。20 世纪 30、40 年代建立的工艺过程中电导率的测定是第一个流动模式的过程分析测量方法。之后，随着检测方法和测量仪器的发展，氧化还原电位、pH、浊度、给定波长吸光度等指标都建立了流动模式的测量方法。这些方法已成为现代过程分析的常用方法。流动模式分析作为化学分析领域的一个独立分支，在过去半个世纪得到了很好的发展并且拥有了大量专门设计的测量仪器，但还有许多具体问题需要解决。这一化学分析领域有很多相关文献，本书不会讨论工艺过程流动分析问题。

　　本书涉及的是实验室流动分析。这种分析与环境、技术条件以及过程、设备的规模有关。简单的处理方法是将得到广泛认可的作者的理论用实际检测方法进行分析。在 20 世纪 50 年代，大型医院的临床分析实验室有大量样本需要分析，所以他们迫切需要一种加快分析过程的方案。第一个发明实验室流动分析系统的是来自美国凯斯西储大学医学院附属医院的生物化学家 Leonard J. Skeggs Jr.，他也是现代人工肾的共同发明者。他设计了第一个用光度法测定血液中尿素氮的实验室流动系统，迅速为新仪器申请了专利，并在三年内由 Technicon Co. 将其推向市场并取得了巨大成功。第一个原型机包含了流通光度计和用于连续记录信号的带状图记录器在内的突破性仪器解决方案。本书将介绍其中的一部分，以说明在该系统构建过程中涉及的许多发明和开创性解决方案。该系统是为分析血液样品而设计的，因此要设计一个旋转进样器来从样品瓶中吸取样品。吸入管路中的样品可能会在流动过程中扩散，但这种扩散可能会受到气泡分割的限制。尿素氮的测定需要去除蛋白质，因此有必要设计流通式膜透析器。液相色谱在早期与各种检测器一同被开发，已知其可以在流体中实现连续检测。开发的空气分段式流动分析系统可以实现众多机械化操作（进样、添加试剂、孵育、透析），这是实验室分析中最重要的突破。

在接下来的 20 年里，基于化学实验室分析机械化概念的仪器在大型临床实验室中占主导地位，但随后关于分析过程机械化的许多其他想法变得越来越具有竞争力，包括离心分析仪、采用固态试剂条的设备，尤其是有各种设计的离散分析仪，在过去的 20 年中已经完全取代了临床流动分析仪。它们更高效、更通用，并且可以对一个样品进行几十次检测。但是，在环境保护、农业分析和食品质检领域的常规分析实验室中仍然广泛使用空气分段式流动分析仪。

在 20 世纪 70 年代中期，通过将少量样品注入流动的载体流或直接注入试剂流的流动分析方法的发明，为实验室流动分析的进一步发展提供了关键的推动力。甚至在几年前，人们可以在流动分析的文献报告中发现，通过引入比实际需求更少的样品可以在气泡分割系统中实现稳态平衡的信号。该发现得出的结论是，在该系统中获得的稳态信号可以用于分析，并且可以提高进样率。根据现有文献，我们注意到流动注射分析的概念来自分析仪器的不同分支。在某种情况下，它被认为是早期开发的空气分段系统的演变，其消除了流动流体的分段，并通过注射端口而不是通过连续抽吸来微量进样。在获得稳态信号的同时减小了管径，该系统可以提供快速分析信号。相同的分析概念源于商业仪器在液相色谱中的应用以及在无分离柱的流动分析中的应用。通过适当的化学条件，可以实现对特定分析物的选择性分析。

在接下来的几年里，人们对这种分析方法的兴趣迅速增加（如果以分析期刊上发表的论文数量来衡量的话，几乎呈指数增长），这在很大程度上归功于 J. Ruzicka、E. H. Hansen 以及他们的研究团队在众多出版物中发表的结论：对于实验室研究，几乎每个分析实验室都可以通过低成本、简单的组件轻松构建流动注射分析系统，而无需大的设备投资。这是实现各种技术设计理念，在此类系统中进行各种化学反应和样品处理操作以及利用各种检测方法的一种方式。这是流动系统中分析过程机械化的一种非常有吸引力的方式，但必须承认这不是测量自动化的一种方式。根据自动化理论，并遵循 IUPAC 术语建议，自动化系统必须配备智能控制系统，使用反馈-循环机制，无需人工操作即可控制和调节测量条件，因此在流动条件下进行分析测量并不意味着测量的自动化。

自 20 世纪 70 年代以来，开发的用于分析的流动进样方法已经在技术上得到了改进，例如最常见的流动系统：将样品和试剂按顺序注射到单管线系统中［称为顺序注射分析（SIA）］，直接注入检测器传感表面的无管系统中的流动测量［称为间歇注入分析（BIA）］，或在具有可移动固体颗粒的流动注入系统中的应用（称为微珠进样分析，其缩写同样为 BIA）。流动注射分析系统发展的另一个方面是流动系统模块的快速小型化以及集成化，例如，通过将一些模块并入进样阀（通常称为阀上实验室的概念），或将它们小型化到微流模式。

通常，流动分析系统可以描述为分析测量设备，其中样品预处理和分析物检测的所有操作都在流体中进行。这似乎是对流动分析的一个非常普遍的理解，但同时我们也可以发现这样的描述有不准确之处。是否可以将带有火焰原子化的简单测量纳入流动分析，其中样品被吸入、雾化，然后传输到火焰进行光学检测？在直接进样的质谱测量中，注入的液体样品被蒸发，分析物被电离（也可以被碎裂），然后传输到检测器，质谱分析能否被认为是流动分析？而最难解决的问题就是液相色谱和流动分析的区别。在柱色谱中，样品中的分析物在柱上被分离，它们有时也可以被衍生化，然后被输送到流通检测器。毛细管电泳也会出现类似的情况。从传统和历史发展的角度来看，更重要的是，从分析化学中的作用来看，将柱色谱包含在流动分析中似乎并不合适。另一方面，在任何类型的典型流动系统中（空气分段连续系统、流动进样系统等），填充反应器常用于样品净化或预浓缩，这些操作就是遵循色谱的常见机制所进行的操作。那么，划分界限在哪里呢？为了本书主题的框架，流动分析是指测量系统中的分析，其中样品处理和检测的所有操作都在流动溶液中进行，但不包含多组分色谱或电泳分离。大多数情况下，它是采用机械化样品预处理的单组分方法，而流动系统中的多组分分析是在具有更复杂歧管的系统中进行的，或者通过使用多组分检测器进行检测。流动分析的动态特性广泛应用于样品处理中，在许多情况下用于改进某些检测方法的参数，目前在多组分测定的设计方面很少涉及。

本书的主要目的是介绍近年来流动分析的成就，这些成就可能有助于确定其在现代化学分析中的地位。尽管自 Skeggs 开创性发明问世的 60 年间发表了数千篇论文，但这种分析方法似乎在常规化学分析的各个领域都被低估了。当然，在 20 世纪 60、70 年代取得的巨大成就，是将带有空气分段系统的商业流动分析仪应用于临床实验室中。经过多年的发展，大量发表的论文和一些用于流动进样方法的商业仪器，并没有将流动进样方法充分引入常规分析实验室。如今，如果在常规分析实验室中使用一些流动分析仪，它们大多是具有空气分段流动并记录稳态平衡信号的连续流动分析仪。

本书所有章节的主题都是我选择的，因为近年来这些流动分析领域取得了巨大的进步。感谢出版商接受我的选择。特别感谢所有接受我邀请为本书出版做出贡献的作者。我相信他们都和我一样希望本书对流动分析的进一步发展和这些化学分析方法的推广有所帮助。

还要感谢所有接受我的邀请并审阅了一些章节的同行：加拿大金斯顿皇后大学 Diane Beauchemin 教授、德国布伦瑞克亥姆霍兹感染研究中心 Ursula Bilitewski 教授、芬兰阿博·阿卡德米大学 Ari Ivaska 教授、瑞典斯德哥尔摩大学 Bo Karlberg 教授、波兰克拉科夫雅盖隆大学 Pawel Koscielniak 教授、捷克布尔诺孟德尔农林大学 Petr Kuban 教授、美国密歇根大学安娜堡分校 Mark E. Meyerhoff 教授、巴西皮拉西卡巴圣保罗大学 CENA Boaventura Reis 教授、捷克赫拉德茨-克拉洛韦查尔斯大学 Petr Solich 教授、马萨诸塞大学阿默斯特分校 Julian Tyson 教授、美国威尔明顿杜邦公司 Bogdan Szostek 博士和英国普利茅斯大学 Paul Worsfold 教授。非常感谢他们在审稿时提供的宝贵帮助。还要感谢参与本书出版的 Wiley-VCH 出版社的所有工作人员，特别是 Manfred Köhl 博士、Waltraud Wüst 博士和 Claudia Nussbeck 女士。

Marek Trojanowicz

目录 CONTENTS

1 流动分析中的发光检测
1.1 引言 ··· 2
1.2 连续流动系统中的发光检测 ··· 3
1.3 趋势和展望 ·· 23
参考文献 ··· 26

2 流动注射分析中的酶
2.1 引言 ··· 46
2.2 酶底物作为分析物 ··· 46
2.3 基于测量酶活性的方法 ·· 58
2.4 结论 ··· 61
参考文献 ··· 61

3 流动电位法
3.1 引言 ··· 70
3.2 背景概念 ·· 70
3.3 电极的开发和检测器池的设计 ···································· 72
3.4 基于电位法的流动分析技术 ·· 77
3.5 趋势与展望 ·· 82
参考文献 ··· 82

4 流动伏安法
4.1 引言 ··· 96
4.2 伏安/安培流动分析 ·· 97
4.3 提高选择性、灵敏度和耐久性的策略 ·························· 105
4.4 趋势与展望 ·· 109

1

参考文献 ·········· 110

5 使用流动分析对蛋白质-蛋白质和蛋白质-配体的相互作用进行亲和相互作用分析

5.1 引言 ·········· 122
5.2 基于质谱流动分析的非共价蛋白质-蛋白质和蛋白质-配体相互作用的分析 ·········· 123
5.3 流动分析和高效液相色谱的集成用于混合物的生物亲和性筛选 ·········· 127
5.4 结论 ·········· 136
参考文献 ·········· 137

6 流动分析中的原子光谱

6.1 引言 ·········· 146
6.2 火焰原子吸收光谱法 ·········· 146
6.3 样品稀释 ·········· 151
6.4 电热原子吸收光谱法 ·········· 152
6.5 原子荧光光谱法 ·········· 157
6.6 结论和展望 ·········· 160
参考文献 ·········· 161

7 流动进样质谱法

7.1 引言 ·········· 174
7.2 FIA-MS 样品引入装置 ·········· 178
7.3 与外部电离源耦合的流动系统-质谱 ·········· 180
7.4 结论 ·········· 190
参考文献 ·········· 191

8 流动分析的环境应用

8.1 引言 ·········· 200
8.2 通过流动方法分析水环境 ·········· 200
8.3 用流动方法分析大气环境 ·········· 210
8.4 用流动方法分析地圈环境 ·········· 214

8.5　环境分析的未来展望 ································· 214
参考文献 ··· 215

9　流动分析方法在医药分析中的应用

9.1　引言 ··· 222
9.2　药物配方分析 ····································· 222
9.3　制药行业的流动程序分析仪 ························· 240
参考文献 ··· 247
拓展阅读 ··· 248

10　连续流动分析的工业和环境应用

10.1　引言 ·· 252
10.2　环境和工业领域概述 ······························ 252
10.3　流动分析方法的应用及其范围 ······················ 258
10.4　流动分析应用程序的开发 ·························· 266
10.5　连续流动分析的趋势 ······························ 268
参考文献 ··· 269

1 流动分析中的发光检测

Antonio Molina-Díaz 和 Juan Francisco García-Reyes

1.1　引言

发光现象（荧光、磷光和化学发光）是广泛使用的分析检测技术的基础。分子荧光（FL）的过程是首先发光体吸收合适能量的光子，从而使电子从基态跃迁到更高能量的空轨道（激发态），然后电子返回到初始基态能级，进而发射光量子。在整个过程中电子自旋保持不变。因此，分子总是处于其基态或激发态（磁场不会分裂这些电子配置的能级）。该过程很快，发光寿命为纳秒级。

有时候，分子的辐射失活会慢得多。这是由于被激发的电子在转变为低能态（称为三重态，因为能级在磁场中分为三个能级）时自旋反转。当电子回到基态能级时，其自旋必须再次反转。该过程称为磷光（PH），比荧光过程慢得多，寿命范围从几十微秒到几毫秒不等。

化学发光（CL）被定义为一种电磁辐射[1]，通过化学反应产生电子激发的中间体或产物进而引发这种电磁辐射，其要么发光（直接化学发光），要么将能量提供给另一个负责发光的分子（间接或敏化化学发光）。当这种发光来自生物体或源自它们的化学系统时，称为生物发光。除了不需要激发源外，该过程与光致发光（荧光或磷光）的过程基本相同，因此，化学发光检测器不会受到背景的干扰。几乎所有的发光检测技术都具有高灵敏度，即使其拥有很低的检测限（即使在亚皮摩尔浓度的情况下）。图1.1所示为不同的发光现象（亚布隆斯基图）。

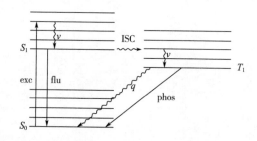

图1.1　与直接激发的室温磷光相关的光物理过程示意图
S_0—单线基态　S_1—最低激发单线态　T_1—最低激发三线态　exc—激发态　flu—荧光
v—振动弛豫　ISC—系统间交叉　q—生物分子猝灭　phos—磷光
（经 Elsevier 许可改编自文献［77］，©2003。）

分子发光现象已广泛用作连续流动分析中的检测系统，包括连续分段流动方法[2]。事实上，在许多早期的流动注射分析（FIA）应用中就采用了荧光检测[3]。发光检测与流动注射分析[4]和最近期开发的流动进样技术相兼容，例如，顺序注射分析（SIA）[5]、多注射器流动进样分析（MSFIA）[6]、多路换向流动系统[7]和多泵流动进样分析（MPFIA）[8]。图1.2显示了流动分析中的发光检测技术方案。本章描述了在流动分析系统中使用的分子发光检测技术，并讨论和展示了近些年的研究在流动分析系统中取得的重要进展。

1.2 连续流动系统中的发光检测

1.2.1 荧光

1.2.1.1 简介

荧光光谱法已成为许多专业应用中的常规技术。它是一种相对简单和快速的分析技术，特别适用于定量分析生物和环境样品中的痕量芳香族或高度不饱和有机分子。荧光检测最显著的特征是高灵敏度和与之相伴的高/中选择性。相比之下，荧光技术的主要缺陷是显示天然荧光的物质数量很少。然而，一些简单的程序/反应极大地扩展了可用荧光团的范围，使该技术能够通过化学标记和衍生程序应用于各种有机化合物和无机化合物。

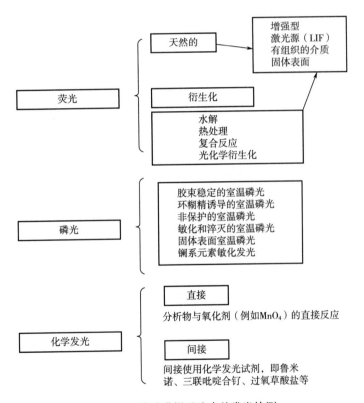

图1.2 流动进样系统中的发光检测

1.2.1.2 流动方法中的荧光检测

荧光测定最直接的分类是基于相关物质是否具有荧光活性或必须通过化学衍生化激活。

（1）基于天然荧光的流动分析方法 基于分析物荧光检测的直接法基本上适用于那些天然带有荧光的分子，其强度足以用于分析。这种具有荧光特性的物质所需的流动注射分析组件很简单：它通常由一个载气管线和一个注射阀组成。含有酸性或碱性基团物质的荧光行为受 pH 影响，因此需要通过载体通道来提供合适的 pH 和介质。天然荧光流动进样方法的一些应用如表 1.1 所示[9-24]。

提高天然荧光方法检测能力的三个主要分析原则：①使用激光源；②胶束或有组织的介质；③固体载体。

表 1.1　　　　　　　　　基于天然荧光分析的选定流动方法[9-24]

分析物	描述	检测限	参考文献
麦穗宁和噻苯达唑	天然荧光，利用顺序注射分析组件并通过多重线性回归（MLR）校准模型对混合物进行分离	LOD：0.04（FBZ），0.08μg/L（TBZ）	[9]
苯呋酯	天然荧光（278/316nm）	0.001～0.5mg/L	[10]
3-吲哚乙酸	天然荧光（280/364nm）	LDR：0.005～0.6mg/L	[11]
黄草灵	天然荧光（258/342nm）	LDR：0.005～15mg/L	[12]
布美他尼	在强碱性介质中的天然荧光（314/370nm）	0.05～10.0μg/mL	[13]
东莨菪碱和伞形酮	在不同 pH 下的天然荧光（350/418nm）	LOD<1μmol/L	[14]
酒石酸麦角胺	顺序注射分析组件，天然荧光（236/390nm）	LOD：10mg/mL	[15]
水杨酸	顺序注射分析组件，天然荧光（299/405nm）	LDR：1～100mg/L	[16]
哌唑嗪	天然荧光（244～389nm）使用顺序注射分析组件	LOD：7ng/mL	[17]
PDT（防晒剂）	在线固相萃取，然后进行天然荧光检测（330/454nm）	LOD：10ng/mL	[18]
PBS（防晒剂）	天然荧光（301～681nm），使用顺序注射分析组件，使用 SAX 迷你柱	LOD：12ng/mL	[19]
17β-雌二醇（E2）	天然荧光；在线固相萃取，通过使用分子印迹聚合物去除基质（281/305nm）	LOD：1.12ng/mL	[20]
增强型天然荧光流动方法			
多菌灵	使用 CTAB 的胶束增强的天然荧光检测	LOD：13nmol/L	[21]
萘普生	萘普生的天然荧光并通过与 β-环糊精形成包合物得到增强，其使用顺序注射分析组件	0.19μmol/L	[22]
华法林	华法林的天然荧光并通过与 β-环糊精形成包合物得到增强，其使用顺序注射分析组件	LOD：0.02μg/mL	[23]
多环芳烃（混合物）	天然荧光，激光 335nm，和多变量校准（SIMCA 和 PLS-1）	—	[24]

注：PDT—苯基二苯并咪唑四磺酸二钠；PBS—2-苯基苯并咪唑-5 磺酸；CTAB—十六烷基三甲基溴化铵。

①流动系统中的激光诱导荧光（LIF）检测。由于仪器灵敏度与激发辐射的强度成正比，激光源显著扩大了荧光检测的潜力。一些利用激光束激发的应用可以检测超痕

量物质，包括罗丹明6G[25]、核黄素[26]、维生素、色氨酸[27]和卟啉[28]。Imasaka等开发了另一种基于FIA-LIF测定乙醇的应用[29]。Amador-Hernández等开发了一种无需事先色谱分离，通过FIA-LIF和多变量校准进行多分析物测定多环芳烃（PAH）的方法，其使用的流动注射分析组件如图1.3所示。

②组织介质中的荧光。一些有机物质的荧光受物理化学环境的影响很大，这促进了组织介质的使用。表面活性剂和环糊精为荧光团提供了一个微环境，该环境可以增强许多溶质的荧光[30]，降低检测限，同时避免了荧光发射强度降低现象的发生。

胶束介质是通过在水或水-有机混合物中加入适量的表面活性剂来制备的。表面活性剂是具有两亲性的化合物。这些分子由两部分组成，头部是极性和亲水的，而尾部是疏水的。另一方面，环糊精（CD）是水溶性环状低聚糖，由单个吡喃葡萄糖单元通过α-1,4-糖苷键连接而成。根据它们的大小，环糊精可以与某些有机分子形成包合物，从而改变其光物理性质。在流动系统中使用有组织的介质可以增强分析物的发光信号进而提高发光检测的灵敏度。表1.1概述了一些采用流动分析方法的应用示例，这些方法通过有组织的介质进行荧光检测。应该注意的是，有组织的介质不仅适用于天然荧光法，还适用于其他衍生化方法，如络合反应或光化学诱导荧光。

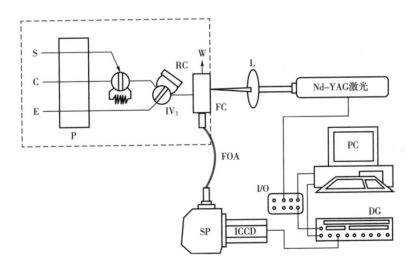

图1.3 使用流动注射分析和激光诱导荧光检测筛选PAH的实验装置
L—镜头 FC—流动池 FOA—光纤组件 SP—光谱仪 ICCD—增强型电荷耦合器件
I/O—多IO组件 DG—延迟发生器 PC—个人电脑 S—样品溶液 C—载体液 E—洗脱液
P—蠕动泵 IV$_2$—进样阀 RC—保留柱 W—废弃物
(经Elsevier许可转载自文献[24]，2001年。)

③固相荧光。该技术通过将分析物吸附或结合到合适载体上测量其发光。1.2.4详细描述了该技术的主要特点和选定的应用。

（2）将分析物衍生化的流动方法 化学衍生化是通过化学试剂或物理处理（即紫外线照射或加热）将非荧光或弱荧光化合物转化为强荧光物质的方法。因此，其使用了各种各样的反应，包括荧光螯合物的形成、荧光配体的置换、离子对的形成、氧化还原过程、酶催化、水解和各种典型的有机合成反应（例如，缩合-环化、标记、抗原-抗体反应）。一些主要的衍生化技术：

①水解反应　水解是将非荧光化合物转化为荧光化合物的最简单的处理方法之一。它通常在强碱性水性介质（NaOH）中完成，在某些情况下，在高温（50~100℃）下形成荧光阴离子。例如，乙酰水杨酸（阿司匹林）具有弱的天然荧光，而水杨酸离子，在310nm 激发时在~400nm 处发出强烈荧光，可用于测定不同样品中的阿司匹林和水杨酸盐。水解反应也已成功应用于测定天然水中的萘草胺，检测限在很低的纳克范围内[31]。使用的流动组件如图1.4所示。

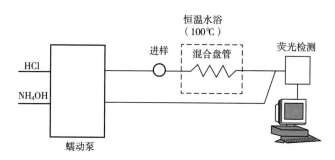

（1）正常流动注射分析配置

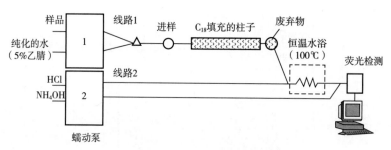

（2）通过固相萃取进行在线预浓缩的配置

图1.4　用于测定萘草胺（基于其水解为1-萘胺）的歧管
线路1是指由第一台泵泵送的污水，线路2是指由第二台泵泵送的污水。
（经 Elsevier 许可转载自文献[31]，1999年。）

②光化学诱导荧光（PIF）。光化学诱导荧光的概念基于紫外线照射下将非荧光分析物转化为强荧光光产物。光化学诱导荧光方法比化学衍生化方法更新，主要的光反应有光环化、光氧化、光异构化、光解和光还原。这种方法的典型例子是荧光测定尿液中的合成代谢剂己烯雌酚和测定不同的农药。

一些参数会影响非荧光化合物向荧光化合物的转化，例如，紫外线照射时间和所用溶剂的性质。氰戊菊酯、除虫脲和溴氰菊酯等农药在质子溶剂中可有效转化为荧光产物（短辐射时间产生高信号），而对于氰戊菊酯和毒死蜱，则选择极性非质子溶剂。

光化学衍生的主要优点如下[32]：a. 不需要使用化学品；b. 反应速度快；c. 所需

设备便宜；d. 适用于流动分析方法；e. 只需可重现的荧光信号，而无需已知荧光化合物的结构。

③络合。络合涉及金属离子与含有供电官能团的弱荧光或非荧光化合物相结合，从而形成强荧光金属螯合物。络合反应已广泛应用于金属离子和有机化合物的荧光测定。

表1.2所示为一些使用不同衍生化技术的流动进样荧光检测方法的应用案例，包括制药、生物医学、食品和环境分析[33-76]。

表1.2 　　选定的包含分析物衍生化的荧光流动分析方法[33-76]

分析物	描述	检测限	参考文献
氧化/还原反应			
硫利达嗪	使用二氧化铅固相反应器（349/429nm）在线氧化硫利达嗪	LOD：5.5ng/mL	[33]
抗坏血酸和半胱氨酸	使用半胱氨酸和/或抗坏血酸等酸性介质（227/419nm）在线还原Tl^{3+}	LDR：0.015~2μg/mL LOD：0.8μmol/L（抗坏血酸） 0.7μmol/L（半胱氨酸）	[34]
硫胺素	将六氰基高铁酸钾（III）固定在阴离子交换树脂（脱氢硫胺素）上进行氧化	LDR：0.1~4mg/L	[35]
青霉胺和硫普罗宁	盐酸介质中Tl^{3+}氧化（227/419nm）	LOD<1μmol/L	[36]
1,4-苯二氮卓类	室温下在乙醇或甲醇介质中先在硫酸的存在下水解，再进行荧光检测	LOD 0.005~0.01mg/mL	[37]
硫胺素和抗坏血酸	Hg^{2+}氧化产生荧光衍生物（硫代硫胺素）和喹喔啉衍生物（抗坏血酸）（356/440nm）	LDR：2~100mg/L（维生素B_1）；5~100mg/L（抗坏血酸）	[38]
氟奋乃静	Ce^{4+}亚砷酸盐作为固态床反应器的强氧化剂	LDR：0.05~100mg/L	[39]
肾上腺素	带有碘的聚氯乙烯盘管	5~25mg/L	[40]
异丙异烟肼和异烟肼	过氧化氢的氧化	LOD：0.008mg/L（异丙异烟肼） 0.005mg/L（异烟肼）	[41]
卡马西平	使用含有二氧化铅的固相反应器在酸性介质中将卡马西平在线氧化成强荧光化合物	LOD：57μmol/L	[42]
水解反应			
吲哚美辛	胶束介质中的碱性水解，使用顺序注射分析系统	16nmol/L	[43]
莠草胺	在酸性介质中水解莠草胺和在碱性介质中测量产物；使用C_{18}柱在线预浓缩分析物	LOD：3μmol/L	[31]
光化学诱导荧光（PIF）			
叶绿醌	紫外光照射后在线还原十二烷基硫酸酯胶束中的叶绿素	LOD：0.05μg/mL	[44]

续表

分析物	描述	检测限	参考文献
苯脲除草剂	在表面活性剂（SDS 和 CTAC）存在下，这些除草剂在紫外光照射下在线光转化为缓冲水溶液中的强荧光光产物	LDR：0.09~45μg/mL LOD：0.33~0.93mg/L	[45]
氯苯氧酸除草剂 吡虫啉	在 CTAC 存在下紫外光照射（PIF），吡虫啉光化学转化为荧光团 1-（6-氯-3-吡啶-甲基）-2-异亚硝基-3,4-双脱氢枞基咪唑啉（334/377nm）	LOD：73.2ng/mL 和 33.5ng/mL LOD：0.3ng/mL	[46] [47]
磺酰脲类除草剂	使用 CTAC 和 SDS 的胶束增强光化学诱导荧光	LOD：0.1~1ng/mL	[48]
氟尿嘧啶	在碱性介质中进行在线光反应，使用多路换向流动组件（247/325nm）	LDR：0.01~4mg/L	[49]
甲胺磷	在紫外线照射下，在过二硫酸盐存在下分解甲胺磷；生成的磷酸盐用于形成磷钼酸，将硫胺素氧化成脱氢硫胺素	LOD：1.7ng/mL	[50]
马拉硫磷	在紫外线照射下，分解马拉硫磷；生成的磷酸盐用于形成磷钼酸，将硫胺素氧化成脱氢硫胺素	LOD：0.02~2mg/mL	[51]
地西泮	在含有 Cu^{2+} 的碱性介质中进行在线光反应（328/382nm）	LDR：0.5~50μg/mL	[52]
抗坏血酸	通过硫氨酸蓝的敏化将抗坏血酸光氧化生成亮氨酸蓝	LDR：0.8~50μmol/L	[53]
噻奈普汀	使用流动注射分析装置在酸性水-乙醇混合物中产生光化学诱导的荧光	LOD：15ng/mL	[54]
氯喹	在碱性介质中进行光化学衍生化，在 355nm 处使用脉冲 Nd：YAG 激光	LOD：8μg/L	[55]
利血平	在丙酮的存在下，利血平的光化学分解形成高度荧光的衍生物	LOD：0.45ng/mL	[56]
阿散酸	在过二硫酸盐的存在下，阿散酸经过紫外光照射在线分解	LOD：10ng/mL	[57]
异丙嗪氨基吩噻嗪	在线光衍生，光化学诱导荧光检测（乙丙嗪、三氟拉嗪、左旋丙哌嗪和硫丙拉嗪）	LOR：0.05~20mg/L LOD：60~90ng/mL	[58] [59]
甲萘醌	紫外光照射下 SDS 胶束中甲萘醌的在线光还原（340/410nm）	LOD：0.18ng/mL	[60]
络合反应			
Al	在表面活性剂 Brij-35 存在下与荧光镓试剂络合。使用填充有 8-羟基喹啉的柱进行在线预浓缩	LOD：0.15nmol/L	[61]

续表

分析物	描述	检测限	参考文献
B	硼酸与变色酸（313/360nm）络合	LOD：3μg/L	[62]
Al	Al^{3+} 和 8-HQSA 形成络合物	LOD：2.8μg/L	[63]
Al	在 triton X-100 中与水杨醛碳腙络合	LOD：2.2ng/mL	[64]
Al	与 8-HQSA 的金属络合；利用 XAD-4 进行在线固相萃取	LOD：0.2ng/mL	[65]
Mg^{2+}	在 CTAC 和 HTAC 的存在下与 HQSA 进行金属络合	LOD：12ng/mL	[66]
Zn	在 triton X-100 中与水杨醛碳腙络合	LOD：5ng/mL	[67]
Zn	在 SDS 中 Zn^{2+} 与 8-（苯磺酰胺）喹啉的络合	LOD：0.2ng/mL	[68]
V^{5+}	V/8-HQ 络合物的浊点萃取	LOD：0.007ng/mL	[69]
铵	和 OPA 以及和亚硫酸铵反应	LOD：7nmol/L	[70]
氰化物	Cu-钙黄绿素络合物和氰化物之间的相互作用	LOQ：0.4mmol/L	[71]
氨基己酸	与 OPA 和 N-乙酰半胱氨酸相互作用产生荧光产物（350/450nm），使用顺序注射分析组件	LOD：0.25μmol/L	[72]
克仑特罗	与 OPA 和巯基乙醇反应	LOD：0.06mg/L	[73]
氯辛	在胶束介质中，氯辛和 Al^{3+} 之间形成络合物	LOD：5nmol/L	[74]
磺胺甲噁唑	与 OPA 和 β-巯基乙醇的荧光反应	LOD：7ng/mL	[75]
雷尼替丁	在 OPA 和 β-巯基乙醇的反应之后进行药物与次氯酸钠的荧光反应	LOD：13ng/mL	[76]

注：CTAC—十六烷基三甲基氯化铵；SDS—十二烷基硫酸钠；β-CD—β-环糊精；HTAC—十六烷基三甲基氯化铵；OPA—邻苯二甲醛；8-HQ—8-羟基喹啉；8-HQSA—8-羟基-喹啉 5-磺酸。

1.2.2 磷光

1.2.2.1 引言

在过去的几十年里，液态的室温磷光（RTP）[77] 已经发展成为分析化学中一种灵敏且通用的工具。室温磷光涵盖了可在流动流体中实施的各种技术，例如：胶束稳定室温磷光（MS-RTP）、环糊精诱导室温磷光（CD-RTP）、非保护室温磷光（NP-RTP）和固体-表面室温磷光（SS-RTP）。虽然这些技术通常以间歇的方式或单独应用，但它们也可以在流动流体中实施，例如 SS-RTP 光传感。

考虑到室温磷光方法内在的缺陷和氧气的存在，液态磷光信号通常太差而无法用于分析，尤其是在使用流动流体时。因此，与其他发光技术相比，例如荧光和化学发光，可用的文献和方法很少。与所有可用的室温磷光方法相比，流动流体采用的更广泛的是 SS-RTP 方法。超过 80% 的所描述的应用都通过利用固体载体来开发流通式室温磷光传感器[78]。在流动流体中，SS-RTP 检测的主要特点和应用将在 1.2.4 详细描述。

1.2.2.2 有序媒介中的室温磷光

Cline-Love 等对水溶液中室温磷光的发展进行了系统研究，探索了十二烷基硫酸钠（SDS）的使用，并表明观察大多数发光体的 MS-RTP 存在三个基本要求[79]：①确保胶束聚集体的存在[表面活性剂浓度高于其临界胶束浓度（CMC）]；②存在重原子或者重物质；③确保氧气清除。

尽管这种方法与流动流体兼容，但在流动注射分析或相关连续流动系统中使用 MS-RTP 作为检测方案的情况很少[80]。Sanz-Medel 及其同事使用 MS-RTP 测定连续流动组件中的铝[81]，该方法已成功应用于测定透析液中的 Al^{3+}。

至于 MS-RTP、CD-RTP 也被描述为液相色谱中的一种检测技术。然而，除非解决色谱中 CD 流动相的低效率问题，否则 CD-RTP 将不能作为 MS-RTP 检测的替代品[82]。尽管已经有 100 多篇关于室温磷光方法（使用环糊精）的论文，但这些论文都没有使用流动进样组件[83]。

1.2.2.3 未受保护的室温磷光

通过对均质室温磷光应用的进一步开发，介质中的保护剂不再成为室温磷光发射的必要条件。1997 年，Li 等报道了一种将硝酸铊作为重原子干扰剂，亚硫酸钠作为脱氧剂来测定水性体系中丹磺酰氯的磷光法[84]。这种在室温下，液体溶液中产生磷光的新概念称为无保护室温磷光[85]或重原子诱导室温磷光（HAI-RTP）[86]。这种方法是一项重要的技术进步，因为许多有机分子，如植物生长调节剂（α-萘氧基乙酸和 β-萘氧基乙酸）、医药（萘甲唑啉、萘普生）和农药（噻苯达唑、甲萘威）、多环芳烃和其他化合物（色胺、色氨酸、吲哚-3-丁酸）可以通过一种非常简单的技术来测定[87]。

Cañabate-Díaz 等[88]开发了一种 FIA-HAI-RTP 方法，该方法使用 1mol/L KI 和 10mmol/L Na_2SO_3 测定药物中的萘甲唑啉，获得的 LOD 为 1.6ng/mL。通过使用 FIA-MS-RTP 和 Amberlite XAD-7 作为固体载体的室温磷光流通光电传感器，该方法比 FIA-RTP 的灵敏度更高。Fernández-González 等[89]开发了一种 FIA-HAI-RTP 方法来测定牛乳中的萘夫西林，LOD 为 0.4μmol/L。使用 HAI-RTP 的流通式光电传感器的开发已被报道，并在 1.2.4 中进行了描述。

1.2.3 化学发光

1.2.3.1 引言

由于化学发光既涉及发光过程又涉及化学反应，因此，观测到的强度取决于化学反应的速率、激发的物质数量及其发光效率。这些反应表现出分析效用，因为发射强度是所涉及的化学物质浓度的函数。从这个意义上说，化学发光观测的信号取决于反应动力学，因此，可以通过使用流动进样系统提供可重现性和适当的混合。

在过去的 25 年中，相关文献描述了数百个基于化学发光检测的流动进样方法和应用[90-92]。本节将涵盖主要的化学发光反应并讨论对分析化学很重要的特征。

流动池设计是化学发光流动系统中的关键的一点。流动池位置和到检测器的距离必须具有高度可重复性。在流动注射分析中最常使用的最合适的是尽可能靠近光电倍增管（PMT）的螺旋几何结构的流动池设计。设计的要求是分析物-试剂混合物在检测器前面应发出最大强度的光。这需要快速混合（特别是在反应快速时）和一个用于测量发射峰的适当体积的流动池。分析信号随着螺旋长度的增加而增加，可接受的最小长度由反应速率决定。样品和试剂流的汇合点与检测池之间的距离（以及体积）需要根据所用反应的动力学进行优化。流动池必须对化学发光发射波长透明，并且对化学反应或溶剂系统呈惰性，通常使用玻璃、石英和聚四氟乙烯管。

图 1.5 显示了用于化学发光测定的基本流动注射分析组件。将样品插入到适当 pH 的载体中，沿不同管线循环的试剂与尽可能靠近混合室的样品载流合并，一旦进入反应室，就会通过光电管监测反应混合物[93]。

（1）样品-试剂反应　流动注射分析系统中的化学发光反应照常发生，即将分别包含样品和试剂的两个流体合并。出于流体动力学原因，混合应在流动池附近进行。化学发光测量可以在混合后几秒钟内完成，这对于监测快速反应的动力学特别有用。涉及固相反应器的多相系统是连续流动化学发光应用中最常见的策略之一。化学发光强度取决于所涉及的反应动力学和样品与试剂接触的方式。

（2）试剂固定化　固定化技术[94]通常基于反应器的使用（试剂固定化包括使用酶或化学发光试剂，通常，还包括使用非极性吸附剂或离子交换剂以保留试剂），这些反应器可放置在化学发光反应发生之前。当使用酶反应器时，可以避免当给定的化学发光试剂与多种化合物产生化学发光时选择性的缺乏。分析物是酶促反应的底物，其中一种产物会敏感地参与化学发光反应。例如，图 1.6 所示为基于固定化葡萄糖氧化酶在六氰基高铁酸钾（Ⅲ）存在下与鲁米诺反应形成过氧化氢时产生化学发光的原理测定葡萄糖的典型歧管示意图。使用鲁米诺反应，可以通过固定吡喃糖氧化酶和过氧化氢测定葡萄糖，并通过固定鲁米诺和 Co^{2+} 测定亚硫酸盐[92]。

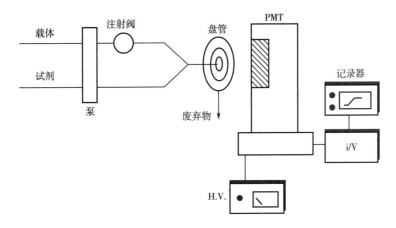

图 1.5　流动进样化学发光计示意图

PMT—光电倍增管　i/V—电流-电压转换器　H.V.—高压

（经 Elsevier 许可转载自文献[93]，2000 年。）

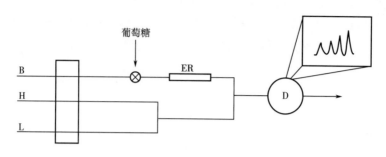

图 1.6 用于分析葡萄糖的 FIA 歧管示意图，该歧管包含基于鲁米诺反应的酶检测反应器

B—缓冲液　H—六氰基铁酸盐　L—鲁米诺溶液　ER—酶反应器　D—检测器

（获得 EDP Sciences 的许可改编自文献[92]，©2000。）

1.2.3.2　化学发光流动方法

在 FI-CL 系统中使用的一系列化学物质中，鲁米诺反应显然是最受欢迎的，但也有大量论文利用高锰酸盐和 Ce^{4+} 的氧化能力。液相化学发光的分析应用可分为涉及众所周知的化学发光反应过程的应用和涉及氧化剂-分析物反应发光过程的应用，氧化剂-分析物是化学发光反应的底物，并且对应于"直接化学发光测定"的概念[95]。

（1）基于试剂的化学发光测定　在第一种方式中，分析物与化学发光反应相互作用，通常作为试剂、催化剂、猝灭剂，甚至增强剂。此类系统的数量有限，其中大多数所采用的试剂包括：鲁米诺、过氧草酸盐、三-（2,2′-联吡啶）钌（Ⅱ）、吖啶酯、Ce^{4+} 等。

①鲁米诺。鲁米诺是目前最常用的液相化学发光材料之一[90-92]。氧化剂包括高锰酸盐、次氯酸盐或碘，但最常用的是过氧化氢。这种化学反应通常需要催化剂，其中包括过渡金属离子（Ti^{4+}、V^{2+}、Cr^{3+}、Mn^{2+}、Fe^{2+}、Fe^{3+}、Co^{2+}、Ni^{2+}、Cu^{2+}）、六氰基铁酸盐（Ⅲ）、血红素和血红素蛋白（血红蛋白、过氧化物酶、过氧化氢酶和细胞色素）。最佳反应 pH8~11。这种化学反应可用于灵敏地测定氧化剂、催化剂或与鲁米诺的衍生产物或相关化合物。

②过氧草酸盐。最常用的试剂是双-（2,4,6-三氯苯基）草酸盐（TCPO）和双-（2,4-二硝基苯基）草酸盐（DNPO）。可灵敏检测荧光化合物（包括蒽、芘、氨基蒽和氨基芘）和适当衍生的分析物｛如胺类、含丹磺酰氯的类固醇；含 N-[4-(2-6-二甲氨基)苯基]马来酰亚胺的硫醇和含荧光胺的儿茶酚胺｝。该试剂的主要缺陷是必须使用有机溶剂。与 FIA-CL 测定相比，这种化学发光试剂更适用于 HPLC-CL[96]。

③三-（2,2′-联吡啶）钌（Ⅱ）。尽管多种化合物可以还原三-（2,2′-联吡啶）钌（Ⅲ），但只有某些特定种类（例如脂肪胺、氨基酸、NADH、某些生物碱、氨基糖苷或四环素抗生素和草酸根离子）会和该试剂发出特征性的橙色光，化学结构的微小差异会对化学发光强度产生巨大的影响[97]。

④Ce^{4+}。该化学发光试剂可以看作是直接化学发光系统或基于试剂的化学发光系统。酸性介质中基于 Ce^{4+} 的化学发光反应已用于测定剂型中的生化物质。少数药物可以直接还原 Ce^{4+} 并产生发光。许多 FI-CL 方法为此类物质而建立，例如萘普生[98]或对

乙酰氨基酚[99]。然而，大多数将 Ce^{4+} 作为氧化剂的测定属于间接类型，这是基于某些分析物对 Ce^{4+}-亚硫酸盐体系、Ce^{4+}-Ru（bpy）$_3^{2+}$ 体系或 Ce^{4+}-Tween 20 等体系化学发光强度的增强效应。这种称为敏化的过程，用于测定还原性化合物，例如氧氟沙星[100]、甲芬那酸[101]或水杨酸[102]。特别是，由 Ce^{4+} 与一些含巯基化合物的化学反应产生的光发射可以通过某些荧光试剂如奎宁[103]、罗丹明 B 或 6G[104]或镧系元素离子[105]增强。

（2）直接（氧化）化学发光方法　这些化学发光方法基于直接化学发光方法，涉及氧化剂（有时是还原剂）和分析物[106]之间的反应。在对具有分析意义的化学发光反应进行表征之前，应进行专门的初步实验研究，该研究基于测试分析物与不同介质中的多种氧化剂之间的化学发光反应。在不同的化学条件下使用强氧化剂（例如 MnO_4^-、ClO^-、Ce^{4+}、H_2O_2、IO_3^-、Br_2 和 N-溴代琥珀酰亚胺）和还原剂，以便从不同的分析物中产生化学发光发射。近年来，带有直接化学发光检测（基于在酸性介质中直接氧化）的流动进样方法在食品、饮料、药物和体液分析方面具有巨大潜力[107]。

流动注射分析是基于高锰酸盐的化学发光系统。已在各种条件下将酸性高锰酸钾作为化学发光试剂从各种分析物引起化学发光。$KMnO_4$ Fl-CL 方法主要基于 $KMnO_4$ 对分析物的直接氧化[108]，或基于分析物对 $KMnO_4^-$-亚硫酸盐体系的化学发光强度的提高[109]，并使用了奎宁、甲醛、乙二醛、甲酸和一些表面活性剂作为一种敏化剂。酸性高锰酸钾可以灵敏地检测含有酚醛、和/或胺部分的分子，因此，它在测定大量重要分析物方面具有相当大的潜力[110]。

CL-FIA 分析的优点包括检测限低和较宽的线性范围，两者通常都可以通过简单、稳定且相对便宜的商业仪器实现。表 1.3 所示为 FIA-CL 方法开发的不同化学发光检测的选定示例，这些示例适用于包括制药、生物医学、环境和食品分析的不同样品[111-163]。

表 1.3　基于直接化学发光或基于试剂的化学发光检测的选定流动方法[111-163]

分析物	样品基质	化学发光反应/试剂	检测限	参考文献
直接化学发光检测				
氯喹	尿	在过氧化氢-亚硝酸盐硫酸介质中氯喹直接化学发光	0.086μmol/L	[111]
盐酸头孢羟氨苄	药物和生物流体	以奎宁作为敏化剂，高锰酸钾-硫酸化学发光	0.05μg/mL	[112]
氯霉素	药物	在线光降解氯霉素，由此产生的光片段通过直接化学发光并使用硫酸介质中的高锰酸钾作为氧化剂进行检测	30ng/mL	[113]
麦角新碱马来酸盐	药物	通过氢氧化钠中的氰化铁（III）直接氧化，通过阳离子表面活性剂（十六烷基氯化吡啶）增强	0.07μg/L	[114]

续表

分析物	样品基质	化学发光反应/试剂	检测限	参考文献
2-乙基-己基-4-(N,N-二甲基氨基)-苯甲酸	防晒剂配方	通过硫酸介质中的高锰酸钾氧化分析物进行直接化学发光	25ng/mL	[115]
异丙烟肼	药物和尿液	室温下，在亚硫酸盐的存在下，在硫酸介质中通过 Ce^{4+} 氧化药物		[116]
纳曲酮酚类化合物	药物	酸性高锰酸钾化学发光	2.5ng/mL	[117]
	葡萄酒	在酸性高锰酸钾的存在下，进行氧化和直接化学发光检测	0.4~700nmol/L	[118]
普萘洛尔	药物	在硫酸介质中，通过高锰酸钾的氧化进行直接化学发光	0.87mg/L	[119]
水杨酰胺	人尿和药物	酸性高锰酸钾产生化学发光	30ng/mL	[120]
酪氨酸	—	高锰酸钾（pH 6.75）和在正磷酸（pH 2）中的酸性高锰酸钾	10~50nmol/L	[121]

基于试剂的化学发光检测

医药和生物应用

分析物	样品基质	化学发光反应/试剂	检测限	参考文献
胆红素	水溶液	N-溴代琥珀酰亚胺或次氯酸钠化学发光	1.75μg/mL	[122]
卡托普利	药物	Ce^{4+}-硫酸化学发光	0.2μmol/L	[123]
儿茶酚胺	血浆	儿茶酚通过咪唑催化分解生成过氧化氢并通过化学发光进行检测	—	[124]
胆碱和乙酰胆碱	大鼠脑组织	使用含有固定在玻璃珠上的乙酰胆碱酯酶和胆碱氧化酶的两个反应器；使用 Co^{2+}-鲁米诺化学发光检测	500~600fmol	[125]
氯丙咪嗪	药物	Ce^{4+} 对亚硫酸盐化学发光氧化的敏化作用	2.5mg/L	[126]
脱氧核糖核酸	—	在酸性介质中通过 DNA 增强罗丹明 B-Ce^{4+} 络合物的化学发光	8.3fg/mL	[127]
D-氨基酸	人血浆	带过氧草酸盐化学发光检测的固定化酶柱反应器	0.4~30pmol（进样 10μL）	[128]
多巴胺	生化样品	咪唑过草酸盐化学发光	10nmol（进样 20μL）	[129]
葡萄糖	兔体液和血液	微透析，然后在固定化葡萄糖氧化酶反应器中反应产生过氧化氢，过氧化氢通过鲁米诺-六氰基高铁酸化学发光进行检测	10μmol/L	[130]

续表

分析物	样品基质	化学发光反应/试剂	检测限	参考文献
三磷酸甘油	水状	固定在可控孔玻璃上的甘油-3-磷酸氧化酶；通过鲁米诺-Co^{2+}化学发光检测	0.5μmol/L	[131]
谷胱甘肽	—	对鲁米诺过氧化氢化学发光系统的增强效应	68nmol/L	[132]
异茴香醚	药物	抑制鲁米诺-过氧化氢-六氰基铁酸盐（Ⅲ）的反应	5mg/L	[133]
甲丙醇酯	药物和人尿	在酸性介质中Ce^{4+}亚硫酸盐化学发光系统的化学发光光敏剂	4.7nmol/L	[134]
萘普生	药物	抑制碱性介质中鲁米诺氧化产生的化学发光		[98]
氧甲唑啉	药物			[135]
吩噻嗪类	药物和生物流体	以罗丹明 B 作为敏化剂，Ce^{4+}-酸化学发光	0.01~0.1μg/mL	[136]
脯氨酸	葡萄酒	硫酸中的 $Ru(bipy)_3^{3+}$	10nmol/L	[137]
蛋白	牛血清白蛋白、人血清白蛋白、γ-球蛋白和卵白蛋白	1,10-菲咯啉-过氧化氢-Cu^{2+}，十六烷基三甲基溴化铵化学发光	0.02μg/mL	[138]
盐酸吡哆醇	片剂	鲁米诺-过氧化氢化学发光	6μg/mL	[139]
芦丁（维生素P）	药物	对鲁米诺-铁氰化物系统的增强作用		
单宁酸	药物	通过单宁酸对鲁米诺-H_2O_2-Mn四磺酸-对酞菁系统化学发光的抑制作用	30ng/mL	[140]
特布他林	药物、血浆和尿液	硫酸特布它林在碱性条件下增强鲁米诺-高锰酸盐系统的化学发光发射	0.8nmol/L	[141]
四环素类	药物制剂	以奎宁作为敏化剂，Ce^{4+}-硫酸化学发光	—	[142]
硫胺素	药物和尿液	鲁米诺和KIO_4固定在阴离子交换柱上；流动注射分析系统中的试剂受控释放技术	0.25~25nmol 1ppm	[143] [144]
三氯生	牙膏样品	在含有荧光素的碱性介质中，三氯生光转化为发光前体，以及与 N-溴代琥珀酰亚胺的化学发光反应	50nmol/L	[145]

译者著：

本书沿用原版书单位。1ppm = 10^{-6}，1ppb = 10^{-9}，1ppt = 10^{-12}。

续表

分析物	样品基质	化学发光反应/试剂	检测限	参考文献
色氨酸	组织	Ce^{4+}-硫酸化学发光，固定于阴离子交换树脂（填充在柱上）上的鲁米诺和高碘酸盐，用作试剂	$0.1\mu g/mL$	[146]
	人尿和血清		1.8ng/mL	[147]
维生素 B_{12}	药物和人类血清	Co^{2+}对流动进样系统中鲁米诺和溶解氧之间化学发光反应的增强作用	50pg/L	[148]
环境应用				
Cu（络合物）	海水	1,10-菲咯啉-过氧化氢化学发光	0.1nmol/L	[149]
Fe 和 Mn（溶解的）	地下水	鲁米诺-高碘酸钾化学发光	3pg/mL 和 5pg/mL	[150]
Fe^{2+} 和 Fe^{3+}	天然水域	Cu 包覆的 Zn 将 Fe^{3+} 还原为 Fe^{2+}；固定在阴离子交换树脂上的鲁米诺；在化学发光检测前用氢氧化钠洗脱	0.4ng/L	[151]
过氧化氢	雨水	固定化的 Co^{2+} 和鲁米诺通过水解洗脱应用于化学发光检测	12nmol/L	[152]
亚硝酸盐	天然水域	亚硝酸盐与过氧化氢反应生成过氧腈，后者与鲁米诺产生化学发光	1nmol/L	[153]
苯酚	水样	苯酚对鲁米诺六氯铁酸盐化学发光系统的增强作用；在线 SPE 提高选择性	4.7ng/mL	[154]
磷酸盐	天然水域	使用鲁米诺-辣根过氧化物酶化学发光检测固定化丙酮酸氧化酶与磷酸盐反应产生的过氧化氢	74nmol/L	[155]
V^{5+}	地球化学和毛发样品	鲁米诺和六氰基铁盐（Ⅱ）均固定在阴离子交换树脂柱上，用磷酸洗脱产生化学发光	5.4ng/mL	[156]
V^{4+}	水样	V^{4+}催化高碘酸盐氧化紫胆素产生化学发光	0.05ng/mL	[157]
食品和饮料检测				
毒死蜱	水果和水样	在阴离子交换柱上使用固定化试剂对鲁米诺-高碘酸盐化学发光反应的抑制作用	0.18ng/mL	[158]
L-苹果酸	葡萄酒	苹果酸脱氢酶/还原型烟酰胺腺嘌呤二核苷酸氧化酶均固定在聚合物微珠上以产生过氧化氢，使用鲁米诺-六氰基高铁酸盐（Ⅲ）化学发光进行检测	$0.08\mu mol/L$	[159]
亚硫酸盐	啤酒和葡萄酒	通过吐温 80 增强的罗丹明 6G（固定在阳离子交换树脂上）敏化进行自氧化	0.03mg/L	[160]
苏丹 I	辣椒酱	鲁米诺-H_2O_2 体系中化学发光的增强作用	3pg/mL	[161]

续表

分析物	样品基质	化学发光反应/试剂	检测限	参考文献
单宁酸	啤酒花颗粒样品	抑制鲁米诺-过氧化氢-Cu^{2+}化学发光	9nmol/L	[162]
四环素类	鱼	四环素对Ce^{4+}和罗丹明B之间反应的增强作用;使用MIP柱的在线SPE	1ng/mL	[163]

(3) 其他 FI-CL 系统　电致化学发光（ECL）。近年来，在酸性介质中电生不稳定氧化剂的化学发光流动系统受到了很多关注，文献中介绍了一些分析应用[164,165]。在 ECL 中，一种或多种试剂是在电解过程中原位生成的。ECL 与化学发光具有许多相同的分析优势，主要是因为不需要昂贵的激发光学器件或复杂的仪器。一些 FIA-ECL 方法已被开发用于分析合适的物质，例如，药物[166-168]、乳酸、葡萄糖、胆固醇和胆碱[169]、氨基酸[170,171]、金属离子[172,173]等。

光致化学发光是一种应用于农药和药物的新技术，其基于在直接化学发光反应之前进行在线光降解，通常使用硫酸介质中的高锰酸钾作为氧化剂[174,175]。该策略已被提议用于测定涕灭威[176]和阿苏拉姆[177]，通过光降解农药，然后进行氧化反应，形成一个全自动的多路换向的流动组件。

1.2.3.3　生物发光

生物发光（BL）是从生物系统中发现的酶催化过程，其中催化蛋白提高了发光反应的效率[178]。生物发光的生物种类繁多，从细菌和真菌到软体动物、鱼类和昆虫。研究最广泛的生物发光有机体是萤火虫（photinus pyralis），因为萤火虫很容易收集，而且它们的发光器官充满了引起发光反应的化学成分。生物发光系统有两个主要优点：由于更高的量子产率而导致灵敏度增加，而选择性因所涉及的酶而得到提高[179]。

生物发光反应的灵敏度和普适性已得到广泛的应用，包括用于酶和底物（葡萄糖、甘油三酯、甘油等）的临床分析、类固醇分析、水生毒性测试、生乳中细菌筛选等。萤火虫荧光素酶、海洋细菌荧光素酶和水母发光蛋白这三种生物发光反应占据了 90% 以上的应用。Hansen[180]描述了一种基于与荧光素酶-荧光素反应的用于分析低浓度 ATP 的流动进样系统，使用了双重进样阀，可以同时注入样品和酶溶液，获得了 0.1nmol/L 的 LOD。此外，Blum 等[181]描述了一种基于酶催化发光反应的光纤传感器，该传感器集成在流动注射分析系统中。

1.2.4　流动流体中基于固相发光的检测

流动注射分析和固体表面发光（SSL）[182]的耦合克服了 SPS 的限制，与离散手动测量相比显现出明显的优势：①样品处理过程（试剂混合、去除干扰、pH 调整等）都可以在线进行；②提高传感器寿命，因为在流动分析中，传感材料暴露在样品中的时间很短，在不同测量间隔内，传感材料处于更温和的基质中；③自动化，这反过来又提供了一系列优势，例如高样品吞吐量（响应时间短）、高精度和准确度、节省试剂和固体传感支撑等。近年来，耦合 FIA-SSL 已成为一个非常有趣的研究领域，并且发光流通传感器已应用于一系列分析物（有机和无机），以及应用于农业、环境、生物医学和药物分析等相关领域。

这种耦合系统使用了配备有市售流动池的常规仪器，传感材料放置在该流动池中，并且采用非常简单的歧管。当分析物保留在流动池中的固体微珠上时，会产生信号。测量结束后，必须重新生成活性固体表面，以备下一次测定使用。

1.2.4.1 固相荧光检测

大多数固体表面荧光检测系统是基于对分析物的天然荧光测量[183]。已描述的系统中，有一大类是基于这一原理，并通过使用弱酸阳离子交换凝胶 CM-Sephadex 作为发光检测中的传感支撑来选择性测定微量的铈、铽、镝或钐。相关文献描述了一系列用于确定药物活性成分的固相天然荧光检测系统[183]。其他应用领域包括环境和食品分析以及 pH 测量和氧传感。表 1.4 所示为如何选择最相关的系统。

反应产物也可以是基于天然荧光信号的系统中检测到的物质，因此，甘油与 NAD^+ 的反应产物（由甘油脱氢酶催化）的天然荧光是间接测定酒中甘油的固相荧光检测系统的基础[184]。最近相关文献描述了顺序注射分析和固相荧光检测的耦合。其中固相荧光检测包含了天然和衍生两种方式[185]。多路换向系统的天然固相荧光检测通过开发单组分和多组分检测来分析评估[186,187]。

表 1.4 基于固相发光检测的分析流动进样方法的代表性示例[203-226]

分析物	样品基质	化学发光反应/试剂	检测限	参考文献
基于荧光的光传感器				
苯海拉明	药物制剂	天然荧光检测；固体载体：Sephadex G-15	0.1~1.2mg/L 奎宁：0.4~20μg/L	[203]
奎宁和奎尼丁	制药、水、洗发水、软饮料	天然荧光检测；固体载体：Sephadex SPC-25	奎尼丁：0.9~20μg/L	[204]
多环芳烃	自来水和矿泉水	天然荧光检测；固体载体：Amberlite XAD-4	—	[205]
α-萘酚和β-萘酚	水	天然荧光检测；二阶导数同步荧光+PLS；固体载体：Sephadex QAE-A25	—	[206]
吡虫啉	辣椒和环境水	光化学诱导荧光检测。固体载体：C_{18} 硅胶微珠	LOD：1.8μg/L	[192]
硫胺素	药物制剂	微珠进样系统；可再生表面传感器；基于硫色素络合物 $Fe(CN)_6^{3-}/OH^-$ (385/433nm) 的 FL 检测；固体载体：C_{18}-PS-DP	LDR：0.06~8mg/L	[202]
氟灭酸	血清、尿液和药物	光化学诱导荧光检测（258/442nm）；多路换向流动组件，固体载体：C_{18} 硅胶微珠	LOD：0.55nmol/L	[207]

续表

分析物	样品基质	化学发光反应/试剂	检测限	参考文献
基于室温磷光的光传感器				
四环素	尿液	镧系元素敏化发光。Eu^{3+}-TC 复合体，(390/622nm)；固体载体：Amberlite XAD-2	LOD：0.25~0.4μg/L	[208]
萘夫西林（β-内酰胺类抗生素）	乳制品	室温磷光检测；283/505nm 固体支持物；萘夫西林印迹溶胶凝胶或印迹有机硅酸盐（有机改性硅烷）	LOD：5.8μmol/L（水乳）33μmol/L（脱脂牛乳）0.1~1.2mg/L	[209-211]
萘甲唑啉	眼霜	基于 HAI-RTP（290/520nm）的光感器，使用 Amberlite XAD-7 作为传感载体。[KI] 1.6mol/L；15mmol/L 亚硫酸钠	LOD：9.4ng/mL	[88, 212]
荧蒽	河水样本	使用基于碘化单体的分子印迹聚合物进行 RTP 光传感	LOD：35ng/L LDR：0~100μg/L	[213]
苯并[a]芘	天然水	使用基于碘化单体的分子印迹聚合物的 RTP 光传感	LOD：10ng/L LDR：0~100μg/L	[214]
苯并[a]芘	水	基于 HAI-RTP（390/690nm）的光感器，使用 Amberlite XAD-7 作为传感载体。[KI] 4mol/L；20mmol/L 亚硫酸钠	LOD：12ng/mL	[215]
2-萘氧基乙酸	水	基于 HAI-RTP（276/516nm）的光感器，使用 Amberlite XAD-7 作为传感载体。[Tl(I)] = 175mmol/L；10mmol/L 亚硫酸钠	LOD：4.9ng/mL	[216]
1-萘基-邻苯二甲酸和代谢物	水	基于 HAI-RTP（276/516 nm）的光感器，使用 Amberlite XAD-7 作为传感载体。[Tl(I)] = 0.2mol/L；15mmol/L 亚硫酸钠	LOD：8.1~11.2ng/mL	[217]
萘乙酸	天然水和苹果	1-NAA 在填充树脂上的在线保留和通过这种天然荧光物质的重原子诱导（HAI）-RTP 发光（290/490nm）来分析测量	LDR：0~500μg/L LOD：1.2μg/L	[218]
基于化学发光的光传感器				
肾上腺素	血清	鲁米诺与亚铁氰化钾之间的化学发光反应；增强保留在固体载体上的肾上腺素 固体载体：MIP 聚合物	LDR：5~100nmol/L LOD：3nmol/L	[219]

续表

分析物	样品基质	化学发光反应/试剂	检测限	参考文献
诺氟沙星	尿样	诺氟沙星印迹聚合物作为识别元件；Ce^{4+}-亚硫酸钠诺氟沙星化学发光反应	LDR：0.1~10μmol/L LOD：30nmol/L	[220]
异烟肼	人尿	异烟肼 MIP 是在异烟肼模板分子存在下，通过甲基丙烯酸（MAA）和乙二醇二甲基丙烯酸酯（EGDMA）的热自由基共聚合成的。固体载体：异烟肼，使用 MIP 识别的 MIP 化学发光检测	LDR：2~200nmol/L	[221]
沙丁胺醇	尿样	鲁米诺与铁氰化钾之间的化学发光反应：沙丁胺醇的增强作用。MIP 作为识别元件	LDR：0.05~10μmol/L LOD：16ng/mL	[222]
肼屈嗪	人尿	鲁米诺-碘之间的化学发光反应：肼苯哒嗪的增强作用，MIP 作为识别元件	LOD：0.6nmol/L LOD：2~800nmol/L	[223]
特布他林	人血清	基于鲁米诺与亚铁氰化物反应的化学发光检测，使用 MIP 作为特布他林的识别材料；芯片上的微流体传感器，注射 12μL 的样品	LDR：8~100ppb LOD：4ppb	[224]
正磷酸盐	矿泉水、地下水、自来水和池塘水	基于鲁米诺反应的化学发光检测	LDR：5~50μg/LP； LOQ：4μg/LP	[225]
水杨酸	药物制剂	固体载体：N-乙烯基-吡咯烷酮/二乙烯基苯-苯吸附剂（Oasis HLB）来自固体载体上的水杨酸与高锰酸盐的直接氧化产生化学发光（Sephadex QAE-A25）	LDR：1~30mg/L LOD：0.30mg/L	[226]

基于分析物化学衍生化的固相荧光检测系统需要更复杂的配置，包括用于合并衍生试剂与注入样品的额外通道，或同时双重注射下（样品和试剂）的载流分叉[188]。

然而，如果荧光试剂（通常是螯合剂）可以永久固定在载体上，则可以实现配置的简化。在这种情况下，每次测量后通过解离荧光复合物并仅洗脱分析物来将传感载体再生[189]。最近，基于使用衍生试剂的顺序注射分析固相荧光检测系统已被应用于测定扑热息痛[190]。在酸性介质中，分析物和亚硝酸钠的反应产物在碱化后插入到系统中。

固相检测也已与光化学诱导荧光检测一起协同实施。该方法 PIF-SPS-FIA 已被应用（包括多路换向方法）于测定药物和农药[191,192]。

只有少数固相荧光多组分检测流动系统（能够检测样品中的多种物质）。实际上，所有这些系统都基于对天然荧光信号的测量。最多可以检测到样品中的三种分析物[193]。另一方面，相关文献提出了一些多路换向荧光固相多组分检测系统[194]。

固相始终被视为固体表面荧光检测系统中的永久成分。测量过后，固体支撑微珠必须进行再生，以便使传感表面为测试下一个样品做好准备，从而使固相可重复使用。然而，当一些特定物质被大量保留时（即离子交换微珠上的多电荷物质），传感表面的再生变得异常难以实现。那么，这时候就需要用到高浓度的盐水和酸溶液来减少传感装置的使用寿命、重现性和基线稳定性[197]。在多次测量后总会出现表面失活的现象。这些缺陷可以通过使用所谓的微珠进样光谱（BIS）或流动进样-可再生固体表面荧光方法[198]来避免，这些设想为第三代流动进样微分析技术[199]。这种 BIS-FIA 荧光测定方法已被用于测定铍、铝[200]和钒[201]。BIS-SIA 荧光测定系统已被用于将硫胺素预先转化为脱氢硫胺素[202]。

1.2.4.2 基于室温磷光的光传感

最近在连续原位监测不同分析物的流动进样系统中引入了不同的 SS-RTP 方法，用以确定分析物，例如氧、金属离子、镧系元素、有机化合物和酶亚基如葡萄糖。

基于发光猝灭的分子氧传感被认为是最典型和应用最广泛的光传感之一。在所有发光猝灭检测方案中，会产生强大而稳定的背景发光信号，该信号在分子氧或其他猝灭剂的存在下进行动态猝灭。用于检测分子氧（基于猝灭室温磷光）[227,228]的实验歧管也可用于间接检测葡萄糖[229,230]和胆固醇[231]等酶底物。就葡萄糖而言，在葡萄糖氧化酶存在的情况下，葡萄糖通过酶促反应消耗氧气形成葡萄糖酸和过氧化氢进而获得室温磷光，这意味着所获的室温磷光可以间接测量葡萄糖浓度。

相关文献描述了通过 SS-RTP 检测金属离子的方法，SS-RTP 则来源于金属与铁的络合物[232]。这些磷光络合物保留在阴离子交换树脂上，树脂通过注入少量强酸性（HCl）溶液进行再生。

基于 SS-RTP 并使用氧清除剂和重原子干扰器的天然磷光特性被用来开发不同的光传感器以确定不同样本中的农药和多环芳烃。近年来，分子印迹材料广泛应用于分离技术，已成为开发更高选择性的新型传感材料的新工具。最近这种识别过程与基于 RTP 的转导一起应用于流动流体[211,213,214]。

使用镧系元素-配体配合物的发光可以检测稀土金属以及配体。使用敏化的镧系元素发光来测定 Eu^{3+}[233]、Gd^{3+}[234]和 Tb^{3+}[235]。在这些例子中，镧系元素和乙酰丙酮之间形成的复合物被瞬时固定在树脂（Chelex 100）上。复合物的固定导致发光强度和寿命增加。使用相同的策略，分析物（用作配体）如抗生素四环素[236-238]和蒽环类抗生素[239]可以通过它们与 Eu^{3+} 的复合物发光来检测。图 1.7 所示为基于铽敏化发光测定药物和生物流体中诺氟沙星的光传感器所采用的歧管。

1.2.4.3 固相化学发光检测

越来越多的分析应用[240,241]采用了流动进样系统中固定化试剂的化学发光检测。在许多情况下，分析物是通过与所溶试剂的化学发光反应进行测定的，这些试剂通过适当的洗脱液从固定的底物上释放出来。在这种情况下，流动池中不使用固体载体，

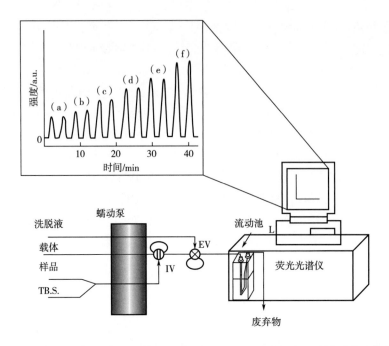

图1.7　基于铽敏化发光的光传感器的流动进样组件，用于测定药物和生物流体中的诺氟沙星
载体为0.1mol/L醋酸盐缓冲液，pH 5.6，样品为在0.1mol/L，pH 5.6的醋酸盐缓冲液中的NFX。
图示为所获得的(a)空白；(b)10ng/mL、(c)30ng/mL、(d)50ng/mL、
(e)70ng/mL和(f)100ng/mL的NFX重复峰

IV—进样阀　EV—洗脱阀　ELUT—洗脱液(0.08mol/L EDTA)　TB.S.—Tb^{3+}(4mmol/L)

(经Elsevier许可，转载自文献[238]，©2005。)

检测是在均质溶液中进行的，因此，这些系统不能被视为真正的流通式化学发光传感器[94]。

(1) 将化学发光试剂固定的化学发光传感器　在化学发光传感器的开发中，将试剂固定到合适的底物上起着重要作用。离子交换树脂广泛用于固定化学发光试剂。通过将鲁米诺和其他化学发光试剂固定在这些类型的载体上开发出一系列用于无机和有机分析物的化学发光传感器。通常，带有固定试剂的树脂被装入一根用作流动池的玻璃管中，并放置在光电倍增管的检测窗口前[242,243]。一种延长这些化学发光传感器使用寿命的非常有趣的方法是固定可重复使用的化学发光试剂：固定在Dowex-50W离子交换树脂上的Ru(bpy)$_3^{2+}$至少可以使用6个月[244]。林等[245]研究了几种固定阴离子试剂的阴离子交换树脂对H_2O_2与高碘酸盐的化学发光反应的影响：包括对Amberlite RA系列、Amberlyst A系列和Muromac系列树脂进行了比较。具有凝胶结构或具有大网状结构的树脂表现出不同的化学发光强度，最强的化学发光信号对应于凝胶丙烯酸型强阴离子交换基树脂(Amberlite IRA 458)。

基于在硫酸介质中分析物对固体氧化剂化学发光氧化亚硫酸盐的敏化作用，几种流通式传感器被开发出来：PbO_2[246]、$NaBiO_3$[247]或MnO_2[248]已被用作吸附在海绵(氯丁二烯)上的氧化剂，并填充到用作流动池的玻璃柱中。化学发光流动传感器(将鲁

米诺固定于合适的聚合物吸附剂上）也已用于气体分析[249,250]。最近相关文献提出了将几种类型的纳米颗粒作为催化剂来开发催化发光气体传感器[251,252]。

（2）基于将酶固定化的化学发光传感器　作为高选择性催化剂，酶的（共）固定是开发化学和生化化学发光传感器最有趣的方法之一。几种化学发光流动传感器将酶（共）固定在异戊二烯化聚乙烯醇微珠上并填充到透明PTFE管（用作螺旋流动池）中：葡萄糖[253]、支链氨基酸（L-亮氨酸、L-缬氨酸和L-异亮氨酸）[254]、血清L-谷氨酸[255]、鱼肉提取物中组胺的测定[256]。

最近，有人建议使用固定有抗体的磁性微珠来测定阴离子表面活性剂的化学发光[257]。使用钕磁铁将磁性微珠吸附在流动池中，在微珠表面发生化学发光反应。在监测到信号后，通过向下移动磁铁来去除磁性微珠。使用溶胶-凝胶来固定酶已成为开发化学发光传感器的常见过程。溶胶-凝胶的一大优势是具有光学透明性和化学稳定性，因而其适用于光学传感器。例如，使用溶胶凝胶固定化血红蛋白作为催化剂证明了这一优势[258]。

（3）将分析物固定在分子印迹聚合物（MIPs）上的化学发光传感器　MIPs已被用作传感材料来设计化学发光流动传感器。开发分子印迹聚合物中具有确定形状和官能团的印迹腔不仅是为了分子识别功能，也可作为一种特殊的化学发光反应介质。这些传感材料已用于测定多种分析物，包括1,10-菲咯啉[259]、盐酸克伦特罗[260]、肾上腺素[219]、诺氟沙星[220]和异烟肼[221]。

在化学发光传感系统的开发过程中还引入了其他流动分析方法，例如多路换向和多注射器流动注射分析。因此，第一台多路化学发光传感器通过使用微型计算机中适当软件控制的四个三通电磁阀来测定水杨酸[226]。类似地，将多重注射器流动注射分析概念与多路换向相结合，开发了一种流通式固相化学发光传感器，用于测定环境水样中的痕量正磷酸盐[225]。

1.3　趋势和展望

人们目前正在研究相关措施以提高流动方法的选择性和灵敏度。其中探索的主要领域是小型化和新材料的开发，小型化旨在降低最小检测限和减少每次分析的样品消耗量，新材料则包括用作特定识别系统的分子印迹聚合物，该聚合物可用于设计感测复杂基质中目标物质的选择性识别系统，或使用具有独特光学特征的量子点进行传感。发光是一种有吸引力且经济高效的检测工具，与这些高级技术完全互补。

1.3.1　小型化

小型化被普遍认为是分析仪器发展的最重要趋势之一，其最终目的是在微观尺度上集成整个分析过程。近年来，基于二氧化硅和玻璃微芯片的微型化全分析系统（μTAS）的研究和开发在化学反应和分析化学领域得到了显著发展，因为其具有便携、试剂消耗量低和分析时间减少的固有优势[261]。

尽管微流体系统主要是为毛细管电泳和微流体注射分析（μFIA）系统等分离装置设计的，但近年来由于其试剂消耗量低、操作简单和适用于小型化系统的特点，微流体系统得到了迅猛发展。从这个意义上说，化学发光是微全分析系统中一种很有前景的检测方法。化学发光测量不需要光源，因此，化学发光的仪器要简单得多，并且可以很容易地集成到芯片上进行检测。

1.3.2 分子印迹材料

另一种通过发光检测提高流动进样方法选择性的策略是使用针对目标物质的新材料和吸附剂。最近，广泛用于分离技术的MIPs（有机聚合物和溶胶-凝胶）已成为一种开发具有高选择性传感材料的新工具。这种识别过程可以在具有发光检测的流通式传感器中实现。从这个意义上说，由于分析物在聚合物结构的特定腔内预浓缩，印迹材料可以进行高度选择性的识别和灵敏的测定。此外，这种识别模式提供了室温磷光所需的刚性。使用印迹材料作为固体载体并结合选择性发光检测模式（如室温磷光），可以设计出高灵敏度、高选择性的光学传感器[78]。

尽管印迹溶胶-凝胶基质拥有一系列优点（例如，可以使用强有效的洗涤步骤，如燃烧，以更好地去除印迹分子），但此类材料（如印迹有机聚合物）尚未在化学应用中取得成功。印迹溶胶凝胶作为活性相（载体）在光学传感领域的用途仍处于早期阶段，需要更多的研究来探索它们的潜力。

1.3.3 量子点

量子点（QDs）是具有独特吸引力的光电特性的纳米结构材料，特别适用于（生物）化学领域的分析应用[262]。量子点通常表现出比传统有机荧光团更高的荧光量子产率，从而可以实现更高的分析灵敏度。随着研究工作致力于扩展这种新型发光体的独特性质，量子点作为用于光学传感的光致发光探针越来越受欢迎。迄今为止，关于量子点应用的大部分工作都仅限于溶液-传感分析。通过将这些量子点固定到合适的固体载体上（为了在流动溶液中制备"活性"固相），从而获得更灵敏和选择性更高的光传感方案[263]。光传感技术可以将量子点的优点与流动分析技术结合起来，量子点可以集成到合适的固体载体中，这个研发过程才刚刚开始，以开发可靠的"活性"相和光传感器并用于流通光学传感或光纤传感。

近期趋势的一个例子是，张及其同事描述了芯片上最先进的化学发光微流动注射分析系统[264]。图1.8所示为其所使用的流动组件方案。利用激光烧蚀技术在聚甲基丙烯酸甲酯（PMMA）芯片上制作微通道。用于特定分子识别的微柱（包括识别分子印迹聚合物、酶和细菌）被引入到流动组件中以提高选择性。这些微流体注射分析系统已应用于临床分析、食品安全检测、体内和实时药物测定以及药代动力学研究[224,265]。

作为这些化学发光微流体传感器中使用的传感方案示例，该传感机制用于通过化学发光测定人血清中的特布他林，如图1.9[221]所示。它以MIP为识别元件。异烟肼MIP是通过甲基丙烯酸（MAA）和乙二醇二甲基丙烯酸酯（EGDMA）在异烟肼模板分

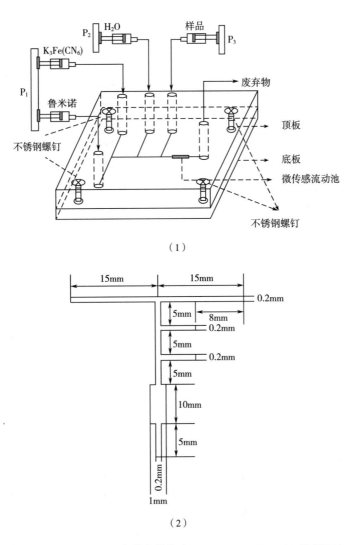

图1.8 （1）用于测定特布他林（P_1、P_2、P_3：Ts2-60注射泵）的芯片上微流体传感器示意图（2）显示微通道尺寸的示意图

（资料来源：经Elsevier许可转载自文献［224］，©2006。）

子存在下进行热自由基共聚反应合成的。MIPs可以选择性地吸附异烟肼，吸附的异烟肼由于对鲁米诺和高碘酸盐（在流动池中混合）之间的弱化学发光反应有很大的增强作用而被感知。传感器是可逆且可重复使用的，它显著提高了化学发光分析的灵敏度和选择性。这些特定的传感方案将在不久的将来有助于开发新的更具选择性和灵敏度的流动进样技术，并有可能通过不需要光源的化学发光系统或使用光纤和发光二极管的荧光或室温磷光方案制造出与发光转导相关的便携式仪器。

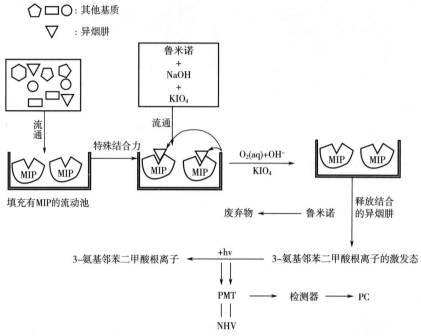

图1.9 基于异烟肼MIP的流动进样化学发光传感器的传感机制

(经Elsevier许可转载自文献[221]，©2007。)

参考文献

[1] Hurtubise, R. J. (1988) in *Molecular Luminescence Spectroscopy: Methods and Applications*, *Part 2* (ed. S. J. Shulman), J. Wiley, New York, Chapter 1.

[2] Valcárcel, M. and Luque de Castro, M. D. (1988) *Automatic Methods of Analysis*, Elsevier, Amsterdam.

[3] Kina, K., Shiraishi, K. and Ishibashi, N. (1978) Ultramicro solvent extraction and fluorimetry based on the flow injection method. *Talanta*, 25, 295-297.

[4] Ruzicka, J. and Hansen, E. H. (1975) Flow injection analyses: Part I. A new concept of fast continuous flow analysis. *Analytica Chimica Acta*, 78, 145-157.

[5] Ruzicka, J. and Marshall, G. (1990) Sequential injection: a new concept for chemical sensors, process analysis and laboratory assays. *Analytica Chimica Acta*, 237, 329-343.

[6] Cerdà, V., Estela, J. M., Forteza, R., Cladera, A., Becerra, E., Altimira, P. and Sitjar, P. (1999) Flow techniques in water analysis. *Talanta*, 50, 695-705.

[7] Reis, B. F., Gine, M. F., Zagatto, E. A. G., Lima, J. L. F. C. and Lapa, R. A. (1994) Multicommutation in flow analysis. Part 1. Binary sampling: concepts, instrumentation and, spectrophotometric determination of iron in plant digests. *Analytica Chimica Acta*, 293, 129-138.

[8] Lapa, R. A. S., Lima, J. L. F. C., Reis, B. F., Santos, J. L. M. and Zagatto, E. A. G. (2002) Multicommutation in flow analysis: concepts, applications and trends. *Analytica Chimica Acta*, 466,

125-132.

[9] De Armas, G., Becerra, E., Cladera, A., Estela, J. M. and Cerdá, V. (2001) Sequential injection analysis for the determination of fuberidazole and thiabendazole by variable-angle scanning fluorescence spectrometry. *Analytica Chimica Acta*, 427, 83-92.

[10] Albert-García, J. R. and Martínez-Calatayud, J. (2006) FIA-fluorimetric determination of the herbricide Benfuresate. *Journal of Flow Injection Analysis*, 23, 19-24.

[11] Martinez-Calatayud, J., Goncalves Ascencao, J. and Albert-García, J. R. (2006) FIA-fluorimetric determination of the Pesticide 3-indolyl acetic Acid. *Journal of Fluorescence*, 16, 61-67.

[12] Subova, I., Assandas, K. A., Catala-Icardo, M. and Martínez-Calatayud, J. (2006) Fluorescence determination of the pesticide asulam by flow injection Analysis. *Analytical Sciences*, 22, 21-24.

[13] Solich, P., Polydorou, C. K., Koupparis, M. A. and Efstathiou, C. E. (2001) Automated flow injection fluorimetric determination and dissolution studies of bumetanide in pharmaceuticals. *Analytica Chimica Acta*, 438, 131-136.

[14] Solich, P., Polasek, M. and Karlicek, R. (1995) Sequential flow injection spectrofluorimetric determination of coumarins using a double-injection single-line system. *Analytica Chimica Acta*, 308, 293-298.

[15] Legnerova, Z., Sklenarova, H. and Solich, P. (2002) Determination of rhodamine 123 by sequential injection technique for pharmacokinetic studies in the rat placenta. *Talanta*, 58, 1151-1155.

[16] Klimundova, J., Mervartova, K., Sklenarova, H. and Solich, P. (2006) Automated sequential injection fluorimetric set-up for multiple release testing of topical formulation. *Analytica Chimica Acta*, 573-574, 366-370.

[17] Legnerova, Z., Huclova, J., Thun, R. and Solich, P. (2004) Sensitive fluorimetric method based on sequential injection analysis technique used for dissolution studies and quality control of prazosin hydrochloride in tablets. *Journal of Pharmaceutical and Biomedical Analysis*, 34, 115-121.

[18] Balaguer, A., Chisvert, A. and Salvador, A. (2006) Sequential-injection determination of traces of disodium phenyl dibenzimidazole tetrasulphonate in urine from users of sunscreens by online solid-phase extraction coupled with a fluorimetric detector. *Journal of Pharmaceutical and Biomedical Analysis*, 40, 922-927.

[19] Vidal, M. T., Chisvert, A. and Salvador, A. (2003) Sensitive sequential-injection system for the determination of 2-phenylbenzimidazole-5-sulphonic acid in human urine samples using on-line solid-phase extraction coupled with fluorimetric detection. *Talanta*, 59, 591-599.

[20] Bravo, J. C., Fernández, P. and Durand, J. S. (2005) Flow injection fluorimetric determination of β-estradiol using a molecularly imprinted polymer. *Analyst*, 130, 1404-1409.

[21] Sacenon, J. F. and De la Guardia, M. (1994) Micellar enhanced fluorimetric determination of carbendazim in natural waters. *Analytica Chimica Acta*, 287, 49-57.

[22] Zisiou, E.-P., Pinto, P. C. A. G., Saravia, M. L. M. F. S., Siquet, C. and Lima, J. L. F. C. (2005) Sensitive sequential injection determination of naproxen based on interaction with β-cyclodextrin. *Talanta*, 68, 226-230.

[23] Tang, L. X. and Rowell, F. J. (1998) Rapid determination of warfarin by sequential injection analysis with cyclodextrin-enhanced fluorescence detection. *Analytical Letters*, 31, 891-901.

[24] Amador-Hernández, J., Fernández-Romero, J. M. and Luque de Castro, M. D. (2001) Flow injection screening and semiquantitative determination of polycyclic aromatic hydrocarbons in water by laser

induced spectrofluorometry—chemometrics. *Analytica Chimica Acta*, 448, 61-69.

[25] Bradley, A. B. and Zare, R. N. (1976) Laser fluorimetry. Sub-part-per-trillion detection of solutes. *Journal of the American Chemical Society*, 98, 620-621.

[26] Richardson, J. H., Tallin, B. W., Jonson, D. C. and Hrubesh, L. W. (1976) Sub-part-per-trillion detection of riboflavin by laser-induced fluorescence. *Analytica Chimica Acta*, 86, 263-267.

[27] Richardson, J. H. (1977) Sensitive assay of biochemicals by laser-induced molecular fluorescence. *Analytical Biochemistry*, 83, 754-762.

[28] Huie, C. W., Airen, J. H. and Williams, W. R. (1991) Rapid screening of porphyrins using flow injection analysis and visible laser fluorimetry. *Analytica Chimica Acta*, 254, 189-196.

[29] Imasaka, T., Higashijima, T. and Ishibashi, N. (1991) Dehydrogenase and ethanol assay based on visible semiconductor laser spectrometry. *Analytica Chimica Acta*, 251, 191-195.

[30] Santana-Rodríguez, J. J., Halko, R., Batancort-Rodríguez, J. R. and Aaron, J. J. (2006) Environmental analysis based on luminescence in organized supramolecular systems. *Analytical and Bioanalytical Chemistry*, 385, 525-545.

[31] Galeano-Díaz, T., Acedo-Valenzuela, M. I. and Salinas, F. (1999) Determination of he pesticide Naptalam, at the ppb level, by FIA with fluorimetric detection and on-line preconcentration by solid-phase extraction on C_{18} modified silica. *Analytica Chimica Acta*, 384, 185-191.

[32] Coly, A. and Aaron, J. J. (1998) Fluorimetric analysis of pesticides: methods, recent developments and applications. *Talanta*, 46, 815-843.

[33] Zhang, Z.-Q., Ma, J., Lei, Y. and Lu, Y.-M. (2007) Flow-injection on-line oxidizing fluorimetry and solid-phase extraction for determination of thioridazine hydrochloride in human plasma. *Talanta*, 71, 2056-2061.

[34] Rezaei, B., Ensafi, A. A. and Nouroozi, S. (2005) Flow-injection determination of ascorbic acid and cysteine simultaneously with spectrofluorometric detection. *Analytical Sciences*, 21, 1067-1071.

[35] Martínez-Calatayud, J., Gómez-Benito, C. and Gaspar-Gimenez, D. (1990) FIA-fluorimetric determination of thiamine. *Journal of Pharmaceutical and Biomedical Analysis*, 8, 667-670.

[36] Pérez-Ruiz, T, Martinez-Lozano, C., Tomás, V. and Sidrach de Cardona, C. (1996) Flow-injection fluorimetric determination of penicillamine and tiopronin in pharmaceutical preparations. *Journal of Pharmaceutical and Biomedical Analysis*, 15, 33-38.

[37] Dolejsova, J., Solich, P., Polydorou, C. K., Koupparis, M. A. and Efstathiou, C. E. (1999) Flow-injection fluorimetric determination of 1, 4-benzodiazepines in pharmaceutical formulations after acid hydrolysis. *Journal of Pharmaceutical and Biomedical Analysis*, 20, 357-362.

[38] Perez-Ruiz, T., Martinez-Lozano, C., Sanz, A. and Guillen, A. (2004) Successive determination of thiamine and ascorbic acid in pharmaceuticals by flow injection analysis. *Journal of Pharmaceutical and Biomedical Analysis*, 34, 551-557.

[39] Laredo Ortiz, S., Gómez-Bendito, C. and Martínez-Calatayud, J. (1993) Determination of fluphenazine hydrochloride in a flow assembly incorporating cerium (IV) arsenite as a solid-bed sector. *Analytica Chimica Acta*, 276, 281-286.

[40] Kojlo, A. and Martínez-Calatayud, J. (1995) Spectrofluorimetric flow injection determination of adrenaline with an iodine solid-phase reactor. *Analytica Chimica Acta*, 308, 334-338.

[41] García-Bautista, J. A., García-Mateo, J. V. and Martínez-Calatayud, J. (1998) Spectrofluorimetric determination of iproniazid and isoniazid in a FIA system provided with a solid-phase

reactor. *Analytical Letters*, 31, 1209-1218.

[42] Zhang, Z.-Q., Liang, G.-X., Ma, J., Lei, Y. and Lu, Y.-M. (2006) A sensitive flow injection fluorimetry for the determination of carbamazepine in human plasma. *Analytical Letters*, 39, 2417-2428.

[43] Pinto, P. C. A. G., Saraiva, M. L. M. F. S., Santos, J. L. M. and Lima, J. L. F. C. (2005) A pulsed sequential injection analysis flow system for the fluorimetric determination of indomethacin in pharmaceutical preparations. *Analytica Chimica Acta*, 539, 173-179.

[44] Pérez-Ruiz, T., Martínez-Lozano, C., Marín, J. and García, M. D. (2006) Automatic determination of phylloquinone in vegetables and fruits using on-line photochemical reduction and fluorescence detection via solid-phase extraction and flow injection. *Analytical and Bioanalytical Chemistry*, 384, 280-285.

[45] Irace-Guigand, S., Leverend, E., Seye, M. D. G. and Aaron, J.-J. (2005) A new online micellar-enhanced photochemically-induced fluorescence method for determination of phenylurea herbicide residues in water. *Luminescence*, 20, 138-142.

[46] García-Campaña, A. M., Aaron, J.-J. and Bosque-Sendra, J. M. (2001) Micellar-enhanced photochemically induced fluorescence detection of chlorophenoxyacid herbicides. Flow injection analysis of mecoprop and 2,4-dichlorophenoxyacetic acid. *Talanta*, 55, 531-539.

[47] Vílchez, J. L., Valencia, M. C., Navalón, A., Molinero-Morales, B. and Capitán-Vallvey, L. F. (2001) Flow injection analysis of the insecticide imidacloprid in water samples with photochemically induced fluorescence detection. *Analytica Chimica Acta*, 439, 299-305.

[48] Coly, A. and Aaron, J.-J. (1999) Sensitive and rapid flow injection analysis of sulfonylurea herbicides in water with micellar-enhanced photochemically induced fluorescence detection. *Analytica Chimica Acta*, 392, 255-264.

[49] Cydzik, I., Albert-García, J. R. and Martinez-Calatayud, J. (2007) Photo-induced fluorescence of fluometuron in a continuous-flow multicommutation assembly. *Journal of Fluorescence*, 17, 29-36.

[50] Pérez-Ruiz, T., Martínez-Lozano, C., Tomas, V. and Martin, J. (2001) Flow injection determination of methamidophos using on-line photooxidation and fluorimetric detection. *Talanta*, 54, 989-995.

[51] Pérez-Ruiz, T., Martínez-Lozano, C., Tomas, V. and Martin, J. (2002) Flow injection spectrofluorimetric determination of malathion in environmental samples using on-line photooxidation. *Analytical Letters*, 35, 1239-1250.

[52] Segarra Guerrero, R., Gómez Benito, C. and Martínez Calatayud, J. (1993) On-line photoreaction and fluorimetric determination of diazepam. *Journal of Pharmaceutical and Biomedical Analysis*, 11, 1357-1360.

[53] Pérez-Ruiz, T., Martínez-Lozano, C., Tomas, V. and Sidrach, C. (1997) Flow injection fluorimetric determination of ascorbic acid based on its photooxidation by thionine Blue. *Analyst*, 122, 115-118.

[54] Nair, M. B., Aaron, J. J., Prognon, P. and Mahuzier, G. (1998) Photochemically induced fluorimetric detection of tianeptine and some of its metabolites. Application to pharmaceutical preparation. *Analyst*, 123, 2267-2270.

[55] Amador-Hernández, J., Fernández-Moreno, J. M. and Luque de Castro, M. D. (2001) Continuous determination of chloroquine in plasma by laser-induced photochemical reaction and fluorescence.

Fresenius' Journal of Analytical Chemistry, 369, 438-441.

[56] Chen, H. and He, Q. (2000) Flow injection spectrofluorimetric determination of reserpine in tablets by on-line acetone sensitized photochemical reaction. Talanta, 53, 463-469.

[57] Pérez-Ruiz, T., Martínez-Lozano, C., Tomas, V. and Martin, J. (2002) Fluorimetric determination of arsanilic acid by flow injection analysis using online photo-oxidation. Analytical and Bioanalytical Chemistiy, 372, 387-390.

[58] Mellado-Romero, A., Gómez-Bendito, C. and Martínez-Calatayud, J. (1992) Photochemical derivatization and fluorimetric determination of promethazine in a FIA Assembly. Analytical Letters, 25, 1289-1308.

[59] Laasis, B. and Aaron, J.-J. (1997) Flow-injection fluorimetric analysis of several aminophenothiazines based on photooxidation. Anahtsis, 25, 183-188.

[60] Pérez-Ruiz, T., Martínez-Lozano, C., Tomás, V. and Martín, J. (2004) Flowinjection fluorimetric determination of menadione using on-line photo-reduction in micellar media. Analytica Chimica Acta, 514, 259-264.

[61] Resing, J. A. and Measures, C. I. (1994) Fluorometric determination of Al in seawater by flow injection analysis with in-line preconcentration. Analytical Chemistiy, 66, 4105-4111.

[62] Economou, A., Themelis, D. G., Bikou, H., Tzanavaras, P. D. and Rigas, P. G. (2004) Determination of boron in water and pharmaceuticals by sequential-injection analysis and fluorimetric detection. Analytica Chimica Acta, 510, 219-224.

[63] Brach-Papa, C., Coulomb, B., Boudenne, J.-L., Cerdá, V. and Theraulaz, F. (2002) Spectrofluorimetric determination of aluminum in drinking waters by sequential injection analysis. Analytica Chimica Acta, 457, 311-318.

[64] Sanchez-Rojas, F., Cristofol Alcaraz, E. and Cano Pavon, J. M. (1994) Determination of aluminium in water by flow injection with fluorimetric detection by using salicylaldehyde carbohydrazone as reagent in a micellar medium. Analyst, 119, 1221-1223.

[65] Brach-Papa, C., Coulomb, B., Branger, C., Margaillan, A., Theraulaz, F., Van Loot, P. and Boudenne, J.-L. (2004) Fluorimetric determination of aluminium in water by sequential injection through column extraction. Analytical and Bioanalytical Chemistry, 378, 1652-1658.

[66] De Armas, G., Cladera, A., Becerra, E., Estela, J. M. and Cerdá, V. (2000) Fluorimetric sequential injection determination of magnesium using 8-hydroxiquinoline-5-sulfonic acid in a micellar medium. Talanta, 52, 77-82.

[67] Gañan-Gutierrez, N., Sánchez-Rojas, F. and Cano-Pavón, J. M. (1991) Determination of zinc by flow injection with fluorimetric detection in a micellar medium. Fresenius' Journal of Analytical Chemistiy, 355, 88-91.

[68] Compañó, R., Ferrer, R., Guitera, J. and Prat, M. D. (1996) Flow injection method for the fluorimetric determination of Zn with 8-(benzenesulphonamido) quinoline. Mikrochimica Acta, 124, 73-79.

[69] Paleologos, E. K., Vlessidis, A. G., Karayannis, M. I. and Veltsistas, P. G. (2001) Nonaqueous catalytic fluorometric trace determination of vanadium based on tire pyronine B-hydrogen peroxide reaction and flow injection after cloud point extraction. Analytical Chemistry, 73, 4428-4433.

[70] Watson, R. J., Butler, E. C. V., Clementson, L. A. and Berry, K. M. (2005) Flow-injection analysis with fluorescence detection for tire determination of trace levels of ammonium in seawater. Journal of Environmental Monitoring, 7, 37-42.

[71] Recalde, D. L., Andres-García, E. and Díaz-Garcia, M. E. (2000) Fluorimetric flow injection and flow-through sensing systems for cyanide control in waste water. *Analyst*, 125, 2100-2105.

[72] Pinto, P. C. A. G., Saraiva, M. L. F. F. S., Santos, J. L. M. and Lima, J. L. F. C. (2006) Fluorimetric determination of aminocaproic acid in pharmaceutical formulations using a sequential injection analysis system. *Talanta*, 68, 857-862.

[73] López-Error, C., Viñas, P., Cerdan, F. J. and Hernández-Córdoba, M. (2000) Determination of clenbuterol in pharmaceutical preparations by reaction with o-phthalaldehyde using a flow injection fluorimetric procedure. *Talanta*, 53, 47-53.

[74] Pérez-Ruiz, T., Martínez-Lozano, C., Tomas, V. and Carpena, J. (1996) Fluorimetric determination of chloroxine using manual and flow injection methods. *Journal of Pharmaceutical and Biomedical Analysis*, 14, 1505-1511.

[75] López-Erroz, C., Viñas, P. and Hernández-Córdoba, M. (1994) Flow-injection fluorimetric analysis of sulfamethoxazole in pharmaceutical preparations and biological fluids. *Talanta*, 41, 2159-2164.

[76] López-Erroz, C., Viñas, P., Campillo, N. and Hernández-Córdoba, M. (1996) Flow injection-fluorimetric method for the determination of ranitidine in pharmaceutical preparations using o-phthalaldehyde. *Analyst*, 121, 1043-1046.

[77] Kuijt, J., Ariese, F., Brinkman, U. A. Th. and Gooijer, C. (2003) Room temperature phosphorescence in the liquid state as a tool in analytical chemistry. *Analytica Chimica Acta*, 488, 135-171.

[78] Sánchez-Barragán, I., Costa-Fernández, J. M., Valledor, M., Campo, J. C. and Sanz Medel, A. (2006) Room-temperature phosphorescence (RTP) for optical sensing. *Trends in Analytical Chemistry*, 25, 958-967.

[79] Cline-Love, L. J., Skrilec, M. and Habarta, J. G. (1980) Analysis of micelle-stabilized room temperature phosphorescence in solution. *Analytical Chemistry*, 52, 754-759.

[80] Díaz-García, M. E. and Sanz Medel, A. (1986) Facile chemical deoxygenation of micellar solutions for room temperature phosphorescence. *Analytical Chemistry*, 58, 1436-1439.

[81] Liu, Y. M., Fernández de la Campa, M. R., Díaz-García, M. E. and Sanz Medel, A. (1990) Phosphorescence detection in flowing systems: selective determination of aluminium by flow injection liquid room-temperature phosphorimetry. *Analytica Chimica Acta*, 234, 233-238.

[82] Weinberger, R., Yarmchuk, P. and Cline Love, L. J. (1982) Liquid chromatographic phosphorescence detection with micellar chromatography and postcolumn reaction modes. *Analytical Chemistry*, 54, 1552-1558.

[83] Muñoz de la Peña, A., Mahedero, M. C. and Bautista-Sánchez, A. (2000) Room temperature phosphorescence in cyclodextrins. *Analytical Applications Analysis*, 28, 670-678.

[84] Li, L., Chen, Y., Zhao, Y. and Tong, A. (1997) Room-temperature phosphorescence of dansyl chloride solution in the absence of protective medium and its medium effect. *Analytica Chimica Acta*, 341, 241-249.

[85] Li, L., Zhao, Y., Yu, Y. and Tong, A. (1998) Non-protected fluid room temperature phosphorescence of several naphthalene derivatives. *Talanta*, 46, 1147-1154.

[86] Segura-Carretero, A., Cruces-Blanco, C., Cañabate-Díaz, B. and Fernández-Gutierrez, A. (1998) An innovative way of obtaining room-temperature phosphorescence signals in solution. *Analytica Chimica Acta*, 361, 217-222.

[87] Segura-Carretero, A., Salinas-Castillo, A. and Fernández-Gutierrez, A. (2005) A review of

heavy-atom-induced room-temperature phosphorescence a straightforward phosphorimetric method. *Critical Reviews in Analytical Chemistry*, 35, 3-14.

[88] Cañabate-Díaz, B., Casado-Terrones, S., Segura-Carretero, A., Costa-Fernández, J. M. and Fernández-Gutierrez, A. (2004) Comparison of three phosphorescent methodologies in solution for the analysis of naphazoline in pharmaceutical preparations. *Analytical and Bioanalytical Chemistry*, 379, 30-34.

[89] Fernández-González, A., Badía, R. and Díaz-García, M. E. (2003) Sensitive flow injection system for nafcillin determination based on non-protected room temperature phosphorescence. *Analytica Chimica Acta*, 498, 69-77.

[90] Fletcher, P., Andrew, K. N., Calokerinos, A. C., Forbes, S. and Worsfold, P. J. (2001) Analytical applications of flow injection with chemiluminescence detection-a review. *Luminescence*, 16, 1-23.

[91] Bowie, A. R., Sanders, M. G. and Worsfold, P. J. (1996) Analytical applications of liquid phase chemiluminescence reactions-a review. *Journal of Bioluminescence and Chemiluminescence*, 11, 61-90.

[92] García-Campaña, A. M. and Baeyens, W. R. G. (2000) Principles and recent analytical applications of chemiluminescence. *Analysis*, 28, 686-698.

[93] Palilis, L. A. and Calokerinos, A. C. (2000) Analytical applications of chemiluminogenic reactions. *Analytica Chimica Acta*, 413, 175-186.

[94] Qin, W. (2002) Flow injection chemiluminescence-based chemical sensors. *Analytical Letters*, 35, 2207-2220.

[95] Lahuerta-Zamora, L., Fuster-Mestre, Y, Duart, M. J., Antón-Fos, G. M., García-Domenech, R., Gálvez Álvarez, J. and Martínez-Calatayud, J. (2001) prediction of the chemiluminescent behavior of pharmaceuticals and pesticides. *Analytical Chemistry*, 73, 4301-4306.

[96] Tsunoda, M. and Imai, K. (2005) Analytical applications of peroxyoxalate chemiluminescence. *Analytica Chimica Acta*, 541, 13-23.

[97] Gerardi, R. D., Barnett, N. W. and Lewis, S. W. (1999) Analytical applications of tris (2,2'-bipyridyl) ruthenium (III) as a chemiluminescent reagent. *Analytica Chimica Acta*, 378, 1-41.

[98] Campiglio, A. (1998) Determination of naproxen with chemiluminescence detection. *Analyst*, 123, 1571-1574.

[99] Koukli, L. L., Calokerinos, A. C. and Hadjiioannou, T. P. (1989) Continuous-flow chemiluminescence determination of acetaminophen by reduction of cerium (IV). *Analyst*, 114, 711-714.

[100] Rao, Y., Tong, Y., Zhang, X. R., Luo, G. A. and Baeyens, W. R. G. (2000) Determination of ofloxacin using a chemiluminescence flow injection method. *Analytica Chimica Acta*, 416, 227-230.

[101] Aly, F. A., Al-Tamimi, S. A. and Alwarthan, A. A. (2000) Determination of flufenamic acid and mefenamic acid in pharmaceutical preparations and biological fluids using flow injection analysis with tris (2,2'-bipyridyl) ruthenium (II) chemiluminescence detection. *Analytica Chimica Acta*, 416, 87-96.

[102] Cui, H., Li, S. F., Li, F., Sun, Y. G. and Lin, X. Q. (2002) A novel chemiluminescent method for the determination of salicylic acid in bactericidal solutions. *Analytical and Bioanalytical Chemistry*, 372, 601-604.

[103] Capitan-Vallvey, L. F., Valencia-Miron, M. C. and Acosta, R. A. (2000) Chemiluminescence determination of sodium 2-mercaptoethane sulfonate by flow injection analysis using cerium (IV) sensitized by quinine. *Talanta*, 51, 1155-1161.

[104] Zhang, Z. D., Baeyens, W. R. G., Zhang, X. R. and Weken, G. V. D. (1996) Chemiluminescence flow injection analysis of captopril applying a sensitized rhodamine 6G method. *Journal of*

Pharmaceutical and Biomedical Analysis, 14, 939-945.

[105] Wang, X., Zhao, H. C., Nie, L. H., Jin, L. P. and Zhang, Z. L. (2001) Europium sensitized chemiluminescense determination of rufloxacin. *Analytica Chimica Acta*, 445, 169-175.

[106] Hindson, B. J. and Barnett, N. W. (2001) Analytical applications of acidic potassium permanganate as a chemiluminescence reagent. *Analytica Chimica Acta*, 445, 1-19.

[107] Chen, J. and Fang, Y. (2007) Flow injection technique for biochemical analysis with chemiluminescence detection in acidic media. *Sensors*, 7, 448-458.

[108] Deftereos, N. T., Grekas, N. and Calokerinos, A. C. (2000) Flow injection chemiluminometric determination of albumin. *Analytica Chimica Acta*, 403, 137-143.

[109] Li, B. X., Zhang, Z. J. and Zhao, L. X. (2002) Flow-injection chemiluminescence detection for studying protein binding for drug with ultrafiltration sampling. *Analytica Chimica Acta*, 468, 65-70.

[110] Barnett, N. W., Hindson, B. J. and Lewis, S. W. (1998) Determination of 5-hydroxytryptamine (serotonin) and related indoles by flow injection analysis with acidic potassium permanganate chemiluminescence detection. *Analytica Chimica Acta*, 362, 131-139.

[111] Liang, Y.-D., Song, J.-F., Yang, Z.-F. and Guo, W. (2004) Flow-injection chemiluminescence determination of chloroquine using peroxynitrous acid as oxidant. *Talanta*, 62, 757-763.

[112] Aly, F. A., Alarfaffj, N. A. and Alwarthan, A. A. (1998) Permanganate-based chemiluminescence analysis of cefadroxil monohydrate in pharmaceutical samples and biological fluids using flow injection. *Talanta*, 48, 471-178.

[113] Catalá-Icardo, M., Misiewicz, M., Ciucu, A., García-Mateo, J. V. and Martínez-Calatayud, J. (2003) FI-on-line photochemical reaction for direct chemiluminescence determination of photodegradated chloramphenicol. *Talanta*, 60, 405-414.

[114] Fuster-Mestre, Y., Fernández-Band, B., Lahuerta-Zamora, L. and Martínez-Calatayud, J. (1999) Flow injection analysis-direct chemiluminescence determination of ergonovine maleate enhanced by hexadecylpyridinium chloride. *Analyst*, 124, 413-416.

[115] Townshend, A., Wheatley, R. A., Chisvert, A. and Salvador, A. (2002) Flow injection-chemiluminescence determination of octyl dimethyl PABA in sunscreen formulations. *Analytica Chimica Acta*, 462, 209-215.

[116] Sanfeliu Alonso, M. C., Lahuerta Zamora, L. and Martínez-Calatayud, J. (2001) Flow-injection with chemiluminescence detection for the determination of iproniazid. *Analytica Chimica Acta*, 437, 225-231.

[117] Campiglio, A. (1998) Chemiluminescence determination of naltrexone based on potassium permanganate oxidation. *Analyst*, 123, 1053-1056.

[118] Costin, J. W., Barnett, N. W, Lewis, S. W. and McGillivery, D. J. (2003) Monitoring the total phenolic/antioxidant levels in wine using flow injection analysis with acidic potassium permanganate chemiluminescence detection. *Analytica Chimica Acta*, 499, 47-56.

[119] Townshend, A., Murillo-Pulgarín, J. A. and Alañon-Pardo, M. T. (2003) Flow injection-chemiluminescence determination of propranolol in pharmaceutical preparations. *Analytica Chimica Acta*, 488, 81-88.

[120] Fuster-Mestre, Y., Lahuerta-Zamora, L. and Martínez-Calatayud, J. (1999) Direct flow injection chemiluminescence determination of salicylamide. *Analytica Chimica Acta*, 394, 159-163.

[121] Costin, J. W., Francis, P. S. and Lewis, S. W. (2003) Selective determination of amino acids

using flow injection analysis coupled with chemiluminescence detection. *Analytica Chimica Acta*, 480, 67-77.

[122] Palilis, L. P., Calokerinos, A. C. and Grekas, N. (1996) Chemiluminescence arising from the oxidation of bilirubin in aqueous media. *Analytica Chimica Acta*, 333, 267-275.

[123] Zhang, Z. D., Baeyens, W. R. G., Zhang, X. R. and VanDerWeken, G. (1996) Chemiluminescence flow injection analysis of captopril applying a sensitized rhodamine 6G method. *Journal of Pharmaceutical and Biomedical Analysis*, 14, 939-945.

[124] Nozaki, O., Iwaeda, T., Moriyama, H. and Kato, Y. (1999) Chemiluminescent detection of catecholamines by generation of hydrogen peroxide with imidazole. *Luminescence*, 14, 123-127.

[125] Fan, W. Z. and Zhang, Z. J. (1996) Determination of acetylcholine and choline in rat brain tissue by FIA with immobilized enzymes and chemiluminescence detection. *Microchemical Journal*, 53, 290-295.

[126] Marques, K. L., Santos, J. L. M. and Lima, J. L. F. C. (2004) Multicommutated flow system for the chemiluminometric determination of clomipramine in pharmaceutical preparations. *Analytica Chimica Acta*, 518, 31-36.

[127] Ma, Y., Zhou, M., Jin, X., Zhang, Z., Teng, X. and Chen, H. (2004) Flowinjection chemiluminescence assay for ultra-trace determination of DNA using rhodamine B-Ce (IV) -DNA ternary system in sulfuric acid media. *Analytica Chimica Acta*, 501, 25-30.

[128] Wada, M., Kuroda, N., Akiyama, S. and Nakashima, K. (1997) A sensititve and rapid FIA with an immobilized enzyme column reactor and peroxyoxalate chemiluminescence detection for the determination of total D-amino acids in human plasma. *Analytical Sciences*, 13, 945-950.

[129] Nozaki, O., Iwaeda, T. and Kato, Y. (1996) Amines for detection of dopamine by generation of hydrogen peroxide and peroxyoxalate chemiluminescence. *Journal of Bioluminescence and Chemiluminescence*, 11, 309-313.

[130] Fang, Q., Shi, X. T., Sun, Y. Q. and Fang, Z. L. (1997) A flow injection microdialysis sampling chemiluminescence system for *in vivo* on-line monitoring of glucose in intravenous and subcutaneous tissue fluid microdialysates. *Analytical Chemistry*, 69, 3570-3577.

[131] Yaqoob, M., Nabi, A. and Masoom, Y. M. (1997) Flow-injection chemiluminescent determination of glycerol-3-phosphate and glycerophosphorylcholine using immobilized enzymes. *Journal of Bioluminescence and Chemiluminescence*, 12, 1-5.

[132] Wang, L., Li, Y., Zhao, D. and Zhu, C. (2003) A Novel Enhancing Flow-Injection Chemiluminescence Method for the Determination of Glutathione Using the Reaction of Luminol with Hydrogen Peroxide. *Microchimica Acta*, 141, 41-45.

[133] Alapont, A. G., Jiménez, E. A., Zamora, L. L. and Martinez-Calatayud, J. (1998) Inhibition of the system luminol-H_2O_2-Fe$(CN)_6^{3-}$ chemiluminescence by the Mn (II) indirect determination of isoniazid in a pharmaceutical formulation. *Journal of Bioluminescence and Chemiluminescence*, 13, 131-137.

[134] Liu, H., Ren, J., Hao, Y., Ding, H., He, P. and Fang, Y. (2006) Determination of metoprolol tartrate in tablets and human urine using flow injection chemiluminescence method. *Journal of Pharmaceutical and Biomedical Analysis*, 42, 384-388.

[135] García-Campaña, A. M., Bosque-Sendra, J. M., Bueno-Vargas, M. P., Baeyens, W. R. G. and Zhang, X. (2004) Flow injection analysis of oxymetazoline hydrochloride with inhibited chemiluminescent detection. *Analytica Chimica Acta*, 516, 245-249.

[136] Aly, F. A., Alarfaj, N. A. and Alwarthan, A. A. (1998) Flow-injection chemiluminometric

determination of some phenothiazines in dosage forms and biological fluids. *Analytica Chimica Acta*, 358, 255–262.

[137] Costin, J. W., Barnett, N. W. and Lewis, S. W. (2004) Determination of proline in wine using flow injection analysis with tris (2,2′-bipyridyl) ruthenium (II) chemiluminescence detection. *Talanta*, 64, 894–898.

[138] Ping, L. Z., An, L. K. and Yang, T. S. (1998) Microdetermination of proteins with the 1,10-phenanthroline–H_2O_2–cetyltrimethylammonium bromide–Cu (II) chemiluminescence system. *Microchemical Journal*, 60, 217–223.

[139] Alwarthan, A. A. and Aly, F. A. (1998) Chemiluminescent determination of pyridoxine hydrochloride in pharmaceutical samples using flow injection. *Talanta*, 45, 1131–1138.

[140] Du, J., Li, Y. and Lu, J. (2001) Flow Injection Chemiluminescence Determination of Rutin based on its enhancing effect on the luminol – ferricyanide/ferrocyanide system. *Analytical Letters*, 34, 1741–1748.

[141] Li, S., Chen, H., Wei, X., Lu, X. and Zhang, L. (2006) Determination of tannic acid by flow injection analysis with inhibited chemiluminescence detection. *Microchimica Acta*, 155, 427–430.

[142] Wang, Z., Zhang, Z., Fu, Z. and Zhang, X. (2004) Sensitive flow injection chemiluminescence determination of terbutaline sulfate based on enhancement of the luminol–permanganate reaction. *Analytical and Bioanalytical Chemistry*, 378, 834–840.

[143] Zhang, X. R., Baeyens, W. R. G., Vandenborre, A., VanDerWeken, G., Calokerinos, A. C. and Schulman, S. G. (1995) Chemiluminescence determination of tetracyclines based on their reaction with hydrogen peroxide catalysed by the copper ion. *Analyst*, 120, 463–466.

[144] Song, Z. and Hou, S. (2002) Determination of picomole amounts of thiamine through flow injection analysis based on the suppression of luminol – KIO_4 chemiluminescence system. *Journal of Pharmaceutical and Biomedical Analysis*, 28, 683–691.

[145] Song, S., Song, Q. J. and Chen, Z. (2007) On – line phototransformation – flow injection chemiluminescence determination of triclosan. *Analytical and Bioanalytical Chemistry*, 387, 2917–2922.

[146] Alwarthan, A. A. (1995) Chemiluminescent determination of tryptophan in a flow injection system. *Analytica Chimica Acta*, 317, 233–237.

[147] Song, Z. and Hou, S. (2002) Chemiluminescence assay for uric acid in human serum and urine using flow injection with immobilized reagents technology. *Analytical and Bioanalytical Chemistry*, 372, 327–332.

[148] Song, Z. and Hou, S. (2003) Subpicogram determination of Vitamin B_{12} in pharmaceuticals and human serum using flow injection with chemiluminescence detection. *Analytica Chimica Acta*, 488, 71–79.

[149] Zamzow, H., Coale, K. H., Johnson, K. S. and Sakamoto, C. M. (1998) Determination of copper complexation in seawater using flow injection analysis with chemiluminescence detection. *Analytica Chimica Acta*, 377, 133–144.

[150] Zhou, Y. X. and Zhu, G. Y. (1997) Rapid automated in–situ monitoring of total dissolved iron and total dissolved manganese in underground water by reverse–flow injection with chemiluminescence detection during the process of water treatment. *Talanta*, 44, 2041–2049.

[151] Qin, W., Zhang, Z. J. and Wang, F. C. (1998) Chemiluminescence flow system for the determination of Fe (II) and Fe (III) in water. *Fresenius' Journal of Analytical Chemistry*, 360, 130–132.

[152] Qin, W., Zhang, Z. J., Li, B. X. and Liu, S. N. (1998) Chemiluminescence flow–sensing

system for hydrogen peroxide with immobilized reagents. *Analytica Chimica Acta*, 372, 357-363.

[153] Mikuska, P., Vecera, Z. and Zdrahal, Z. (1995) Flow – injection chemiluminescence determination of ultra low concentrations of nitrite in water. *Analytica Chimica Acta*, 316, 261-268.

[154] Qi, H., Lv, J. and Li, B. (2007) Determination of phenol at ng l^{-1} level by flow injection chemiluminescence combined with on-line solid-phase extraction. *Spectrochimica Acta Part A*, 66, 874-878.

[155] Ikebuko, K., Wakamura, H., Karube, I., Kubo, I., Inagawa, M., Sugawara, T., Arikawa, Y., Suzuki, M. and Takeuchi, T. (1996) Phosphate sensing system using pyruvate oxidase and chemiluminescence detection. *Biosensors & Bioelectronics*, 11, 959-965.

[156] Qin, W., Zhang, Z. J. and Zhang, C. J. (1997) Chemiluminescence flow system for vanadium (v) With immobilized reagents. *Analyst*, 122, 685-688.

[157] Nakano, S., Sakamoto, K., Takenobu, A. and Kawashima, T. (2002) Flow – injection chemiluminescent determination of vanadium (IV) and total vanadium by means of catalysis on the periodate oxidation of purpurogallin. *Talanta*, 58, 1263-1270.

[158] Song, Z., Hou, S. and Zhang, N. (2002) A new green analytical procedure for monitoring sub-nanogram amounts of chlorpyrifos on fruits using flow injection chemiluminescence with immobilized reagents. *Journal of Agricultural and Food Chemistry*, 50, 4468-4474.

[159] Kiba, N., Inagaki, J. and Furusawa, M. (1995) Chemiluminometric flow injection method for determination of free L-malate in wine with co-immobilized malate dehydrogenase/NADH oxidase. *Talanta*, 42, 1751-1755.

[160] Huang, Y. M., Zhang, C., Zhang, X. R. and Zhang, Z. J. (1999) Chemiluminescence of sulfite based on auto-oxidation sensitized by rhodamine 6G. *Analytica Chimica Acta*, 391, 95-100.

[161] Liu, Y, Song, Z., Dong, F. and Zhang, L. (2007) Flow injection chemiluminescence determination of Sudan I in hot chilli sauce. *Journal of Agricultural and Food Chemistry*, 55, 614-617.

[162] Cui, H., Li, Q., Meng, R., Zhao, H. Z. and He, C. X. (1998) Flow injection analysis of tannic acid with inhibited chemiluminescent detection. *Analytica Chimica Acta*, 362, 151-155.

[163] Xiong, Y., Zhou, H., Zhang, Z., He, D. and He, C. (2006) Molecularly imprinted on-line solid-phase extraction combined with flow injection chemiluminescence for the determination of tetracycline. *Analyst*, 131, 829-834.

[164] Zheng, X. W. and Zhang, Z. J. (2002) Flowinjection chemiluminescence detecting sulfite with *in situ* electrogenerated Mn^{3+} as the oxidant. *Sensors and Actuators B*, 84, 142-147.

[165] Zhang, C., Huang, J., Zhang, Z. and Aizawa, M. (1998) Flow injection chemiluminescence determination of catecholamines with electrogenerated hypochlorite. *Analytica Chimica Acta*, 374, 105-110.

[166] Zhu, L, Li, Y. and Zhu, G. (2002) Flow injection determination of dopamine based on inhibited electrochemiluminescence of luminol. *Analytical Letters*, 35, 2527-2537.

[167] Tomita, I. N. and Bulhoes, L. O. S. (2001) Electrogenerated chemiluminescence determination of cefadroxil antibiotic. *Analytica Chimica Acta*, 442, 201-206.

[168] Zheng, X., Guo, Z. and Zhang, Z. (2001) Flow-injection electrogenerated chemiluminescence determination of epinephrine using luminol. *Analytica Chimica Acta*, 441, 81-86.

[169] Marquette, C. A., Leca, B. D. and Blum, L. J. (2001) Electrogenerated chemiluminescence of luminol for oxidase-based fibre-optic biosensors. *Luminescence*, 16, 159-165.

[170] Zhu, L., Li, Y. and Zhu, G. (2003) Electrochemiluminescent determination of L-cysteine with a flow injection analysis system. *Analytical Sciences*, 19, 575-578.

[171] Waseem, A., Yaqoob, M., Nabi, A. and Greenway, G. M. (2007) Determination of thyroxine using tris (2, 2′ - bipyridyl) ruthenium (iii) - NADH enhanced electrochemiluminescence detection. *Analytical Letters*, 40, 1071-1083.

[172] Whitchurch, C. and Andrews, A. (2000) Development and characterization of a novel electrochemiluminescent reaction involving cadmium. *Analyst*, 125, 2065-2070.

[173] Lv, J., Zhang, Z. and Luo, L. (2003) An Online Galvanic Cell - generated Electrochemiluminescence and Flow Injection Determination of Calcium in Milk and Vegetables. *Analytical Sciences*, 19, 883-886.

[174] Gómez-Taylor, B., Palomeque, M., García-Mateo, J. V. and Martínez-Calatayud, J. (2006) Photoinduced chemiluminescence of pharmaceuticals. *Journal of Pharmaceutical and Biomedical Analysis*, 41, 347-357.

[175] Sahuquillo-Ricart, I., Antón-Fos, G. M., Duart, M. J., García-Mateo, J. V., Lahuerta-Zamora, L. and Martínez-Calatayud, J. (2007) Theoretical prediction of the photoinduced chemiluminescence of pesticides. *Talanta*, 72, 378-386.

[176] Palomeque, M., García - Bautista, J. A., Catalá - Icardó, M., García - Mateo, J. V. and Martínez - Calatayud, J. (2004) Photochemical - chemiluminometric determination of aldicarb in a fully automated multicommutation based flow-assembly. *Analytica Chimica Acta*, 512, 149-156.

[177] Chivulescu, A., Catalá-Icardó, M., García-Mateo, J. V. and Martínez-Calatayud, J. (2004) New flow-multicommutation method for the photo-chemiluminometric determination of the carbamate pesticide asulam. *Analytica Chimica Acta*, 519, 113-120.

[178] Roda, A., Pasini, P., Guardigli, M., Baraldini, M., Musiani, M. and Mirasoli, M. (2000) Bio-and chemiluminescence in bioanalysis. *Fresenius' Journal of Analytical Chemistry*, 366, 752-759.

[179] Roda, A., Guardigli, M., Michelini, E., Mirasoli, M. and Pasini, P. (2003) Analytical Bioluminescence and Chemiluminescence. *Analytical Chemistry*, 75, 462A-470A.

[180] Gamborg, G. and Hansen, E. H. (1994) Flow-injection bioluminescent determination of ATP based on the use of the luciferin-luciferase system. *Analytica Chimica Acta*, 285, 321-328.

[181] Blum, L. J., Gautier, S. M. and Coulet, P. R. (1993) Design of bioluminescence-based fiber optic sensors for flow injection analysis. *Journal of Biotechnology*, 31, 357-368.

[182] Yoshimura, K. (1987) Implementation of ion - exchanged absorptiometric detection in flow analysis systems. *Analytical Chemistry*, 59, 2922-2924.

[183] Molina-Díaz, A. Ruiz-Medina, A. Fernández-de-Córdova, M. L. (2002) The potential of flow-through optosensors in pharmaceutical analysis. *Journal of Pharmaceutical and Biomedical Analysis*, 29, 399-419.

[184] Cañizares, P. and Luque de Castro, M. D. (1995) Flow-through spectrofluorimetric sensor for the determination of glycerol in wine. *Analyst*, 120, 2837-2840.

[185] Llorent - Martínez, E. J., Satinsky, D. and Solich, P. (2007) Fluorescence optosensing implemented with sequential injection analysis: a novel strategy for the determination of labetalol. *Analytical and Bioanalytical Chemistry*, 387, 2065-2069.

[186] García Reyes, J. F., Ortega Barrales, P. and Molina Diaz, A. (2005) Rapid determination of diphenylamine residues in apples and pears with a single multicommuted fluorometric optosensor. *Journal of Agricultural and Food Chemistry*, 53, 9874-9878.

[187] García Reyes, J. F., Ortega Barrales, P. and Molina Díaz, A. (2007) Multicommuted

fluorometric multiparameter sensor for simultaneous determination of naproxen and salicylic acid in biological fluids. *Analytical Sciences*, 23, 423-427.

[188] Chen, D. , Luque de Castro, M. D. and Valcárcel, M. (1990) Flow-through sensor for the fluorimetric determination of cyanide. *Talanta*, 37, 1049-1055.

[189] Torre, M. , Fernández Gámez, F. , Lázaro, F. , Luque de Castro, M. D. and Valcárcel, M. (1991) Spectrofluorimetric flow-through sensor for the determination of beryllium in alloys. *Analyst*, 116, 81-83.

[190] Llorent-Martínez, E. J. , Šatínský, D. , Solich, P. , Ortega-Barrales, P. and Molina-Díaz, A. (2007) Fluorimetric SIA optosensing in pharmaceutical analysis: determination of paracetamol. *Journal of Pharmaceutical and Biomedical Analysis*, 45, 318-321.

[191] López-Flores, J. , Fernández de Córdova, M. L. and Molina-Díaz, A. (2005) Implementation of flow-through solid phase spectroscopic transduction with photochemically induced fluorescence: determination of thiamine. *Analytica Chimica Acta*, 535, 161-168.

[192] López Flores, J. , Molina Díaz, A. and Fernández de Córdova, M. L. (2007) Development of a photochemically induced fluorescence-based optosensor for the determination of imidacloprid in peppers and environmental waters. *Talanta*, 72, 991-997.

[193] García Reyes, J. F. , Llorent-Martínez, E. J. , Ortega Barrales, P. and Molina Díaz, A. (2004) Multiwavelength fluorescence based optosensor for simultaneous determination of fuberidazole, carbaryl and benomyl. *Talanta*, 62, 742-749.

[194] Llorent Martínez, E. J. , García Reyes, J. F. , Ortega Barrales, P. and Molina Díaz, A. (2006) A multicommuted fluorescence-based sensing system for simultaneous determination of Vitamins B_2 and B_6. *Analytica Chimica Acta*, 555, 128-133.

[195] Yoshimura, K. , Matsuoka, S. , Inokura, Y. and Hase, U. (1992) Flow analysis for trace amounts of copper by ion-exchanger phase absorptiometry with 4, 7-diphenyl 2, 9-dimethyl-1, 10-phenanthroline disulphonate and its application to the study of karst groundwater storm runoff. *Analytica Chimica Acta*, 268, 225-233.

[196] Matsuoka, S. , Yoshimura, K. and Tateda, A. (1995) Application of ion-exchanger phase visible light absorption to flow analysis. Determination of vanadium in natural water and rock. *Analytica Chimica Acta*, 317, 207-213.

[197] Ruedas Rama, M. J. , Ruiz Medina, A. and Molina Díaz, A. (2003) Bead injection spectroscopic flow-through renewable surface sensors with commercial flow cells as an alternative to reusable flow-through sensors. *Analytica Chimica Acta*, 482, 209-217.

[198] Ruzicka, J. (1994) Discovering flow injection: journey from sample to a live cell and from solution to suspension. *Analyst*, 119, 1925-1934.

[199] Ruzicka, J. and Scampavia, L. (1999) Bead injection-a novel approach. *Analytical Chemistry*, 71, 257A-263A.

[200] Ruedas Rama, M. J. , Ruiz Medina, A. and Molina Díaz, A. (2004) Implementation of flow-through multisensors with bead injection spectroscopy: fluorimetric renewable surface biparameter sensor for determination of berillium and aluminum. *Talanta*, 62, 879-886.

[201] Ruedas Rama, M. J. , Ruiz Medina, A. and Molina Díaz, A. (2005) A flow injection renewable surface sensor for the fluorimetric determination of vanadium (V) with Alizarin Red S. *Talanta*, 66, 1333-1339.

[202] Zhu, H., Chen, H. and Zhou, Y. (2003) Determination of thiamine in pharmaceutical preparations by sequential injection renewable surface solid-phase spectrofluorometry. *Analytical Sciences*, 19, 289-294.

[203] Pascual-Reguera, I., Guardia-Rubio, M. and Molina-Díaz, A. (2004) Native fluorescence flow-through optosensor for the fast determination of diphenhydramine in pharmaceuticals. *Analytical Sciences*, 20, 799-803.

[204] Ortega-Algar, S., Ramos-Martos, N. and Molina-Díaz, A. (2004) Fluorimetric flow-through sensing of quinine and quinidine. *Microchimica Acta*, 147, 211-214.

[205] Fernández-Sánchez, J. F., Segura-Carretero, A., Costa-Fernández, J. M., Bordel, N., Pereiro, R., Cruces-Blanco, C., Sanz-Medel, A. and Fernández-Gutierrez, A. (2003) Fluorescence optosensors based on different transducers for the determination of polycyclic aromatic hydrocarbons in water. *Analytical and Bioanalytical Chemistry*, 377, 614-623.

[206] Ortega-Algar, S., Ramos-Martos, N. and Molina-Díaz, A. (2003) A flow-through fluorimetric sensing device for determination of α- and β-naphthol mixtures using a partial least-squares multivariate calibration approach. *Talanta*, 60, 313-323.

[207] López-Flores, J., Fernández de Córdova, M. L. and Molina-Díaz, A. (2007) Multicommutated flow-through optosensors implemented with photochemically induced fluorescence: determination of flufenamic acid. *Analytical Biochemistry*, 361, 280-286.

[208] Álava-Moreno, F., Díaz-García, M. E. and Sanz-Medel, A. (1993) Room temperature phosphorescence optosensor for tetracyclines. *Analytica Chimica Acta*, 281, 637-644.

[209] Guardia, L., Badía, R. and Díaz-García, M. E. (2007) Molecularly imprinted solgels for nafcillin determination in milkbased Products. *Journal of Agricultural and Food Chemistry*, 55, 566-570.

[210] Fernández-González, A., Badía-Laíño, R., Díaz-García, M. E., Guardia, L. and Viale, A. (2004) Assessment of molecularly imprinted sol-gel materials for selective room temperature phosphorescence recognition of nafcillin. *Journal of Chromatography B*, 804, 247-254.

[211] Guardia, L., Badía, R. and Díaz-García, M. E. (2006) Molecular imprinted ormosils for nafcillin recognition by room temperature phosphorescence optosensing. *Biosensors and Bioelectronics*, 21, 1822-1829.

[212] Casado-Terrones, S., Fernández-Sánchez, J. F., Cañabate-Díaz, B., Segura-Carretero, A. and Fernández-Gutiérrez, A. (2005) A fluorescence optosensor for analyzing naphazoline in pharmaceutical preparations: comparison with other sensors. *Journal of Pharmaceutical and Biomedical Analysis*, 38, 785-789.

[213] Sánchez-Barragán, I., Costa-Fernández, J. M., Pereiro, R., Sanz-Medel, A., Salinas, A., Segura, A., Fernández-Gutiérrez, A., Ballesteros, A. and González, J. M. (2005) Molecularly imprinted polymers based on iodinated monomers for selective room-temperature phosphorescence optosensing of fluoranthene in water. *Analytical Chemistry*, 77, 7005-7011.

[214] Traviesa-Alvarez, J. M., Sánchez-Barragán, I., Costa-Fernández, J. M., Pereiro, R. and Sanz-Medel, A. (2007) Room temperature phosphorescence optosensing of benzo[a]pyrene in water using halogenated molecularly imprinted polymers. *Analyst*, 132, 218-223.

[215] Salinas-Castillo, A., Fernández-Sánchez, J. F., Segura-Carretero, A. and Fernández-Gutiérrez, A. (2005) Solid-surface phosphorescence characterization of polycyclic aromatic hydrocarbons and selective determination of benzo(a)pyrene in water samples. *Analytica Chimica Acta*, 550, 53-60.

[216] Casado-Terrones, S., Fernández-Sánchez, J. F., Segura-Carretero, A. and Fernández-Gutiérrez, A. (2005) The development and comparison of a fluorescence and a phosphorescence optosensors for determining the plant growth regulator 2-naphthoxyacetic acid. *Sensors and Actuators B*, 107, 929-935.

[217] Salinas-Castillo, A., Fernández-Sánchez, J. F., Segura-Carretero, A. and Fernández-Gutiérrez, A. (2004) A facile flow-through phosphorimetric sensing device for simultaneous determination of naptalam and its metabolite 1-naphthylamine. *Analytica Chimica Acta*, 522, 19-24.

[218] Fernández-Argüelles, M. T., Cañabate, B., Segura-Carretero, A., Costa-Fernández, J. M., Pereiro, R., Sanz-Medel, A. and Fernández-Gutiérrez, A. (2005) Flow-through optosensing of 1-naphthaleneacetic acid in water and apples by heavy atom induced-room temperature phosphorescence measurements. *Talanta*, 66, 696-702.

[219] Du, J., Sehn, L. and Lu, J. (2003) Flow injection chemiluminescence determination of epinephrine using epinephrine-imprinted polymer as recognition material. *Analytica Chimica Acta*, 489, 183-189.

[220] He, Y., Lu, J., Zhang, H. and Du, J. (2005) Molecular Imprinting-chemiluminescence determination of norfloxacin using a norfloxacin-imprinted polymer as the recognition material. *Microchimica Acta*, 149, 239-244.

[221] Xiong, Y., Zhou, H., Zhang, Z., He, D. and He, C. (2007) Flow-injection chemiluminescence sensor for determination of isoniazid in urine sample based on molecularly imprinted polymer. *Spectrochimica Acta Part A*, 66, 341-346.

[222] Zhou, H., Zhang, Z., He, D. and Xiong, Y. (2005) Flow through chemiluminescence sensor using molecularly imprinted polymer as recognition elements for detection of salbutamol. *Sensors and Actuators B*, 107, 798-804.

[223] Xiong, Y., Zhou, H., Zhang, Z., He, D. and He, C. (2006) Determination of hydralazine with flow injection chemiluminescence sensor using molecularly imprinted polymer as recognition element. *Journal of Pharmaceutical and Biomedical Analysis*, 41, 694-700.

[224] He, D., Zhang, Z., Zhou, H. and Huang, Y. (2006) Micro flow sensor on a chip for the determination of terbutaline in human serum based on chemiluminescence and a molecularly imprinted polymer. *Talanta*, 69, 1215-1220.

[225] Morais, I. P. A., Miró, M., Manera, M., Estela, J. M., Cerdá, V., Souto, M. R. S. and Rangel, A. O. S. S. (2004) Flow-through solid-phase based optical sensor for the multisyringe flow injection trace determination of orthophosphate in waters with chemiluminescence detection. *Analytica Chimica Acta*, 506, 17-24.

[226] Llorent-Martínez, E. J., Ortega-Barrales, P. and Molina-Díaz, A. (2006) Chemiluminescence optosensing implemented with multicommutation: determination of salicylic acid. *Analytica Chimica Acta*, 580, 149-154.

[227] Álava-Moreno, F., Valencia-González, M. J., Sanz Medel, A. and Díaz García, M. E. (1997) Oxygen sensing based on the room temperature phosphorescence intensity quenching of some lead-8-hydroxyquinoline complexes. *Analyst*, 122, 807-810.

[228] Costa-Fernández, J. M., Díaz-García, M. E. and Sanz-Medel, A. (1998) Sol-gel immobilized room-temperature phosphorescent metal-chelate as luminescent oxygen sensing material. *Analytica Chimica Acta*, 360, 17-26.

[229] Valencia-González, M. J., Liu, Y. M., Díaz-García, M. E. and Sanz Medel, A. (1993)

Optosensing of D-glucose with an immobilized glucose oxidase minireactor and an oxygen roomtemperature phosphorescence transducer. *Analytica Chimica Acta*, 283, 439–446.

[230] Paprovsky, D. B., Savitsky, A. P. and Yaropolov, A. I. (1990) Oxygen and glucose optical biosensors based on phosphorescence quenching of metalloporphyrins. *Journal of Analytical Chemistry*, 45, 1441–1445.

[231] Valencia-González, M. J. and Díaz-García, M. E. (1994) Enzymic reactor/room-temperature phosphorescence sensor system for cholesterol determination in organic solvents. *Analytical Chemistry*, 66, 2726–2731.

[232] Pererio García, R., Liu, Y. M., Díaz García, M. E. and Sanz Medel, A. (1991) Solid-surface room-temperature phosphorescence optosensing in continuous flow systems: an approach for ultratrace metal ion determination. *Analytical Chemistry*, 63, 1759–1763.

[233] Lu, J. and Zhang, Z. (1995) Determination of europium with solid-surface room-temperature phosphorescence optosensing. *Analyst*, 120, 2585–2588.

[234] Gong, Z. and Zhang, Z. (1997) Room temperature phosphorescence optosensing for gadolinium. *Microchimica Acta*, 126, 117–121.

[235] Gong, Z., Zhang, Z. and Zhang, A. (1996) Room Temperature Phosphorescence Optosensor for Terbium. *Analytical Letters*, 29, 515–527.

[236] Cuenca-Trujillo, R. M., Ayora-Cañada, M. J. and Molina-Díaz, A. (2002) Determination of ciprofloxacin with a room-temperature phosphorescence flow-through sensor based on lanthanide-sensitized luminescence. *Journal of AOAC International*, 85, 1268–1272.

[237] Traviesa-Alvarez, J. M., Costa-Fernández, J. M., Pereiro, R. and Sanz Medel, A. (2007) Direct screening of tetracyclines in water and bovine milk using room temperature phosphorescence detection. *Analytica Chimica Acta*, 589, 51–58.

[238] Llorent-Martínez, E. J., García-Reyes, J. F., Ortega-Barrales, P. and Molina-Díaz, A. (2005) Terbium-sensitized luminescence optosensor for the determination of norfloxacin in biological fluids. *Analytica Chimica Acta*, 532, 159–164.

[239] Álava-Moreno, F., Valencia-González, M. J. and Díaz-García, M. E. (1998) Room temperature phosphorescence optosensor for anthracyclines. *Analyst*, 123, 151–154.

[240] Aboul-Enein, H. Y., Stefan, R. I. and van Staden, J. F. (1999) Chemiluminescence-Based (Bio) Sensors—an overview. *Critical Reviews in Analytical Chemistry*, 29, 323–331.

[241] Aboul-Enein, H. Y., Stefan, R. I., van Staden, J. F., Zhang, X. R., García-Campaña, A. M. and Baeyens, W. R. G. (2000) Recent developments and applications of chemiluminescence sensors. *Critical Reviews in Analytical Chemistry*, 30, 271–289.

[242] Huang, Y. M., Zhang, C., Zhang, X. R. and Zhang, Z. J. (1999) A novel chemiluminescence flow-through sensor for the determination of analgin. *Fresenius' Journal of Analytical Chemistry*, 365, 381–383.

[243] Zhang, S. and Li, H. (2001) Flow-injection chemiluminescence sensor for the determination of isoniazid. *Analytica Chimica Acta*, 444, 287–294.

[244] Lin, J. M., Qu, F. and Yamada, M. (2002) Chemiluminescent investigation of tris (2,2′-bipyridyl) ruthenium (II) immobilized on a cationic ion-exchange resin and its application to analysis. *Analytical and Bioanalytical Chemistry*, 374, 1159–1164.

[245] Lin, J. M., Sato, K. and Yamada, A. (2001) Hydrogen peroxide chemiluminescent flow-

through sensor based on the oxidation with periodate immobilized on ion-exchange resin. *Microchemical Journal*, 69, 73-80.

[246] Li, B., Zhang, Z., Zhao, L. and Xu, C. (2002) Chemiluminescence flow-through sensor for ofloxacin using solid-phase PbO_2 as an oxidant. *Talanta*, 57, 765-771.

[247] Li, B., Zhang, Z., Zhao, L. and Xu, C. (2002) Chemiluminescence flow-through sensor for pipemidic acid using solid sodium bismuthate as an oxidant. *Analytica Chimica Acta*, 459, 19-24.

[248] Zhao, L., Li, B., Zhang, Z. and Lin, J. M. (2004) Chemiluminescent flow-through sensor for automated dissolution testing of analgin tablets using manganese dioxide as oxidate. *Sensors and Actuators B*, 97, 266-271.

[249] Collins, G. E. and Rose-Pehrsson, S. L. (1995) Chemiluminescent chemical sensors for oxygen and nitrogen dioxide. *Analytical Chemistry*, 67, 2224-2230.

[250] Okabayashi, T, Fujimoto, T., Yamamoto, I., Utsomomiya, K., Wada, T., Yamashita, Y., Yamashita, N. and Nakagawa, M. (2000) High sensitive hydrocarbon gas sensor utilizing cataluminescence of $\gamma-Al_2O_3$ activated with Dy^{3+}. *Sensors and Actuators B*, 64, 54-58.

[251] Zhu, Y., Shi, J., Zhang, Z., Zhang, C. and Zhang, X. R. (2002) Development of a gas sensor utilizing chemiluminescence on nanosized titanium dioxide. *Analytical Chemistry*, 74, 120-124.

[252] Zhou, K., Ji, X., Zhang, N. and Zhang, X. (2006) On-line monitoring of formaldehyde in air by cataluminescence-based gas sensor. *Sensors and Actuators B*, 119, 392-397.

[253] Kiba, N., Itagaki, A., Fukumura, S., Saegusa, K. and Furusawa, M. (1997) Highly sensitive flow injection determination of glucose in plasma using an immobilized pyranose oxidase and a chemiluminometric peroxidase sensor. *Analytica Chimica Acta*, 354, 205-210.

[254] Kiba, N., Tachibana, M., Tani, K. and Miwa, T. (1998) Chemiluminometric branched chain amino acids determination with immobilized enzymes by flow injection analysis. *Analytica Chimica Acta*, 375, 65-70.

[255] Kiba, N., Ito, S., Tachinaba, M., Tani, K. and Koizime, H. (2001) Flow-through chemiluminescence sensor using immobilized oxidases for the selective determination of L-glutamate in a flow injection system. *Analytical Sciences*, 17, 929-933.

[256] Sekiguchi, Y, Nishikawa, A., Makita, H., Yamamura, A., Matsumoto, K. and Kiba, N. (2001) Flow-through chemiluminescence sensor using immobilized histamine oxidase from *Arthrobacter crystallopoietes* KAIT-B-007 and peroxidase for selective determination of histamine. *Analytical Sciences*, 17, 1161-1164.

[257] Zhang, R., Hirakawa, K., Seto, D., Soh, N., Nakano, K., Masadome, T., Nagata, K., Sakamoto, K. and Imato, T. (2005) Sequential injection chemiluminescence immunoassay for anionic surfactants using magnetic microbeads immobilized with an antibody. *Talanta*, 68, 231-238.

[258] Li, B. X., Zhang, Z. J. and Zhao, L. X. (2001) Chemiluminescent flow-through sensor for hydrogen peroxide based on sol-gel immobilized hemoglobin as catalyst. *Analytica Chimica Acta*, 445, 161-167.

[259] Lin, J. M. and Yamada, M. (2001) Chemiluminescent flow-through sensor for 1,10-phenanthroline based on the combination of molecular imprinting and chemiluminescence. *Analyst*, 126, 810-815.

[260] Zhou, H., Zhang, Z., He, D., Hu, Y, Huang, Y. and Chen, D. (2004) Flow chemiluminescence sensor for determination of clenbuterol based on molecularly imprinted polymer. *Analytica*

Chimica Acta, 523, 237-242.

[261] Manz, A., Graber, N. and Widmer, H. M. (1990) Miniaturized total chemical analysis systems: a novel concept for chemical sensing. *Sensors and Actuators B*, 1, 244-248.

[262] Alivisatos, A. P. (1996) Semiconductor clusters, nanocrystals, and quantum dots. *Science*, 271, 933-937.

[263] Costa-Fernández, J. M., Pereiro, R. and Sanz-Medel, A. (2006) Tire use of luminescent quantum dots for optical sensing. *Trends in Analytical Chemistry*, 25, 207-218.

[264] Zhang, Z., He, D., Liu, W. and Lv, Y. (2005) Chemiluminescence micro-flow injection analysis on a chip. *Luminescence*, 20, 377-381.

[265] He, D., Zhang, Z., Huang, Y. and Hu, Y. (2007) Chemiluminescence microflow injection analysis system on a chip for the determination of nitrite in food. *Food Chemistry*, 101, 667-672.

2 流动注射分析中的酶

Robert Koncki、Lukasz Tymecki 和 Beata Rozum

2.1 引言

将生物催化过程纳入分析程序有两个主要原因。首先是使传统分析方法的可用分析物范围得到显著扩展。例如，无电化学活性的分析物可以通过一些酶促反应转化为电活性物质。其次，在许多情况下，由于生物催化过程的高度特异性，这些方法的选择性得到了显著提高。另一个生物分析目的是需要确定一些酶的活性。这些分析方法可进一步用于间接检测对酶活性有影响的化合物，如抑制剂、激活剂、辅酶等。由于此类相互作用的生物特异性，这些测定也具有高度选择性。

在流动注射分析系统中实施酶促的方法利用了该技术通常的优势，包括自动化分析的可能性、减少（生物）试剂消耗、缩短分析时间等。显然，这种方法在多次和频繁重复分析（大规模、常规分析）以及在线实时监测的情况下是合理的，但没有必要将流动注射分析开发用于单次和偶尔测量。这一分析领域似乎是为低成本和易于使用的一次性设备［例如，条带测试、单次（生物）传感器等］或配备高度专业化仪器的大型中央分析实验室保留的。

酶促分析方法本质上具有动力学特征。生物催化转化的动力学是试剂浓度、酶活性、时间、反应条件等的函数。显然，为了准确测定底物浓度或酶活性，所有这些参数都应严格控制。流动注射分析系统提供了这种可能性。反应时间和传输条件由流动注射分析歧管的配置（反应盘管的长度、流速等）定义。这些条件的高度重现性是此类系统的一个明显优势。此外，该过程的"化学"条件（pH、缓冲容量、底物浓度或酶活性、辅助因子、活化剂的存在等）由所用载体的组成明确定义。应该指出的是，流动注射分析系统使分析程序的所有步骤（不仅是酶促反应）自动化。此外，每一个步骤都可以单独优化。这意味着精心设计的流动注射分析系统可为采样、去除/屏蔽干扰物、（生物）化学转化和最终检测提供最佳条件。这些可能性都提高了分析人员的工作舒适度以及分析的质量、准确性、精确性、再现性等。

本章的主要目标是报道酶在流动注射分析系统中的不同使用方式以及设计的检测方案在实际分析中的应用。描述了不同的分析方法：从最常见的通过测定酶活性测定底物（酶作为生物催化剂），到更复杂的方案，如抑制剂和辅因子的间接测定。如果没有流动分析，一些酶促应用将变得非常困难甚至不可能执行，相关实例得到了证实。本章还提供了一些在流动分析系统中使用酶的例子，其中可溶性和固定化酶作为生物传感器或生物反应器是流动系统的一部分。在上面的例子中，作者专注于光学和电化学检测方法。最后，作者回顾了临床、环境和食品分析领域的一些实际应用。本章涵盖了过去五年（2003—2007 年）流动分析相关文献的报道工作。

2.2 酶底物作为分析物

酶的分析用途主要与相应底物的检测有关。此类应用需要生物催化系统的尽可能高的活性，以提供分析物到检测产物的尽可能高的转化率。酶活性也需要恒定，以确保反应过程中产生的产物量仅取决于底物浓度而不取决于任何其他条件。在这些生物检测方案中，酶既不被消耗也不会失活，因此它们可以重复使用。从严格的经济和实用角度来看，流动注射分析系统中的酶主要以其固定形式应用。文献中很少报道利用可溶性酶的流动注射分析系统，并且该系统仅针对那些容易获得的廉价的高活性酶（例如葡萄糖氧化酶）。此外，通过合适的系统配置（即在停流模式下工作的微系统或顺序注射分析系统），可以尽可能地减少酶的消耗[1,2]。在这种方法中，单次分析所需的酶活性显著降低。通常，将固定化酶并入流动注射分析系统的方式主要有两种：该酶可以作为系统的单独部分，以流通式生物反应器的形式使用；或者，它可以与检测器集成在一起，作为其不可分割的一部分，致力于扮演着生物分子识别过程受体的角色。

2.2.1 基于生物传感器的流动注射分析系统

酶与检测器的集成简化了流动注射分析系统。带有生物传感器的歧管中的分散显著减少，这有助于提高灵敏度，即使分析物的酶促转化不是定量的。此类系统的灵敏度很高，因为负责产生分析信号的生物试剂的浓度变化会直接显示在检测器表面上。最后，一些生物传感方案在任何其他流动注射分析配置中都不可能实现，例如，在pH检测的情况下，因为生物催化反应的产物具有质子迁移特性。全段样品的酶促反应引起的pH变化主要取决于其缓冲能力。因此，由此产生的pH变化与检测到的底物浓度并不相关。对于基于pH的生物传感器，样品的缓冲能力会影响响应大小，但信号仍然是特异性的。

遗憾的是，在某些情况下，流动注射分析与生物传感器的集成也限制了生物传感器优化的可能性。众所周知，在某些情况下，检测器和酶的最佳操作条件是不同的，因此无法在两个生物传感器组件的最佳条件下进行测量，需要折中，这显然会降低生物检测系统的效率。另一个显著缺陷是由生物传感器动力学引起的系统响应变慢（特别是生物传感器的"记忆效应"导致返回基线信号的时间过长）。通过选择合适的固定化方法可以尽可能减少这种不便，比如提高生物催化层的渗透性以及加快层内的物质传输。另一方面，应用的固定化方法应延长生物催化层的使用寿命，以便生物传感器可以在很长一段时间内进行许多分析。最后一个缺陷是在生物受体层破坏后，整个流动注射分析的检测系统元件都必须更换。

毫无疑问，电流型酶电极构成了流动注射分析系统中的主要生物传感器组。表2.1汇总了许多采用不同内部电极的示例，这些电极能够消除氧化还原物质的干扰或有效地固定酶。

表2.1　使用电流型生物传感器作为检测器的选定流动注射分析系统

分析物	酶	生物传感器的构建	样品	参考文献
		（单酶电极）		
过氧化氢	辣根过氧化物酶	在3-巯基丙酸自组装单层修饰的金电极上，酶与使用戊二醛的介质（四硫代富瓦烯）相交联	染发剂、雨水	[3]

续表

分析物	酶	生物传感器的构建	样品	参考文献
过氧化氢	辣根过氧化物酶	酶包埋在大量丝网印刷石墨电极中	—	[4]
过氧化氢	辣根过氧化物酶	酶包埋在掺杂二茂铁—甲酸-BSA 偶联物和多壁碳纳米管的有机改性硅酸盐复合材料中	—	[5]
葡萄糖	葡萄糖氧化酶	酶与全氟磺酸沉积在掺杂 RhO_2 的石墨丝网印刷电极上	速溶茶、蜂蜜	[6]
葡萄糖	葡萄糖氧化酶	酶与沉积在铂电极上的普鲁士蓝取代的聚吡咯复合膜共价结合	尿液、血清、输液、软饮料、葡萄酒	[7]
葡萄糖	葡萄糖氧化酶	酶固定在普鲁士蓝电化学修饰的玻碳电极上，全氟磺酸薄膜作为保护层	速溶咖啡	[8]
葡萄糖	葡萄糖氧化酶	酶包埋在玻碳电极上的多孔二氧化钛溶胶-凝胶基质中，形成酶/二氧化钛/全氟磺酸；全氟磺酸薄膜作为保护层	血清标样	[9]
葡萄糖	葡萄糖氧化酶	酶包埋在铂针电极上的电聚合间苯二胺中	大鼠脑透析液	[10]
葡萄糖	葡萄糖氧化酶	通过与包埋在氧硅烷溶胶-凝胶基质中的戊二醛交联，用 BSA 固定化酶	葡萄汁	[11]
半乳糖	半乳糖氧化酶	酶与分散在壳聚糖中的单壁碳纳米管共价结合并沉积在玻璃碳盘电极上；全氟磺酸薄膜作为保护层或/和中间层	血浆	[12]
甘油	甘油脱氢酶	酶与吩嗪硫酸甲酯共固定或使用聚（乙二醇）二缩水甘油醚交联到 Os-复合物改性的聚（乙烯基咪唑）氧化还原聚合物上	葡萄酒	[13]
乙醇	酒精氧化酶	在过氧化聚吡咯修饰的金电极上电聚合吡咯中的酶	红酒	[14]
乙醇	酒精氧化酶	在过氧化聚吡咯修饰的金电极上与戊二醛交联的酶和 BSA	红酒	[14]
胆碱	胆碱氧化酶	在过氧化聚吡咯修饰的铂电极上与戊二醛交联的酶和 BSA	乳粉、牛乳、大豆卵磷脂水解物	[15]
赖氨酸	L-赖氨酸 α-氧化酶	在电聚合间苯二胺修饰的金电极上与戊二醛交联的酶和 BSA	牛乳、豆浆、面粉、面食	[16]
谷氨酸、β-ODAP	L-谷氨酸氧化酶	酶、BSA 和 Tween-20 在普鲁士蓝修饰的玻碳电极上与戊二醛交联	—	[17]
NADH	NADH 氧化酶	在掺杂普鲁士蓝的丝网印刷石墨电极上固定全氟磺酸、戊二醛和酶	红葡萄酒和白葡萄酒	[18]

续表

分析物	酶	生物传感器的构建	样品	参考文献
L-乳酸	L-乳酸氧化酶	在位于铂电极上的 Anylon 膜上,酶与聚乙烯亚胺和戊二醛交联	细胞培养基	[19]
葡萄糖	D-葡萄糖脱氢酶	酶与奥芬二酮和 NAD^+ 包埋在碳糊电极中	果汁、乳制品果汁、乳制品	[20]
果糖	D-果糖脱氢酶	酶与 $Os(bpy)_2Cl_2$ 和聚乙烯亚胺包埋在碳糊电极中	—	[20]
乳糖、纤维二糖	纤维二糖脱氢酶	酶吸附在石墨电极上	—	[21]
葡萄糖酸盐	葡萄糖脱氢酶	在3-巯基丙酸自组装单层修饰的金电极上,酶与使用戊二醛的介质(四硫代富瓦烯)相交联	葡萄酒,新葡萄汁	[22]
组胺	组胺脱氢酶酪氨酸酶	酶在玻碳电极上与聚(乙二醇)二缩水甘油醚和 PVI-dmeos 交联	金枪鱼肌肉提取物	[23]
酚类衍生物	酪氨酸酶	酶与戊二醛在3-巯基丙酸自组装单层修饰的金电极上交联	炼油厂废水	[24,25]
苯酚	漆酶	酶在玻碳电极上与戊二醛交联	葡萄酒	[26]
酚类衍生物	漆酶	吸附在石墨电极上的酶	—	[27]
三甲胺	含黄素单加氧酶	酶与含有苯乙烯基团的聚乙烯醇光在与克拉克型氧电极耦合的透析膜上进行光交联	鱼提取物(鲭鱼)	[28]
有机磷农药	有机磷水解酶	在光刻微结构金电极上,使用戊二醛的单层自组装胱胺与酶相结合	—	[29]
(双酶电极)				
L-乳酸丙酮酸	L-乳酸氧化酶、丙酮酸氧化酶	酶和明胶在涂有电聚合1,2-二氨基苯的铂电极上与戊二醛交联	大鼠脑透析液、人血清标样	[30]
L-乳酸	L-乳酸脱氢酶、心肌黄酶	酶和明胶在涂有电聚合1,2-二氨基苯的铂电极上与戊二醛交联	啤酒、清酒、红酒	[31]
D-乳酸	D-乳酸脱氢酶、心肌黄酶	酶和明胶在涂有电聚合1,2-二氨基苯的铂电极上与戊二醛交联	啤酒、清酒、红酒	[31]
甘油	NADH 氧化酶、甘油脱氢酶	在掺杂普鲁士蓝的丝网印刷石墨电极上固定酶和全氟磺酸以及戊二醛酶	—	[18]
甘油	甘油脱氢酶、心肌黄酶	酶包埋在玻碳电极的聚氨基甲酰基磺酸盐中	厌氧发酵液	[32]
L-谷氨酸	L-谷氨酸氧化酶、L-谷氨酸脱氢酶	与氧电极传感元件相连的聚碳酸酯膜上,酶与戊二醛交联	酱油和番茄酱、泰式鸡汤、辣椒鸡肉汁	[33]

续表

分析物	酶	生物传感器的构建	样品	参考文献
（三酶电极）				
醋酸盐	醋酸激酶、丙酮酸激酶、丙酮酸氧化酶	酶包埋在含苯乙烯基团的聚乙烯醇的膜中；在聚（二甲基硅氧烷）改性的铂电极上进行膜光交联	红葡萄酒	[34]
醋酸盐	乙酸激酶、丙酮酸激酶、乳酸脱氢酶	酶与介体修饰的（亮甲酚蓝）聚（乙二醇）二缩水甘油醚碳棒电极交联	葡萄酒、醋	[35]
乳糖	β-半乳糖苷酶、变旋酶、D-葡萄糖脱氢酶	酶与奥芬二酮和NAD$^+$包埋在碳糊电极中	果汁、乳制品	[20]
蔗糖	转化酶、变旋酶、D-葡萄糖脱氢酶	酶与奥芬二酮和NAD$^+$包埋在碳糊电极中	果汁、乳制品	[20]
葡萄糖、L-乳酸丙酮酸	葡萄糖氧化酶、L-乳酸氧化酶、丙酮酸氧化酶	酶和明胶在通过电聚合1，2-二氨基苯改性的铂电极上与戊二醛交联	大鼠脑透析液人血清清标样	[36]
甘油	甘油激酶、甘油-3-磷酸氧化酶、辣根过氧化物酶	酶包埋在玻碳电极上的聚氨基甲酰基磺酸盐中	厌氧发酵液	[32]
正磷酸盐	麦芽糖磷酸化酶、变旋酶、葡萄糖氧化酶	酶在通过电聚合1，2-二氨基苯改性的铂电极上与戊二醛交联	—	[37]

单酶电极主要通过氧化酶[3-19]、脱氢酶[20-23]或加氧酶[24-28]进行修饰。多酶连续生物传感器很少报道。这些生物传感器包括一个生物催化层，其中一些酶依次催化两个或多个转化过程，进而从检测的底物中生成电活性产物（例如甘油[18,32]、乳酸盐[30,31,36]、丙酮酸盐[30,36]和乙酸盐[34,36]的生物传感器）。另一种有趣的多酶敏化方法由双酶谷氨酸生物传感器验证，其中生物传感层通过分析物的氧化还原循环放大分析信号[33]。最后，值得一提的是流动注射分析系统，其中的生物传感器是一个复杂的装置，由两个或三个对不同生物分析物敏感的单酶电极组成。这些所谓的双酶电极[30,31]和三重酶电极[36]能够同时进行多个分析物的生物检测。图2.1说明了在生物传感层中包含两种酶的生物传感器的三种不同传感方案。

在流动注射分析条件下很少使用电位生物传感器，最常用的是对尿素[38-40]和肌酐[40-42]敏感的传感器。以完整的生物电化学流动池形式存在的尿素生物传感器可以仅通过丝网印刷制造[38,39]。在这种微型装置中，生物传感器的内部电极和参比电极均为基于厚膜且对pH敏感的二氧化钌电极。这种带状生物流动池在专门用于分析尿液和血液透析后液体的流动注射分析歧管中得到了很好的应用。另一个在差分电位滴定法条件下工作的生物传感系统例子是用于氨基酸检测的流动注射分析系统，该系统采用了L-赖氨酸氧化酶改性的固态铵离子选择性电极[43]。通过肌酸酐二亚胺酶或尿素酶与铵离子选择性电极的聚合物膜的共价结合，得到了非常稳定的肌酐和尿素生物传感

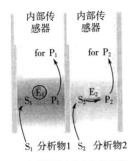 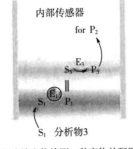 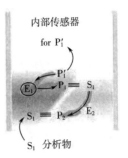

(1)同时检测两种底物的配置（双生物传感器）　　(2)连续生物检测一种底物的配置（双酶生物传感器）　　(3)用于将一种底物信号生物放大的配置（双酶生物传感器）

图 2.1　受体部分包含两种酶的生物传感器

器[40,41]。基于此类生物传感器的流动注射分析系统已成功应用于尿液、血清和透析后液体的分析。在流动注射分析条件下工作的电位生物传感器的另一个例子是用碱性磷酸酶[44]改性的 pH 膜电极[29]。用于构建生物传感器的酶可以催化该磷酸有机化合物的水解。酸性反应产物会导致内部 pH 传感器的电位发生变化。由于反应过程中产生的对硝基苯酚是一种电活性产物，也可以对其进行电流检测，因此，开发的流动注射分析系统能够对选定的有机磷农药进行双重（电位和电流）生物检测。

2.2.2　基于生物反应器的流动注射分析系统

与生物反应器集成的流动注射分析系统不存在大多数上述生物传感器的缺陷。在这样的系统中，检测器和生物催化组件的操作条件可以分别优化，并且不需要相互的折中，就像在生物传感器中一样。当不止一种酶参与分析物生物转化时，这一点尤其重要；然后，流动注射分析系统中的酶可以被分离并放入连续的生物反应器中，每个生物反应器都可以在最佳条件下工作。第二个显著优势是可以设计成可更换生物反应器的流动注射分析系统，这些生物反应器是根据目标分析物进行选择的。此外，在这样的系统中，可以平行放置不同的生物反应器，从而有机会用相同的检测器同时测定两种或更多的分析物。类似地，通过使用生物传感器阵列可以实现此目的，但不幸的是，它需要应用多通道测量仪器。最后，由于生物反应器是流动注射分析系统的独立部分，因此无需更换检测器即可将生物反应器更换。

生物反应器的另一个优势使得其在流动注射分析中有更广泛的应用（与生物传感器相比）。即生物反应器可与非传感检测器在流动注射分析系统中结合使用，例如，基于对转化样品的吸光度、荧光或化学发光的测量。如上所述，这样的检测系统可以独立于先前的酶促反应进行优化，并且可以在不利于酶工作的条件下检测。表 2.2 收集了过去五年内文献报道的基于生物反应器的流动注射分析系统的一些示例，其中应用了各种电化学[19,37,45-54]和光学[55-74]检测系统。内置生物反应器的流动注射分析系统的主要缺陷是样品分散性太强，尤其是在填充有颗粒的微柱存在的情况下。为了克服这一缺陷，人们构建了具有高生物催化活性并且能够快速和有效地转化分析物的生物反应器。或者，可以通过使用"带涂层的管状生物反应器"显著减少分散。样品分散延长了测定时间，降低了流动注射分析系统的样品吞吐量。最后，在流动注射分析系统中引入任何替代组件都会增加系统的复杂性，并且还会增加发生故障的频率。

2.2.3 流动注射分析的其他优势

表2.1和表2.2中汇编的几个示例清楚地表明了在流动注射分析系统中可以应用不同的酶促生物传感器和生物反应器。除了引言中所提方法具有的明显优势外，还应记住的是，流动注射分析概念涉及分析过程的所有步骤，而不仅仅是分析物的转化和检测步骤。基于流动注射分析的分析程序的质量改进也与采样或降低干扰等过程能否自动化有关。

表2.2 带有生物反应器的流动注射分析系统

分析物	酶	生物反应器	检测	样品	参考文献
葡萄糖	葡萄糖氧化酶	酶与熔融石英柱的功能化壁共价结合	铂电极上的电流法（H_2O_2）检测	人血清	[45]
葡萄糖	葡萄糖氧化酶、辣根过氧化物酶	酶分别固定在用tresyl功能化的硅微珠上，该硅微珠填充于HPLC柱上	玻璃碳电极上的电流法（H_2O_2）检测	人脑透析液	[46]
乳酸	乳酸氧化酶、辣根过氧化物酶	酶分别固定在用tresyl功能化的硅微珠上，该硅微珠填充于HPLC柱上	玻璃碳电极上的电流法（H_2O_2）检测	人脑透析液	[46]
L-乳酸、葡萄糖	乳酸氧化酶、葡萄糖氧化酶	通过戊二醛将酶固定在可控微孔玻璃珠（CPG）上，每个酶均填充到单独的聚合物柱中	聚（1,2-二氨基苯）改性铂电极上的电流法（H_2O_2）检测	大鼠脑透析液、人血清标准品	[47]
L-苹果酸	苹果酸酶	酶固定在戊二醛功能化的氨丙基玻璃珠上，酶包被的微珠填充到聚合物柱中	铂电极上的电流法（H_2O_2）检测	细胞培养基	[19]
磷酸盐	酸性磷酸酶	通过戊二醛将酶固定在CPG珠上，每个酶均填充到单独的聚合物柱中	铂电极上的电流法（H_2O_2）检测	—	[37]
果糖基肽、果糖基氨基酸	果糖基-氨基酸氧化酶、果糖基-肽氧化酶	酶固定在戊二醛功能化的单向传递的C微珠上，每种类型的微珠分别填充到单独的柱中	玻璃碳电极上的电流法（H_2O_2）检测	蛋白酶消化的血细胞	[48, 49]

续表

分析物	酶	生物反应器	检测	样品	参考文献
甘油、三酰甘油	甘油激酶、甘油-3-磷酸氧化酶、脂肪酶	酶与功能化壁共价结合；甘油激酶和甘油-3-磷酸氧化酶共固定化，脂肪酶存在于单独的玻璃柱中	铂电极上的电流法（H_2O_2）检测	三油酸甘油酯标样	[50]
酚类	漆酶	使用碳二亚胺法将酶固定在 ECH 琼脂糖树脂上；凝胶填充在聚合物柱中	玻碳电极上的电流法（1,4-苯醌）检测	橄榄油厂废水	[51]
尿素	脲酶	酶固定在 CPG、硅胶或 Poraver 上，用 3-氨基-丙基-三乙氧基硅烷和戊二醛载体进行功能化并填充到聚合物柱上	电导测定法	血清	[52]
异柠檬酸	异柠檬酸脱氢酶	将酶固定在戊二醛功能化的氨丙基玻璃微珠上；将酶包被的微珠填充 Teflon 柱中	带离子选择性电极的电位（CO_3^{2-}）法检测	果汁	[53]
柠檬酸盐	柠檬酸裂解酶、草酰乙酸脱羧酶	将酶固定在戊二醛功能化的氨丙基玻璃微珠上；将酶包被的微珠填充 Teflon 柱中	带离子选择性电极的电位（CO_3^{2-}）法检测	果汁	[54]
H_2O_2	过氧化物酶	通过戊二醛将酶与 AmberliteIRA-173 树脂偶联；酶和树脂填充到聚合物柱中	光度法（H_2O_2）作为抗吡咯醌亚胺染料（505nm）	雨水	[55]
葡萄糖	葡萄糖氧化酶	通过戊二醛将酶与 AmberliteIRA-173 树脂偶联，树脂与酶填充到聚合物柱中	作为抗吡咯醌亚胺（505nm）的光度法（H_2O_2）检测	全人血	[56]
葡萄糖	葡萄糖氧化酶、辣根过氧化物酶	使用戊二醛将酶与功能化的 CPG 珠上交联，载体填充到玻璃柱中	使用 Trinder 方法的光度（H_2O_2）（490nm）检测	细胞培养基	[57,58]
半乳糖	半乳糖氧化酶、辣根过氧化物酶	通过戊二醛使酶在功能化的 CPG 珠上交联，载体填充到玻璃柱中	使用 Trinder 方法的光度（H_2O_2）（490nm）检测	细胞培养基	[58]
乙醇	酒精氧化酶、辣根过氧化物酶	通过戊二醛使酶在功能化的 CPG 珠上交联，载体填充到玻璃柱中	使用 Trinder 方法的光度（H_2O_2）（490nm）检测	细胞培养基	[58]

续表

分析物	酶	生物反应器	检测	样品	参考文献
乳酸	乳酸氧化酶、辣根过氧化物酶	通过戊二醛使酶在功能化的 CPG 珠上交联，载体填充到玻璃柱中	使用 Trinder 方法的光度（H_2O_2）（490nm）检测	细胞培养基	[58]
氨基酸	L-氨基酸氧化酶、辣根过氧化物酶	通过戊二醛使酶在功能化的 CPG 珠上交联，载体填充到玻璃柱中	使用 Trinder 方法的光度（H_2O_2）（490nm）检测	细胞培养基	[58]
氰酸盐	氰酶	酶固定在用 3-氨基丙基三乙氧基硅烷和戊二醛功能化的载体（CPG）上，然后将载体填充到聚合物柱中	使用 Berthelot 反应的光度（NH_4^+）（700nm）检测	废电镀液	[59]
正磷酸盐、植酸盐	植酸酶	酶与硅烷化孔径可控的硅微珠通过共价键结合，微珠填充到玻璃柱中	钼蓝（650nm）的荧光法（PO_4^{3-}）	玉米、牛乳、大豆	[60]
尿素	脲酶	酶共价固定在 CPG 珠上并填充到聚合物柱中	荧光法（异吲哚衍生物，485nm）	含酒精的饮料	[61]
D-苹果酸、L-苹果酸	D-苹果酸脱氢酶、L-苹果酸脱氢酶	酶与氨丙基 CPG 珠相结合，每种酶填充到单独的玻璃柱中	荧光法（NADH 或 NADPH，455nm）	果汁、软饮料	[62]
丙酮酸	丙酮酸脱羧酶、乙醛脱氢酶	通过戊二醛将酶与氨丙基 CPG 相结合，微珠填充到玻璃柱中	荧光法（NADH，455nm）	—	[63]
丙酮酸、L-乳酸	L-乳酸氧化酶	通过戊二醛将酶与氨丙基 CPG 相结合，微珠填充到玻璃柱中	荧光法（NADH，455nm）	—	[63]
丙酮酸盐、醋酸盐	乙酸激酶、丙酮酸激酶	通过戊二醛将酶与氨丙基 CPG 相结合，微珠填充到玻璃柱中	荧光法（NADH，455nm）	—	[63]
丙酮酸盐、柠檬酸盐	柠檬酸裂解酶、草酰乙酸脱羧酶	通过戊二醛将酶与氨丙基 CPG 相结合，微珠填充到玻璃柱中	荧光法（NADH，455nm）	—	[63]
D-葡萄糖酸盐	葡萄糖酸激酶、6-磷酸葡萄糖酸脱氢酶	通过戊二醛将酶与氨丙基 CPG 相结合，微珠填充到玻璃柱中	荧光法（NADH，455nm）	蜂蜜、醋、贵腐酒	[64]
延胡索酸	延胡索酸酶、苹果酸脱氢酶	酶与环氧载体微珠通过共价键结合，微珠一层一层地填充到塑料柱中	荧光法（NADH，455nm）	细胞培养基	[65]

续表

分析物	酶	生物反应器	检测	样品	参考文献
琥珀酸	异柠檬酸裂解酶、异柠檬酸脱氢酶	酶与环氧树脂载体微珠通过共价键结合,微珠一层一层地填充到塑料柱中	荧光法（NADH,440nm）	细胞培养基	[66]
GABA 谷氨酸	γ-氨基丁酸谷氨酸转氨酶、L-谷氨酸氧化酶、过氧化氢酶	酶与氨丙基CPG珠相结合,将L-谷氨酸氧化酶/过氧化氢酶和γ-氨基丁酸谷氨酸氨基转移酶共固定到单独的玻璃柱上	荧光法（NADPH,455nm）	细胞培养基	[67]
葡萄糖	葡萄糖氧化酶、辣根过氧化物酶	通过戊二醛将葡萄糖氧化酶固定在氨基二基玻璃微珠上；微珠填充到聚合物柱中；载体溶液中含有辣根过氧化物酶	化学发光（鲁米诺+H_2O_2,Co^{2+}催化）	人尿、低糖饮料	[68]
葡萄糖	葡萄糖氧化酶、辣根过氧化物酶	酶与硅湿蚀刻的微芯片表面（由聚乙烯亚胺和戊二醛或戊二醛和3-氨基丙基三乙氧基硅烷改性）结合	化学发光（鲁米诺+H_2O_2）	—	[69]
乙醇	酒精氧化酶、辣根过氧化物酶	酶与硅湿蚀刻的微芯片表面（由聚乙烯亚胺和戊二醛或戊二醛和3-氨基丙基三乙氧基硅烷改性）结合	化学发光（鲁米诺+H_2O_2）	—	[69]
甘油	甘油激酶、3-磷酸甘油氧化酶	通过戊二醛将酶与CPG微珠（用3-氨基丙基-三乙氧基硅烷改性）相结合；微珠填充玻璃柱中	化学发光（鲁米诺+H_2O_2）	血清中的甘油三酯（用脂肪酶预处理）	[70]
胆碱、乙酰胆碱	胆碱氧化酶、过氧化物酶、乙酰胆碱酯酶	酶单独固定,胆碱氧化酶与过氧化物酶共固定在乙烯基聚合物珠的甲苯基化表面；微珠粒逐层填充到聚合物柱中	化学发光法（鲁米诺+H_2O_2）	兔脑匀浆	[71]
磷酸胆碱	磷脂酶C、碱性磷酸酶、胆碱氧化酶	酶共固定到戊二醛功能化的氨丙基-CPG珠上并填充到聚合物柱中	化学发光法（鲁米诺+H_2O_2,Co^{2+}催化）	沉积物	[72]
PO_4^{3-}	麦芽糖磷酸化酶、变旋酶、葡萄糖氧化酶	酶固定在NHS多孔纤维素微珠上,微珠填充到不锈钢柱中	化学发光法（鲁米诺+H_2O_2）	淡水	[73]

续表

分析物	酶	生物反应器	检测	样品	参考文献
PO_4^{3-}	丙酮酸氧化酶	通过戊二醛将酶与CPG珠（用3-氨基丙基三乙氧基硅烷改性）相结合，微珠填充到玻璃柱中	化学发光法（鲁米诺+H_2O_2）	水	[74]

可以采取三种消除干扰的策略：①抗干扰屏障的使用；②干扰掩蔽；③干扰去除。第一种方法的常见例子是构建涂有全氟磺酸膜的电流生物传感器，可排除来自乙酰氨基酚、抗坏血酸和尿酸等电活性样品成分的影响。另一个例子是开发合适的电极材料为生物转化的反应产物提供低氧化还原反应电位。干扰剂的掩蔽可以通过在流动注射分析系统中选择合适的载体来实现。这种方法的一个例子是在载体溶液中加入螯合剂，如EDTA，掩蔽了在流动注射分析条件下抑制酶系统的重金属离子。另一个例子是使用能很好控制干扰物质的载体。然而，这种方法仅适用于目标分析物含量相对较高的样品。最后，干扰去除，通常是传统分析程序的费力步骤，有时在流动注射分析条件下很容易执行。一个很好的例子是去除内源性铵和碱性阳离子，这两者基于低选择性的内部铵离子选择性电极[40,41]强烈地影响着电位生物传感器对肌酐和尿素的响应。应用系统的方案如图2.2所示。可以在生物传感器之前使用阳离子交换微柱从样品中去除干扰离子。重要的是，在设计得当的流动注射分析系统中实行微柱不仅自动化地去除干扰物，而且可以使微柱中的离子交换剂连续再生[40]。

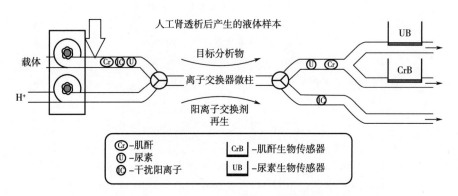

图2.2 带有干扰物去除装置的流动注射分析系统
带有离子交换剂的微柱保留了影响尿素和肌酐生物传感器响应的阳离子。
流动注射分析系统用于血液透析治疗的双参数监测。
(资料来源：文献[40]，2004年。)

自动化采样是流动注射分析生物检测的另一个重要优势。一个有趣的例子是通过透析进行微量采样。这种流通式采样器可以以浸入液体样品中的微透析采样管[31]的形式出现，也可以作为微透析探针直接位于分析环境中[10,30,36,46,47]。图2.3显示了在基于生物反应器的流动注射分析系统中应用这种微量采样探针并用于在手术过程中在线监测人脑中的葡萄糖和乳酸水平[46]。

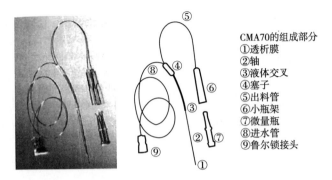

图 2.3 使用在线微采样单元的流动注射分析生物分析系统，用于监测脑中的葡萄糖和乳酸
（来自 CMA 微透析的带金尖端的脑微透析导管，www.microdialysis.se。相关细节请参照文献[46]。）

2.2.4 流动注射分析系统作为监测器的分析应用

需要强调的是，开发流动注射分析的主要目标是将其应用于实际样品的分析。本章回顾的生物分析系统满足了这一需求。酶底物识别的性质定义了实际分析化学的领域，在这些领域中，此类系统可以得到广泛应用。从表 2.1 和表 2.2 可以看出，系统最适用的主要领域是：临床/生物医学[7,9,10,12,19,30,36,45-49,52,56,68,70,71]、食品[4-8,14-16,18,20,26,28,31,33-35,54,60-62,64,68]和环境分析[3,24,25,51,55,59,72-74]。此时有必要强调流动注射分析概念的更多优势及其实际效果。流动注射分析系统适用于大量样品的连续分析，并且可以实时进行测量。这意味着流动注射分析系统可以同时扮演分析器和监测器的角色。这个属性对于离散监测实际过程的流动注射分析系统的发展有很大影响，相关分析文献广泛报道了几个用于监测临床和生物技术过程的流动注射分析系统的例子。

已经提到的带有微透析采样并且基于生物传感器[10,30,36]和生物反应器[47]的流动注射分析系统，被用于监测葡萄糖、乳酸和丙酮酸——主要的脑营养物质及其代谢物。来自大鼠脑的纹状体透析液被自动取样。类似的生物分析流动注射分析系统[46]具有两个平行的生物反应器和电流检测器，已应用于在手术过程中和头部外伤患者神经重症监护期间对脑葡萄糖和乳酸的连续监测。使用酶促流动注射分析系统进行生物医学监测的另一个例子是对血液透析治疗的控制。开发的基于厚膜尿素生物传感器[39]或酶促敏化尿素和肌酐[40]的铵离子选择性电极的系统，已成功应用于监测人工肾产生的废液中这两种尿毒症标志物的水平。此类监测有助于确定治疗是否有效，从此类在线测量中获得的尿素和肌酐谱图可用于定量描述治疗以及对其进行充分性评估。

基于酶的流动注射分析系统还可以应用于监测先进生物技术过程。基于电流型 L-乳酸生物传感器和苹果酸酶生物反应器（与电流型 H_2O_2 检测器偶联）的两个流动注射分析系统[19]已应用于监测由乳酸菌引起的红葡萄酒中的微苹果酸乳酸发酵（通过降低苹果酸水平来提高葡萄酒质量的生物技术过程）。带有固定化延胡索酶和苹果酸脱氢酶的生物反应器的荧光流动注射分析系统已被用于在线监测延胡索酸产生菌（*Rhizopus oryzae*）的培养[65]。一个类似的流动注射分析系统带有共固定化异柠檬酸裂解酶和脱氢酶的双酶反应器，其已被用于监测固定化大肠杆菌生产琥珀酸的生物技术过程[66]。几种基于固定化辣根过氧化物酶和相应氧化酶的双酶反应器的荧光流动注射分析系统已被用于监测大肠杆菌发酵液中的葡萄糖[57,58]、乙醇[58]、乳酸[58]和L-亮氨酸[58]以及酿酒酵母发酵液中的葡萄糖[57,58]、半乳糖[58]和乙醇[58]。这些生物分析系统还用于监测大鼠朗格汉斯胰岛细胞（生物人工胰腺装置）培养基中的葡萄糖[57]以及监测间充质干细胞培养基中的乳酸和亮氨酸水平[58]。为环境需要而开发的基于酶的流动注射分析设备的一个很好的例子是带有固定化氰酶的荧光系统，用于监测珠宝行业废水生物修复过程中的氰酸盐消耗量[59]。

2.3 基于测量酶活性的方法

上一节中提到的酶促生物检测底物的方案可以适用于所选酶的检测。简而言之，通过使用相同的检测系统并且选择浓度恒定的适当底物，在流动注射分析系统中获得的分析信号，经过一定的酶反应时间，应该是（根据 Michaelis-Menten 理论）与测定的酶的活性成比例。以酶的检测为导向的生物分析程序，可以很容易地通过流动注射分析实现。此外，应该指出的是，通过酶活性检测可以间接检测出影响酶活性的相关物质。对流动注射分析酶活性检测系统的调查清楚地表明了其应用的主要领域。第一大领域是控制工业酶的生产或其在消费品中的活性水平，第二个领域是用于临床诊断的生理液体（主要是血清）中酶活性的分析。基于酶抑制的方法主要应用于环境分析，很少有系统适用于酶辅因子的间接检测。这种检测的一般方案如图 2.4 所示。下面简要回顾一些实际应用的例子。

2.3.1 酶活性检测

角质酶（脂解酯酶）是应用于各种化学和生物技术中（乳制品和油脂化学工业）的有前景的催化剂，因为它们能够在水性和非水性环境中运行。能够检测角质酶的常见显色底物是对硝基苯丁酸盐，它很容易水解成黄色的对硝基苯酚。基于此反应和检测的荧光流动注射分析系统已通过额外调整生成该底物的胶束[75]。该流动注射分析系统已应用于监测来自发酵液的酿酒酵母（*Saccharomyces cerevisiae*）菌株分泌的角质酶的生物生产、预浓缩和纯化[75,76]。工业枯草杆菌蛋白酶类蛋白酶如 savinase 和 Purafect 被大量生产并广泛用于洗涤剂行业。为检测这些酶的活性而开发的流动注射分析系统的主要部分是含有固定在溶胶-凝胶颗粒表面的明胶-德州红共轭物的柱子[77]。将带有固定荧光团标记的蛋白质底物的生物反应器暴露于枯草杆菌蛋白酶样品中，导致形成荧光峰。

另一种重要的工业酶是 α-淀粉酶，广泛用于通过转糖基化的淀粉降解和寡糖生产。此外，从临床角度来看，血清和尿液中 α-淀粉酶活性的测定对于急性胰腺炎的诊断是有参考价值的。淀粉酶活性测定基于对所选寡糖（如麦芽五糖）通过酶促反应生成的麦芽糖的检测。用于此类目的的流动注射分析系统基于麦芽糖双酶电极[78]或基于与电流型葡萄糖生物传感器连接的葡萄糖苷酶生物反应器[79]。后一种系统已被用于唾

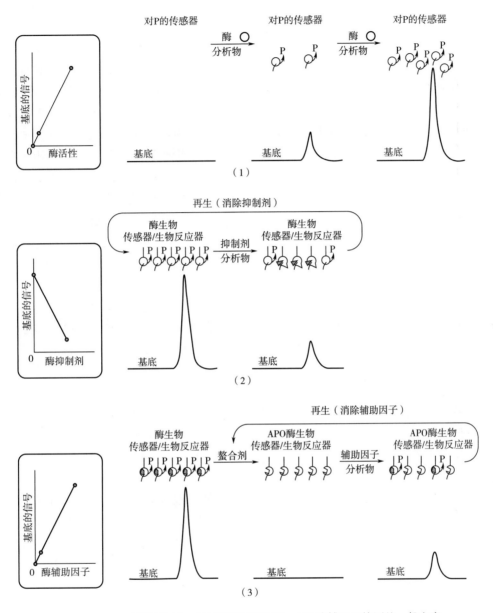

图2.4 （1）酶活性检测 （2）酶抑制剂检测 （3）酶辅因子检测的一般方案

液淀粉酶的检测，唾液淀粉酶被认为是与应激反应相关的一个因素。

已经开发了两种不同的 FIA 系统来检测人血清中的碱性磷酸酶活性。这种酶被认为是骨病和肝脏疾病的指示物。半自动流动/微珠进样系统中的生物分析基于对酶促反应过程中磷酸盐释放的对硝基苯酚的荧光检测[80]。使用包覆小麦胚芽的微珠可以对骨碱性磷酸酶进行生物特异性分离和检测。基于氟离子选择性电极的更简单的电位流动注射分析系统利用了酶催化单氟磷酸盐水解的能力[81]。反应过程中从底物产生的氟离子的量与酶活性成正比。该流动注射分析系统可用于测定血清碱性磷酸酶的生理和病理水平以及血清样品热预处理后骨碱性磷酸酶和肝碱性磷酸酶的定量区分。

2.3.2 酶抑制剂检测

用于酶活性测定的生物传感方案和流动注射分析系统同样适用于间接检测相应的酶抑制剂，最深入的研究致力于检测乙酰胆碱酯酶（AChE）抑制剂。常见的目标分析物是有机磷物质。这些分析物使得此类系统在环境、农业和军事分析中具有潜在效用，用于 AChE 抑制剂测定的流动注射分析系统在酶的应用形式（生物传感器、生物反应器或可溶性酶）以及检测方法方面有所不同。前人已经通过逐层静电自组装方法制造了电流式生物传感器并建立了夹心状结构，其中酶位于聚电解质层之间[82]。通过纳米管改性的玻碳电极作为内部传感器，开发的基于生物传感器的流动注射分析系统已被提议用于监测有机磷农药和神经毒剂。对于氨基甲酸酯的检测，流动注射分析系统基于半一次性反应器，其中 AChE 通过共价键结合固定在硅胶上[83]。该检测基于 pH 和电导率的测量，开发的方法用于检测水中的甲萘威，无需任何样品预浓缩。使用 pH 敏感型探针和高灵敏度荧光检测器开发了一种通过抑制可溶性 AChE 来测定有机磷农药和氨基甲酸酯农药的类似方法[84]。另一种利用流动注射分析系统的有机磷杀虫剂分光光度分析法使用了填充有 AChE 的反应器，其中 AChE 固定在胺化的玻璃微珠上。使用埃尔曼试剂对硫代乙酰胆碱酶水解过程中形成的硫代胆碱进行光学检测，该检测方法用于测定残留酶的活性。生物反应器在每次使用双复磷抑制后在线再生[85]。该方法已应用于分析实际水样，最近发表的综述中报道了许多早期关于使用基于 AChE 的流动系统测定农药的出版物[86,87]。

除乙酰胆碱酯酶外，其他酶很少用于在流动注射分析条件下间接测定抑制剂。最近，碱性磷酸酶（以可溶形式）被用于测定一些选定的抑制剂[88]。单氟磷酸盐和氟离子选择性电极分别用作底物和检测器。应该强调的是，由于较慢抑制过程的动力学差异，该流动注射分析系统提高了这种抑制检测的选择性。由于这种方法，该系统可以检测 ppb 范围内的铍和钒酸盐离子，而不受其他抑制剂（包括重金属离子）的影响。另一个开发的流动注射分析系统用于测定黄嘌呤氧化酶抑制剂[89]。所有测量均使用可溶性酶并应用克拉克型氧电极作为检测器。酶促反应和抑制作用在流动注射分析系统中在线进行。该系统被用于槲皮素的抑制性检测，槲皮素是一种在抑制许多与活性氧相关的疾病中起重要作用的化合物。在优化条件下，流动注射分析可以在亚毫摩尔浓度范围内测定槲皮素。

2.3.3 酶辅因子检测

一些酶是金属蛋白，在它们的生物活性中心具有起辅助因子作用的金属离子。金属离子的去除会导致形成脱辅基酶（没有生物催化活性）。相反，将辅助因子添加到脱辅基酶可以恢复酶的活性。这种可逆转化构成了间接测定辅助因子、离子（再生剂）或螯合剂（灭活剂）的生物传感平台。这两种可逆操作（失活和再生）在流动注射分析条件下似乎特别容易重复执行。虽然这种方法具有很大的潜力，但关于流动注射分析条件下脱辅基酶形成和再生的出版物少得令人惊讶。

使用 α-淀粉酶证明了基于脱辅基酶的识别可以检测钙离子[90]。流动注射分析系统

已用于评估固定化 α-淀粉酶的活性，这是一种生物识别元件。酶被固定在流通微型柱中。淀粉用作酶促反应的底物。使用包含额外双层柱的流动注射分析歧管可以进行测定，该柱由固定化的 α-葡萄糖苷酶和吡喃糖氧化酶组成，最后是电流检测器。该系统能够检测毫摩尔水平的 Ca^{2+}。

另一种更简单的流动注射分析系统基于用碱性磷酸酶共价结合的单分子层改性的 pH 电极[44]。这种酶的辅因子作用源自锌离子。半胱氨酸用作螯合剂，用于生成固定化脱辅基酶。分析过程的所有步骤，包括酶失活（生物受体形成）、活性恢复（分析物识别）、活性检测（底物的信号测量）都是通过载体成分的交替变化在线重复进行。报道的电位法流动注射分析系统能够选择性检测微摩尔范围内的锌离子。

2.4 结论

长期以来，酶一直被用于分析。如今，酶在与流动注射分析相结合时发挥着重要作用。它们被认为是分析物，并且有大量的流动注射分析程序来确定它们的活性。此外，酶不仅负责底物还包括抑制剂和辅助因子的生物分子识别。考虑到生物分子识别过程的多样性，酶在不同测定中的应用数量并不奇怪。使用酶的新生物检测方案仍在开发中。此外，对于已经开发的分析系统，人们正在探索新型材料、技术和仪器概念，如纳米颗粒应用、薄膜和厚膜微器件，以及用于顺序注射分析和微全分析系统的新型平台。除了这些发展之外，人们还关注如何实现更低的检测限和降低试剂与样品消耗量以及缩短分析时间。所有这些需求都可以通过各种基于酶的流动注射分析系统来满足。最后，应该记住，流动注射分析不可避免地与实际分析化学相关。正如本章所示，基于酶的流动注射分析系统完美地证实了这一说法。

最后应该提到的是，酶的活性检测在本章未报道的一大类生物测定中也起着重要作用。一些酶，如碱性磷酸酶、辣根过氧化物酶、葡萄糖氧化酶和脲酶，常被用作各种免疫化学分析方法中的生物催化标记物。使用酶标抗体和免疫识别系统的其他组件是一种常见和流行的做法，因为酶标记物能够显著放大分析信号，从而降低检测限。值得注意的是，用酶标记生物分子具有与基因分析相同的优势。所有这些基于生物亲和力的分析物识别以及酶活性检测都可以在流动注射分析中实现，显著提高了此类生物的测定质量。

参考文献

[1] Panoutsou, P. and Economou, A. (2005) Rapid enzymatic chemiluminescent assay of glucose by means of a hybrid flow injection/sequential-injection method. *Talanta*, 67, 603–609.

[2] Economou, A., Panoutsou, P. and Themelis, D.G. (2006) Enzymatic chemiluminescent assay of glucose by sequential-injection analysis with soluble enzyme and on-line sample dilution. *Analytica Chimica Acta*, 572, 140–147.

[3] Campuzano, S., Pedrero, M. and Pingarrón, J.M. (2005) A peroxidasetetrathiafulvalene biosensor based on selfassembled monolayer modified Au electrodes for the flow injection determination of hydrogen

peroxide. *Talanta*, 66, 1310–1319.

[4] Ledru, S., Ruillé, N. and Boujtita, M. (2006) One-step screen-printed electrode modified in its bulk with HRP based on direct electron transfer for hydrogen peroxide detection in flow injection mode. *Biosensors & Bioelectronics*, 21, 1591–1598.

[5] Tripathi, V. S., Kandimalla, V. B. and Ju, H. (2006) Amperometric biosensor for hydrogen peroxide based on ferrocene-bovine serum albumin and multiwall carbon nanotube modified ormosil composite. *Biosensors & Bioelectronics*, 21, 1529–1535.

[6] Kotzian, P., Brázdilová, P., Řezková, S., Kalcher, K. and Vytřas, K. (2006) Amperometric glucose biosensor based on rhodium dioxide-modified carbon ink. *Electroanalysis*, 18, 1499–1504.

[7] Derwińska, K., Miecznikowski, K., Koncki, R., Kulesza, P. J., Głąb, S. and Malik, M. (2003) Application of prussian blue based composite film with organic polymer to construction of enzymatic glucose biosensor. *Electroanalysis*, 15, 1843–1849.

[8] De Mattos, I. L. and Da Cunha Areias, M. C. (2005) Automated determination of glucose in soluble coffee using Prussian Blue–glucose oxidase–Nafion® modified electrode. *Talanta*, 66, 1281–1286.

[9] Yu, J., Liu, S. and Ju, H. (2003) Glucose sensor for flow injection analysis of serum glucose based on immobilization of glucose oxidase in titania sol-gel membrane. *Biosensors & Bioelectronics*, 19, 401–409.

[10] Osborne, P. G. and Hashimoto, M. (2004) Chemical polymerization of mphenylenediamine, in the presence of glucose oxidase, produces an enzyme – retaining electrooxidisable polymer used to produce a biosensor for amperometric detection of glucose from brain dialysate. *Analyst*, 129, 759–765.

[11] Barsan, M. M., Klinčar, J., Batič, M. and Brett, C. M. A. (2007) Design and application of a flow cell for carbon-film based electrochemical enzyme biosensors. *Talanta*, 71, 1893–1900.

[12] Tkac, J., Whittaker, J. W. and Ruzgas, T. (2007) The use of single walled carbon nanotubes dispersed in a chitosan matrix for preparation of a galactose biosensor. *Biosensors & Bioelectronics*, 22, 1820–1824.

[13] Niculescu, M., Sigina, S. and Csöregi, E. (2003) Glycerol dehydrogenase based amperometric biosensor for monitoring of glycerol in alcoholic beverages. *Analytical Letters*, 36, 1721–1737.

[14] Carelli, D., Centonze, D., De Giglio, A., Quinto, M. and Zambonin, P. G. (2006) An interference-free first generation alcohol biosensor based on a gold electrode modified by an overoxidised nonconducting polypyrrole film. *Analytica Chimica Acta*, 565, 27–35.

[15] Pati, S., Quinto, M., Palmisano, F. and Zambonin, P. G. (2004) Determination of choline in milk, milk powder, and soy lecithin hydrolysates by flow injection analysis and amperometric detection with a choline oxidase based biosensor. *Journal of Agricultural and Food Chemistry*, 52, 4638–4642.

[16] Divritsioti, M. H., Karalemas, I. D., Georgiou, C. A. and Papastathopoulos, D. S. (2003) Flow injection analysis system for L – lysine estimation in foodstuffs using a biosensor based on lysine oxidase immobilization on a gold-poly (m-phenylenediamine) electrode. *Analytical Letters*, 36, 1939–1963.

[17] Varma, S., Yigzaw, Y. and Gorton, L. (2006) Prussian blue-glutamate oxidase modified glassy carbon electrode: A sensitive L-glutamate and b-N-oxalyl-a, b-diaminopropionic acid (β-ODAP) sensor. *Analytica Chimica Acta*, 556, 319–325.

[18] Radoi, A., Compagnone, D., Devic, E. and Palleschi, G. (2007) Low potential detection of NADH with Prussian Blue bulk modified screen – printed electrodes and recombinant NADH oxidase from *Thermus thermophilus*. *Sensors and Actuators B*, 121, 501–506.

[19] Esti, M., Volpe, G., Micheli, L., Delibato, E., Compagnone, D., Moscone, D. and Palleschi, G. (2004) Electrochemical biosensors for monitoring malolactic fermentation in red wine using two strains of Oenococcus oeni. *Analytica Chimica Acta*, 513, 357-364.

[20] Maestre, E., Katakis, L., Narváez, A. and Domínguez, E. (2005) A multianalyte flow electrochemical cell: Application to the simultaneous determination of carbohydrates based on bioelectrocatalytic detection. *Biosensors & Bioelectronics*, 21, 774-781.

[21] Harreither, W., Coman, V., Ludwig, R., Haltrich, D. and Gorton, L. (2007) Investigation of graphite electrodes modified with cellobiose dehydrogenase from the Ascomycete Myriococcum thermophilum. *Electroanalysis*, 19, 172-180.

[22] Campuzano, S., Gamella, M., Serra, B., Reviejo, A. J. and Pingarrón, J. M. (2007) Integrated electrochemical gluconic acid biosensor based on self-assembled monolayer-modified gold electrodes. Application to the analysis of gluconic acid in musts and wines. *Journal of Agricultural and Food Chemistry*, 55, 2109-2114.

[23] Takagi, K. and Shikata, S. (2004) Flow injection determination of histamine with a histamine dehydrogenase-based electrode. *Analytica Chimica Acta*, 505, 189-193.

[24] Campuzano, S., Serra, B., Pedrero, M., De Villena, F. J. M. and Pingarrón, J. M. (2003) Amperometric flow injection determination of phenolic compounds at self-assembled monolayer-based tyrosinase biosensors. *Analytica Chimica Acta*, 494, 187-197.

[25] Serra, B., Reviejo, A. J. and Pingarrón, J. M. (2003) Flow injection amperometric detection of phenolic compounds at enzyme composite biosensors application to their monitoring during industrial waste waters purification processes. *Analytical Letters*, 36, 1965-1986.

[26] Gamella, M., Campuzano, S., Reviejo, A. J. and Pingarrón, J. M. (2006) Electrochemical estimation of the polyphenol index in wines using a laccase biosensor. *Journal of Agricultural and Food Chemistry*, 54, 7960-7967.

[27] Haghighi, B., Jarosz-Wilkołązka, A., Ruzgas, T., Gorton, L. and Leonowicz, A. (2005) Characterization of graphite electrodes modified with laccases from Trametes hirsuta and Cerrena unicolor and their use for flow injection amperometric determination of some phenolic compounds. *International Journal of Environmental Analytical Chemistry*, 85, 753-770.

[28] Mitsubayashi, K., Kubotera, Y., Yano, K., Hashimoto, Y., Kon, T., Nakakura, S., Nishi, Y. and Endo, H. (2004) Trimethylamine biosensor with flavin-containing monooxygenase type 3 (FMO3) for fish-freshness analysis. *Sensors and Actuators B*, 103, 463-467.

[29] Schöning, M. J., Krause, R., Block, K., Musahmeh, M., Mulchandani, A. and Wang, J. (2003) A dual amperometric/potentiometric FIA-based biosensor for the distinctive detection of organophosphorus pesticides. *Sensors and Actuators B*, 95, 291-296.

[30] Yao, T. and Yano, T. (2004) On-line microdialysis assay of L-lactate and pyruvate *in vitro* and *in vivo* by a flow injection system with a dual enzyme electrode. *Talanta*, 63, 771-775.

[31] Nanjo, Y., Yano, T, Hayashi, R. and Yao, T. (2006) Optically specific detection of D-and L-lactic acids by a flow injection dual biosensor system with on-line microdialysis sampling. *Analytical Sciences*, 22, 1135-1138.

[32] Katrlík, J., Mastihuba, V., Voštiar, I., Šefčovičova, J., Štefuca, V. and Gemeiner, P. (2006) Amperometric biosensors based on two different enzyme systems and their use for glycerol determination in samples from biotechnological fermentation process. *Analytica Chimica Acta*, 566, 11-18.

[33] Basu, A. K., Chattopadhyay, P., Roychudhuri, U. and Chakraborty, R. (2006) A biosensor based on coimmobilized L-glutamate oxidase and L-glutamate dehydrogenase for analysis of monosodium glutamate in food. *Biosensors & Bioelectronics*, 21, 1968-1972.

[34] Mizutani, F., Hirata, Y., Yabuki, S. and Iijima, S. (2003) Flow injection analysis of acetic acid in food samples by using trienzyme/poly (dimethylsiloxane) -bilayer membrane-based electrode as the detector. *Sensors and Actuators B*, 91, 195-198.

[35] Mieliauskiene, R., Nistor, M., Laurinavicius, V. and Csoregi, E. (2006) Amperometric determination of acetate with a tri-enzyme based sensor. *Sensors and Actuators B*, 113, 671-676.

[36] Yao, T., Yano, T. and Nishino, H. (2004) Simultaneous *in vivo* monitoring of glucose, L-lactate, and pyruvate concentrations in rat brain by a flow injection biosensor system with an on-line microdialysis sampling. *Analytica Chimica Acta*, 510, 53-59.

[37] Yao, T., Takashima, K. and Nanjyo, Y. (2003) Simultaneous determination of orthophosphate and total phosphates (inorganic phosphates plus purine nucleotides) using a bioamperometric flow injection system made up by a 16-way switching valve. *Talanta*, 60, 845-851.

[38] Tymecki, Ł., Zwierkowska, E. and Koncki, R. (2005) Strip bioelectrochemical cell for potentiometric measurements fabricated by screen-printing. *Analytica Chimica Acta*, 538, 251-256.

[39] Tymecki, Ł. and Koncki, R. (2006) Thickfilm potentiometric biosensor for bloodless monitoring of hemodialysis. *Sensors and Actuators B*, 113, 782-786.

[40] Radomska, A., Koncki, R., Pyrzyńska, K. and Głąb, S. (2004) Bioanalytical system for control of hemodialysis treatment based on potentiometric biosensors for urea and creatinine. *Analytica Chimica Acta*, 523, 193-200.

[41] Radomska, A., Bodenszac, E., Głąb, S. and Koncki, R. (2004) Creatinine biosensor based on ammonium ion selective electrode and its application in flow injection analysis. *Talanta*, 64, 603-608.

[42] Rasmussen, C. D., Andersen, J. E. T. and Zachau-Christiansen, B. (2007) Improved performance of the potentiometric biosensor for the determination of creatinine. *Analytical Letters*, 40, 39-52.

[43] García-Villar, N., Saurina, J. and Hernández-Cassou, S. (2003) Flow injection differential potentiometric determination of lysine by using a lysine biosensor. *Analytica Chimica Acta*, 477, 315-324.

[44] Rozum, B., Koncki, R. and Tymecki, Ł. (2007) The potentialities of pH-electrode modified with alkaline phosphatase. *Sensors and Actuators B*, 127, 632-636.

[45] Ho, J., Wu, L., Fan, N. C., Lee, M. S., Kuo, H. Y. and Yang, C. S. (2007) Development of a long-life capillary enzyme bioreactor for the determination of blood glucose. *Talanta*, 71, 391-396.

[46] Jones, D. A., Parkin, M. C., Langemann, H., Landolt, H., Hopwood, S. E., Strong, A. J. and Boutelle, M. G. (2002) On-line monitoring in neurointensive care: Enzyme-based electrochemical assay for simultaneous, continuous monitoring of glucose and lactate from critical care patients. *Journal of Electroanalytical Chemistry*, 538-539, 243-252.

[47] Yao, T, Yano, T., Nanjyo, Y. and Nishino, H. (2003) Simultaneous determination of glucose and L-lactate in rat brain by an electrochemical *in vivo* flow injection system with an on-line microdialysis sampling. *Analytical Sciences*, 19, 61-65.

[48] Nanjo, Y, Hayashi, R. and Yao, T. (2006) Determination of fructosyl amino acids and fructosyl peptides in protease-digested blood sample by a flow injection system with an enzyme reactor. *Analytical Sciences*, 22, 1139-1143.

[49] Nanjo, Y., Hayashi, R. and Yao, T. (2007) An enzymatic method for the rapid measurement of

the hemoglobin A_{1C} by a flow injection system comprised of an electrochemical detector with a specific enzyme-reactor and a spectrophotometer. *Analytica Chimica Acta*, 583, 45-54.

[50] Wu, L. C. and Cheng, C. M. (2005) Flowinjection enzymatic analysis for glycerol and triacylglycerol. *Analytical Biochemistry*, 346, 234-240.

[51] Vianello, F., Ragusa, S., Cambria, M. T. and Rigo, A. (2006) A high sensitivity amperometric biosensor using laccase as biorecognition element. *Biosensors & Bioelectronics*, 21, 2155-2160.

[52] Limbut, W., Thavarungkul, P., Kanatharana, P., Asawatreratanakul, P., Limsakul, C. and Wongkittisuksa, B. (2004) Comparative study of controlled pore glass, silica gel and Poraver for the immobilization of urease to determine urea in a flow injection conductimetric biosensor system. *Biosensors & Bioelectronics*, 19, 813-821.

[53] Kim, M. and Kim, M. J. (2003) Isocitrate analysis using a potentiometric biosensor with immobilized enzyme in a FIA system. *Food Research International*, 36, 223-230.

[54] Kim, M. (2006) Determining citrate in fruit juices using a biosensor with citrate lyase and oxaloacetate decarboxylase in a flow injection analysis system. *Food Chemistry*, 99, 851-857.

[55] Matos, R. C., Coelho, E. O., de Souza, C. F., Guedes, F. A. and Matos, M. A. C. (2006) Peroxidase immobilized on Amberlite IRA-743 resin for on-line spectrophotometric detection of hydrogen peroxide in rainwater. *Talanta*, 69, 1208-1214.

[56] De Oliveira, A. C. A., Assis, V. C., Matos, M. A. C. and Matos, R. C. (2005) Flowinjection system with glucose oxidase immobilized on a tubular reactor for determination of glucose in blood samples. *Analytica Chimica Acta*, 535, 213-217.

[57] Vojinovič, V., Calado, C. R., Silva, A. I., Mateus, M., Cabral, J. M. S. and Fonseca, L. P. (2005) Micro-analytical GO/HRP bioreactor for glucose determination and bioprocess monitoring. *Biosensors & Bioelectronics*, 20, 1955-1961.

[58] Vojinovič, V., Esteves, F. M. F., Cabral, J. M. S. and Fonseca, L. P. (2006) Bienzymatic analytical microreactors for glucose, lactate, ethanol, galactose and L-amino acid monitoring in cell culture media. *Analytica Chimica Acta*, 565, 240-249.

[59] Luque-Almagro, V. M., Blasco, R., Fernandez-Romero, J. M. and Lucue de Castro, M. D. L. (2003) Flow-injection spectrophotometric determination of cyanate in bioremediation processes by use of immobilised inducible cyanase. *Analytical and Bioanalytical Chemistry*, 377, 1071-1078.

[60] Carvalho Vieira, E. and Nogueira, A. R. A. (2004) Orthophosphate, phytate, and total phosphorus determination in cereals by flow injection analysis. *Journal of Agricultural and Food Chemistry*, 52, 1800-1803.

[61] Iida, Y., Ikeda, M., Aoto, M. and Satoh, I. (2004) Fluorometric determination of urea in alcoholic beverages by using an acid urease column-FIA system. *Talanta*, 64, 1278-1282.

[62] Tsukatani, T. and Matsumoto, K. (2005) Sequential fluorometric quantification of malic acid enantiomers by a single line flow injection system using immobilized-enzyme reactors. *Talanta*, 65, 396-401.

[63] Tsukatani, T. and Matsumoto, K. (2006) Flow-injection fluorometric quantification of pyruvate using co-immobilized pyruvate decarboxylase and aldehyde dehydrogenase reactor: Application to measurement of acetate, citrate and l-lactate. *Talanta*, 69, 637-642.

[64] Tsukatani, T. and Matsumoto, K. (2005) Fluorometric quantification of total D-gluconate by a flow injection system using an immobilized-enzyme reactor. *Analytica Chimica Acta*, 530, 221-225.

[65] Rhee, J. I. and Sohn, O. J. (2003) Flow injection system for on-line monitoring of fumaric acid in

biological processes. *Analytica Chimica Acta*, 499, 71–80.

[66] Sohn, O. J. , Han, K. A. and Rhee, J. I. (2005) Flow injection analysis system for monitoring of succinic acid in biotechnological processes. *Talanta*, 65, 185–191.

[67] Tsukatani, T. and Matsumoto, K. (2005) Sequential fluorometric quantification of γ–aminobutyrate and L–glutamate using a single line flow injection system with immobilized–enzyme reactors. *Analytica Chimica Acta*, 546, 154–160.

[68] Manera, M. , Miró, M. , Estela, J. M. and Cerdá, V. (2004) A multisyringe flow injection system with immobilized glucose oxidase based on homogeneous chemiluminescence detection. *Analytica Chimica Acta*, 508, 23–30.

[69] Davidsson, R. , Genin, F. , Bengtsson, M. , Laurell, T. and Emnéus, J. (2004) Microfluidic biosensing systems part I. Development and optimisation of enzymatic chemiluminescent μ–biosensors based on silicon microchips. *Lab on a Chip*, 4, 481–487.

[70] Yaqoob, M. and Nabi, A. (2003) Flow injection chemiluminescent assays for glycerol and triglycerides using a co–immobilized enzyme reactor. *Luminescence*, 18, 67–71.

[71] Kiba, N. , Ito, S. , Tachibana, M. , Tani, K. and Koizumi, H. (2003) Simultaneous determination of choline and acetylcholine based on a trienzyme chemiluminometric biosensor in a single line flow injection system. *Analytical Sciences*, 19, 1647–1651.

[72] Amini, N. and McKelvie, I. (2005) An enzymatic flow analysis method for the determination of phosphatidylcholine in sediment pore waters and extracts. *Talanta*, 66, 445–452.

[73] Nakamura, H. , Hasegawa, M. , Nomura, Y. , Ikebukuro, K. , Arikawa, Y. and Karube, I. (2003) Improvement of a CL–FIA system using maltose phosphorylase for the determination of phosphate–ion in freshwater. *Analytical Letters*, 36, 1805–1817.

[74] Yaqoob, M. , Anwar, M. and Nabi, A. (2005) Determination of phosphate in fresh waters by flow injection with immobilized enzyme and chemiluminescence detection. *International Journal of Environmental Analytical Chemistry*, 85, 451–459.

[75] Almeida, C. F. , Calado, C. R. C. , Bernardino, S. A. , Cabral, J. M. S. and Fonseca, L. P. (2006) A flow injection analysis system for on–line monitoring of cutinase activity at outlet of an expanded bed adsorption column almost in real time. *Journal of Chemical Technology and Biotechnology*, 81, 1678–1684.

[76] Almeida, C. F. , Cabral, J. M. S. and Fonseca, L. P. (2004) Flow injection analysis system for on–line cutinase activity assay. *Analytica Chimica Acta*, 502, 115–124.

[77] Theaker, B. J. and Rowell, F. J. (2003) A rapid and sensitive fluorometric flow injection assay for subtilisin–type enzymes utilising sol–gel particles directly coated with gelatin–Texas Red substrate. *Analyst*, 128, 1043–1047.

[78] Zajoncová, L. , Jílek, M. , Beranová, V. and Peč, P. (2004) A biosensor for the determination of amylase activity. *Biosensors & Bioelectronics*, 20, 240–245.

[79] Yamaguchi, M. , Kanemaru, M. , Kanemori, T. and Mizuno, Y. (2003) Flow–injection–type biosensor system for salivary amylase activity. *Biosensors & Bioelectronics*, 18, 835–840.

[80] Hartwell, S. K. , Somprayoon, D. , Kongtawelert, P. , Ongchai, S. , Arppornchayanon, O. , Ganranoo, L. , Lapanantnoppakhun, S. and Grudpan, K. (2007) Online assay of bone specific alkaline phosphatase with a flow injectionbead injection. *Analytica Chimica Acta*, 600, 188–193.

[81] Ogończyk, D. and Koncki, R. (2007) Potentiometric flow injection system for determination of alkaline phosphatase in human serum. *Analytica Chimica Acta*, 600, 194–198.

[82] Liu, G. and Lin, Y. (2006) Biosensor-based on self-assembling acetylcholinesterase on carbon nanotubes for flow injection/ amperometric detection of organophosphate pesticides and nerve agents. *Analytical Chemistry*, 78, 835-843.

[83] Suwansa-Ard, S., Kanatharana, P., Asawatreratanakul, P., Limsakul, C., Wongkittisuksa, B. and Thavarungkul, P. (2005) Semi disposable reactor biosensors for detecting carbamate pesticides in water. *Biosensors & Bioelectronics*, 21, 445-454.

[84] Jin, S., Xu, Z., Chen, J., Liang, X., Wu, Y. and Qian, X. (2004) Determination of organophosphate and carbamate pesticides based on enzyme inhibition using a pH-sensitive fluorescence probe. *Analytica Chimica Acta*, 523, 117-123.

[85] Dǎneţ, A. F., Bucur, B., Cheregi, M. C., Badea, M. andS erban, S. (2003) Spectrophotometric determination of organophosphoric insecticides in a FIA system based on AChE inhibition. *Analytical Letters*, 36, 59-73.

[86] Prieto-Simón, B., Campás, M., Andreescu, S. and Marty, J. L. (2006) Trends in flow-based biosensing systems for pesticide assessment. *Sensors*, 6, 1161-1186.

[87] Solé, S., Merkoci, A. and Alegret, S. (2003) Determination of toxic substances based on enzyme inhibition. Part II. Electrochemical biosensors for the determination of pesticides, using flow systems. *Critical Reviews in Analytical Chemistry*, 33, 127-143.

[88] Koncki, R., Rudnicka, K. and Tymecki, Ł. (2006) Flow injection system for potentiometric determination of alkaline phosphatase inhibitors. *Analytica Chimica Acta*, 577, 134-139.

[89] Lam, L. H., Sakaguchi, K., Ukeda, H. and Sawamura, M. (2006) Flow injection determination of xanthine oxidase inhibitory activity and its application to food samples. *Analytical Sciences*, 22, 105-109.

[90] Iida, Y., Sato, Y. and Satoh, I. (2003) Novel detection system for calcium (II) ions based on an apoenzyme reactivation method using an amylase column as a recognition element. *Electrochemistry*, 71, 449-452.

3 流动电位法

M. Conceicão B. S. M. Montenegro 和 Alberto N. Araújo

3.1 引言

分析化学部门协助解决了许多与人类福祉和可持续发展相关领域的问题。然而，可用于它们活动的财政资源有限，因此，方法和程序的自动化可以尽可能提高测定的数量和性质与其单一成本之间的比率。自动化显著减少了劳动密集型任务，并可以在短时间内产生高质量的结果。在受控流体动力学条件下运行的液体流动技术是当今自动化方案的一个重要组成部分。当需要实现其他操作单元的自动化时，可以在样品分离或连续流动分析系统中使用色谱法或电迁移技术。在最近的发展进程中，这些工具与检测器设备一起被集成和小型化，以更环保的方式减少试剂和样品的消耗。与更常见的光谱分析仪器一样，使用离子选择性电极（ISEs）的电位计已被证明十分适用于连续流动系统。除了可以在广泛的分析物浓度范围内以非破坏性方式检测外，ISEs 还具有能耗低、小型化但不会显著降低灵敏度和成本低的特点。本章重点回顾了连续流动系统中电位检测的主要内容：最新进展、耦合配置和实际应用，以及趋势和未来发展前景。

3.2 背景概念

常见的本科教科书将电位法定义为一种分析技术，其中为了达到电化学流动池内化学平衡而产生的自由能变化是在没有电流流动的情况下，通过电位测量的。这一经典的定义符合 Luigi Galvani（1737—1798 年）和 Walter Nernst 在 1888 年提出的初步实验，当时它们试图解释将金属丝浸入其自身离子溶液或将惰性金属丝浸入氧化还原离子对溶液时产生电势的原因[1]。实验结果是，Nernst 开发了第一个能够测量溶液酸度的电极，尽管其应用由于结构复杂而受到限制[2]。仅仅九年后，Cremer 观察到原电池的电势取决于膜两侧质子浓度的差异，其中原电池的一半组成由薄玻璃膜隔开[3]。因此，玻璃电极实际上是在近一个世纪前被提出的第一个离子选择性电极[4]，但人们错误地认为其响应是由透过玻璃膜的质子扩散控制的[5]。尽管在早期取得了这些成就，非传统电极直到 20 世纪下半叶才普遍被人们使用。这是由于理论方面的巩固[6,7]、高阻抗电压表的出现[8,9]、对相关种类电极的构思（带有对氟化物有选择性的结晶膜[10]或包含对钙[11]和钾[12,13]具有选择性的高黏性液体膜）以及这些构思的快速实现[14]。由于这些电极在 20 世纪 90 年代被大规模引入常规临床化学，人们因此能够测定钠、钾、氯、钙和质子。这主要是因为它们的操作几乎不需要维护，并且能够在全血、血浆、血清和尿中进行直接测量[15,16]。如今，寻找新的电活性物质以提高选择性和/或扩展新的应用、朝着小型化和与自动化设备耦合所作的努力正在成为全球科学界的热点。

电位检测是通过将指示电极和参比电极同时浸入待测溶液中来实现的。离子选择性电极通常用作指示电极，而在某些情况下使用基于金属表面的常规电极会得到令人满意的结果。电池电位的大小取决于所用电极的性质以及样品和参照溶液连接处的离子迁移率。因此，观察到的不同样品之间电池电位的变化主要是由于与指示电极表面或膜接触的被监测物质的活性发生了变化,积累的关于 ISE 性能的知识已经证明样品离子在样品/膜界面处的分配(该分配是为了达到化学平衡)决定了电位响应[17,18]。这种相界电位模型基于两点假设可以用来预测分析响应:一是假设与水溶液样品接触的有机相处于化学平衡,二是膜内的扩散电位可以忽略[19]。根据模型,在 ISE 的有机/水相边界中产生的电势由下式给出:

$$E = \frac{RT}{z_i F}\ln k_1 + \frac{RT}{z_i F}\ln \frac{a_i(\mathrm{aq})}{a_i(\mathrm{org})}$$

其中 $a_i(\mathrm{aq})$ 和 $a_i(\mathrm{org})$ 分别是水样和有机相接触边界中未复合的监测离子(带电荷 z_i)的活性。对于常数 $a_i(\mathrm{org})$,预计会出现类似于经典电极的能斯特响应:

$$E_i = E_i^0 + \frac{RT}{z_i F}\ln a_I$$

方程中的第一项表示一次的标准电位,k_1 是样品相和膜相中离子溶剂化相对自由能的函数。在这种情况下,人们可以理解提出新化学实体的根本原因,这些新化学实体能够萃取膜中的目标离子并能保持这些离子的活性(最好是低活性)。效果比较好的膜可以通过以下两种方式制备:一种是通过疏水带电离子交换剂的组合,该交换剂将目标离子而不是反离子拉入膜(膜的选择渗透特性),另一种是通过选择性疏水络合剂(称为阴离子载体)来缓冲其活性[20](图 3.1)。

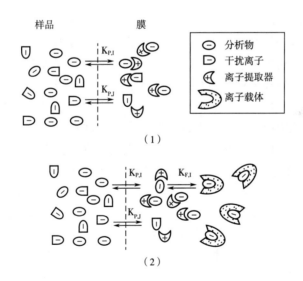

图 3.1 基于使用(1)离子提取器的离子选择性膜的平衡机制;(2)离子提取器加离子载体。
样品溶液侧反离子以及膜侧游离和键合的主要离子或干扰离子间的平衡并未显示

膜的定性和定量成分以及膜结构决定了电极的最终特性,例如耐用性、分析范围、斜率、响应时间和选择性。在分析复杂样品时,具有相同信号电荷或更亲脂性的其他离子[分别具有不同的 $a_i(\mathrm{org})$ 和 k_1 值,见上文]可能会在一定程度上干扰最终电位读数。在这些条件下,ISE 电位可以通过半经验的 Nikolskii-Eisenmann 方程[6,7]更好地描述,其对样品中目标离子能斯特响应的三个或四个活性数据几十年有效:

$$E_i = E_i^0 + S\lg\left(a_i + \sum K_j a_j^{z_i/z_j}\right)$$

其中 E_i 是被监测离子的活性 a_i 和其他（干扰）离子的活性 a_j 所贡献的电极电位。每个干扰离子的贡献取决于其在样品中的实际活性、电荷的比率和电位选择性系数 K_j。根据 IUPAC 的建议[21]确定的系数高于或低于电极的系数，因而赋予目标离子更低或更高的电位响应（与干扰离子相比）。最后，S 表示电极电位和主要离子活性之间半对数关系的实际斜率。在25℃时，对于带有单个正电荷或负电荷的主要离子，S 的正或负值为59.2mV/decade。对于双电荷离子，该值的净值降至29.6mV/decade。

尽管由于选择性特性和宽线性动态范围，通常需要简单的样品预处理，但如果在流动条件下进行基于电位的测量，该情形通常会得到改善[22,23]。长久的液体流体对 ISE 膜具有调节作用，从而获得更好的电位稳定性。与使用传统方法获得的电位相比，可重复的样品体积和传输时间有助于减少分析时间并提高电位读数的重现性和灵敏度。此外，与传统方法相比，选择性电位系数有所提高。一般来说，该方法对干扰离子（对膜系统具有更强亲和力）结果的提高比主要离子更明显，因为瞬态浓度和非热力学平衡通常被用于分析测定[24]。

3.3　电极的开发和检测器池的设计

无论电位测定的实际应用模式是什么，是间歇还是连续系统，其灵敏度和选择性主要取决于指示电极的特性和检测池的设计。传统的金属电极由于选择性不足而使开发过程受到限制，这也阻碍了其应用于痕量分析或具有复杂基质的样品分析，相比之下，聚合物膜离子选择性电极可以根据最终用途定制选择性分布范围，该类型的电极正成为一种提高电位应用分析的更具吸引力的方式[25]。它们通常可以描述为嵌有离子交换剂的高黏性膜相或所谓的"离子载体"或"离子载体"的亲脂性络合剂，其决定了 ISE 最终的选择性表现。这些术语仍然源自最初使用的能够通过亲脂相（例如生物细胞膜或人工模型膜）传输离子的试剂[26,27]。这种类型电极的理论建模是最新的[28-31]，并且已经能够预测如何改善响应特性。虽然电位计应当在几个分析领域中具有吸引力（由于数十年的活性或浓度范围），但是它的应用受到普遍报道的微摩尔检测限的影响。这种行为要么来自主离子的反扩散（样品中存在干扰离子），要么来自内部参比溶液的跨膜浸出。这些现象增加了 ESI/样品界面中的主要离子活性，影响了稀释样品的测量。在内部参比溶液中使用高浓度的干扰离子和低浓度的初级离子可以提高 ESI 检测限并保证热力学选择性系数的准确性。该理论强调，检测限在皮摩尔范围内的 Pb^{2+} 选择性电极[32]最初被提出，随后是对银[33]、钙[34]、镉[35]和碘化物[36]敏感的新电极，从而开辟了它们在环境分析中的应用[37]。未来的改进将以更深入理解基本原理和显著减少实验偏差为指导，而不是基于与其他分析技术类似的技术成就[38,39]。

实验室中电位测定法的使用不仅仅实现了节省成本、简易化或改善检测限，也许更重要的是提供准确的结果，这取决于 ISE 膜对分析物与样品中其他离子的区分能力。如前所述，这种能力是通过建立主要离子对每个干扰离子的电位选择系数来评估的。通过亲脂性高分子可以很容易地制备膜，这些亲脂性分子负责样品溶液进行跨膜界面的离子交换。分析物的季铵盐或鏻盐主要用于检测阴离子物质和四烷基硼酸盐衍生物或用于阳离子敏感膜的杂多酸[40-42]。然而，人们认为与 Hofmeister 系列类似的一般选择性模式与离子交换器和被测离子的疏水特性有关。许多研究人员已经表明，选择性模式可以通过额外加入离子载体来改变和改善，尽管其他膜成分的性质和相对含量仍然在很大程度上决定了它们最终的电分析性能[43,44]。离子载体是在超分子化学领域发展起来的具有阳离子或阴离子选择性识别能力的有机物质。根据新理论，电位膜中的主要离子活性被离子载体缓冲，并与之形成高度稳定的复合物[45]。依据分析物的性质和所带电荷，人们对离子载体的类别进行了分类。例如，含有庞大亚基或噻唑[46-49]的冠醚，在刚性分子上具有氢键官能团（例如，尿素、硫脲和胍）的阴离子选择性物质框架[50-54]，在下缘或上缘带有识别离子的环芳烃[55-57]，吡咯和聚吡咯类[58,59]，金属卟啉和金属酞菁[60]，手性阴离子识别环糊精和脱氧胆酸类[61]。然而，这是一个非常开放的研究领域，在未来几年所涉及的知识会大量增加，对于阴离子而言尤其如此，这是因为阴离子的尺寸、形状、水合作用和 pH 依赖性变量比阳离子大得多。回到 ISE 的传感膜上来，对其描述最多的缺陷之一是使用其嵌入成分进行渐进浸出，这会造成选择性的损失和灵敏度的下降。为了克服这一缺陷，人们提出了新的膜配方（不包含增塑剂溶剂）[62-66]和离子载体与膜聚合物之间的共价连接[63,65,67]，以及非常规聚合物的使用，例如，胺化或羧化 PVC、聚（乙烯-共-乙烯基）醋酸酯（EVA）[68]、聚甲基丙烯酸甲酯或聚苯乙烯[69]以及固体接触电极中电活性聚合物的混合物[58]。

尽管之前主要报道的是用于间歇分析的电极，但其中一些考虑了不连续流动条件，在该条件下 ISE 内部参比溶液的去除，如稍后所述，具有特别重要的意义。大多数工作都集中在通过有机聚合物膜或化学改性碳糊电极的离子交换剂（例如杂多酸）配制的电极来测定阳离子物质[70-90]（表 3.1）。后者具有容易制备/再生的优点，并且具有低电位漂移和欧姆电阻。所报道的对选择性分布的改进[81]似乎与流动条件所提供的动力学识别更相关。我们的团队开发了新的 ISE，其中包含具有特定金属-配体相互作用（金属卟啉）的离子载体或可以与脂肪族和芳香族物质（环糊精）形成质子桥接的包合物的离子载体。除了可以提高选择性之外，它们在测定氯化物[91]、特定脂肪族羧酸盐和芳香族物质（如双氯芬酸）[92]时产生了可重现的响应。考虑到电极的使用寿命会由于膜上的连续液体流而变短，以同样的方式，通过将 PVC 基质膜聚合物改为 EVA[93]或通过使用二氧化硅无机溶胶-凝胶载体[94]，可以延长电极的使用寿命。通过主客体识别的大环离子载体与光固化聚合物的结合[95-97]，不仅可以极大提升在导体载体上的附着力，还提高了电位响应的稳定性。在需要电位 pH 测量的连续流动应用领域也取得了重大进展。先前的方法表明由于电极设计（在玻璃电极条件下非常明显）、狭窄的 pH 范围和系统耦合脆弱性而难以实施连续流动。在这种情况下，建议使用玻璃膜[98]、PVC 或环氧丙烯酸酯[99,100]、金属/金属氧化物[101,102]、ISFET[103]和不锈钢电极[104]来实现连续流动设置。这两种基于醌氢醌[105,106]和 Fe_2O_3 石墨-环氧树脂管状电极（通过在聚氨酯树脂表面填充混合物[107]制备）的复合电极似乎具有与玻璃电极相当的特性，但其更易于实施和维护以适应流动安装和乳液中的 pH 测量。

表 3.1　最近（2000—2007 年）在流动进样系统中用于测定阳离子有机物质的电极膜组件

主离子	传感器固定化	膜组成 离子载体/增塑剂	构造类型	分析应用	参考文献
苯丙醇胺	聚合	PT/DOP 和 DBP	常规带内参比溶液	制药	[70]
特布它林	聚合	PT/DOP	常规带内参比溶液	制药	[71]
吡贝地尔	化学改性	ST、SM、PT	常规无内参比溶液	药物和生物液体（尿液）	[72]
莨菪碱	碳糊聚合	TPB、PT/DOP	常规带内参比溶液	制药	[73]
氯丙嗪	聚合	TPB/o-NPOE	管状的	制药	[74]
西地那非	聚合	TP、Re/o-NPOE	常规带内参比溶液	药物和生物液体（人血清）	[75]
肌酐	聚合	PT、MP、PC/o-NPOE	包覆线材和管材	生物液体（人血清）药物	[76]
双环胺	聚合	ST、SM、PT、MP、TPB/DBP	常规带内参比溶液	生物液体（血清和尿液）牛乳	[77]
美维柏林	聚合	ST、SM、PT、MP/DBP	常规带内参比溶液	药物	[78]
阿米替林	聚合	PT、MP 和 PT/MP/DOP 的混合物	常规带内参比溶液	生物液体（血清和尿液）药物	[79]
四环素	聚合（PVC 和 EVA）	t-TCPB/BEHS	管状	制药	[80]
哌醋甲酯	聚合	PT、MP 和 PT/MP/DOP 的混合物	常规带内参比溶液	制药	[81]
烷基酚	聚合	TPB/o-NPOE	管状	—	[82]
聚乙氧基化物双嘧达莫	聚合	TPB、Re/o-NPOE	常规带内参比溶液	制药	[83]
氨溴索	聚合	TPB/DOP	常规和带涂层石墨	制药	[84]
氯化铵	聚合	TPB/o-NPOE、BEHS、DBP	管状	河流水域	[85]
半胱氨酸	聚合	TPB、BTPPA/BEHS、o-NPOE	管状	制药	[86]
沙丁胺醇	聚合	PT、MP 和 PT/MP/DOP 的混合物	常规带内参比溶液	制药	[87]
雷尼替丁	聚合	TPB、PT/DOP、DBP	常规带内参比溶液	制药	[88]
屈它维林	聚合	ST、SM、PT、MPTPB/DBP、DOP、DOS、TCP、DINP	常规带内参比溶液	制药	[89]
氯氮卓	聚合	PT、MP/DBP	包覆线	制药	[90]

注：PT—磷钨酸盐；ST—硅钨酸盐；SM—硅钼酸盐；TPB—四苯基硼酸盐；Re—赖内克酸盐；MP—磷酸钼；PC—苦杏仁酸；t-TCPB—四（4-氯苯基）硼酸钾；BTPPA—双（三苯基磷+佛兰烯）铵；DOP—邻苯二甲酸二辛酯；DBP—邻苯二甲酸二丁酯；o-NPOE—邻硝基苯辛醚；BEHS—葵二酸二辛酯；DOS—癸二酸二辛酯；TCP—磷酸三酯；DINP—邻苯二甲酸二异壬酯；PVC—聚氯乙烯；EVA—乙烯（醋酸乙烯酯）。

基于对 ISE 性能积累的知识和与液体管状导管的充分耦合，电位检测已成为在不同连续流动模式以及在联用技术（使用 HPLC 和毛细管电泳[108,109]）和小型化流动系统，如芯片实验室[110]和阀上实验室[111]中增强使用的备选检测技术。与液膜内溶液 ISEs（本书第一卷图 3.2）相比，通过将膜沉积在铂或玻璃碳基底上以制备的包覆涂层的线状电极甚至是实心电线电极（图 3.2）更受人们欢迎。它们易于组装，灵敏度高，并且通过优化分析物离子（该离子穿过膜接触界面）化学势的差异，可以获得一到两秒的响应时间[112,113]。它们的性能优于电流检测，因为响应与采用的流动方案和速率几乎无关，这是新型超高流速 HPLC 系统中的一个重要问题。在该特定联用技术中，通过分子识别方法（主客体化学）优化分析物和缓冲液之间相互作用能的差异。通过这种方式，有机胺的分析灵敏度比使用分光光度检测法获得的灵敏度高 20 倍[114]，并且该程序可以进一步扩展到低分子质量氨基醇和烷基胺的测定[115]。此外，首次使用定量构效关系（QSAR）方法根据亲脂性和极化率预测非选择性膜电极对 β-肾上腺素药物和脂肪胺的电位响应和检测限[116,117]。基于铵盐的电极与基于醚脲衍生物（带有胺官能团或大环多胺）的电极相比，由于膜组分很少浸出进入洗脱液，后者可以更好地对羧酸进行色谱分析（检测限更低）。将基于大环离子载体的电位检测器与 HPLC 相连被证明具有极低的检测限并且有利于分析具有生化重要性的有机酸[113,118-121]、胺类[116]和多离子单核苷酸和寡核苷酸[122]。

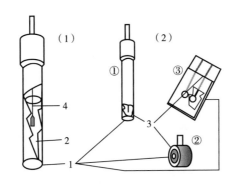

图 3.2 离子选择性电极配置

（1）液膜 （2）①固体接触②管状固体接触和③丝网印刷

1—聚合物敏感膜 2—内参比溶液 3—导电环氧树脂，金属表面或导电墨水 4—内参比电极

在流通条件下的电位检测包括通过电子线路设备监测、记录和处理电分析检测池中产生的信号。如果打算直接测量，则检测器位于采样点下游较近的距离，如果还需要进行在线样品预处理，则检测器应位于较远的距离。无论考虑何种特定系统，流通检测池都会监测流动流体中分析物的浓度（质量）-时间曲线。我们可以发现大量流通检测池都通过池中通道的长度、直径和几何构型区分，从而确定样品传输到膜电极界面的主要传输机制。检测池的构造需要确保分析物向指示电极的界面有效传输，以充分利用电极的响应特性。根据检测器探测分散样品区的方式，检测池测量壁面射流配置中通道横截面处的平均浓度或测量管状和薄池配置（其保持切向流动条件[123]）中传

感探头表面的局部浓度（图 3.3）。因此，指示电极可位于检测通道的中心或其壁面处。电位检测器可被定性为表面检测装置，这意味着采用了可再现的径向浓度分布。无论如何，流体动力学分析方法都需要坚固的电极。构建坚固 ISE 表面所需的材料从液体膜和导电聚合物到陶瓷材料和金属材料不等。通过电化学家之间的几次讨论来决定使用这些材料，讨论的内容包括分析电势的产生究竟是靠离子的传输（透过膜），还是仅限于通过膜界面对离子的分隔。根据第一种方法，人们提出了检测流动池，其中液膜电极带有可以插入的已知浓度的内部溶液，流动的分析物样品以壁面射流检测模式冲洗膜的另一侧。在过去几年中，新的工作模式似乎支持后一种方法[27,28,44,124]。因此，使用指示电极也获得了良好的电位信号，其中省略了内部溶液并且液膜覆盖导电表面，例如金属，即形成所谓的带涂层的线电极。如前所述，带涂层的线 ISEs 经常与连续流动程序相关联，其优点是具有不同的构造，即管状或平面结构，并且易于小型化可以在液相色谱、毛细管电泳和微型化全分析系统中使用[113]。大多数带涂层的线电极，有时称为固体接触电极和固态电极，是通过直接在固体导电表面上施加 PVC 膜实现的。然而，附着力差是一个普遍遇到的问题，由于载体在膜表面连续流动导致膜分离。或者，当电极涂有 PVC 膜时，可以使用电活性共轭聚合物作为离子电子换能器，或直接掺杂离子识别物质。在几个应用示例中，它们主要是与平面微电极配置中的流动配件耦合[125-127]，显著提高了耐用性和潜在稳定性。管状配置也被成功地使用，其中膜覆盖在导电支撑件中纵向孔的内壁。这种类型的 ISE 配置适合作为流动配件中管状导管的延伸，从而减少检测池的内部死体积并能够在不破坏循环样品塞的流体动力学传输条件的情况下进行测量。其主要优点之一是可以通过其他检测技术对下游样品作进一步处理从而进行形态分析或多重测定，该优点在流通平面阵列传感器或壁面射流电池中会被削弱[128]。

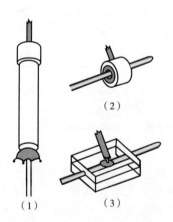

图 3.3　连续流系统中 ISE 的一般耦合模式

（1）在壁面射流检测模式下，流动的样品区垂直撞击电极表面
（2）在管状电极中，样品与电极表面作相切流动　（3）在薄池电极中，样品与电极表面作相切流动

电位测定法是一种有价值的分析工具，适用于实验室中远程进行的各种医疗和环境方面的应用，满足低成本、设计简单、尺寸小甚至一次性模的要求。相关文献描述

了各种平面技术来开发用于流动方法的固体接触传感器[129,130]。薄膜和厚膜技术被证明非常适合生产一次性传感器。由于大规模生产的简单性和高效率，丝网印刷特别有用[127]。通过这种制造过程可以实现电极的小型化，从而实现具有传感器阵列形式的多传感器系统[131]。与场效应晶体管（FET）相比，该制备方法具有更高的耐用性，并且它们在微流体系统中的集成很简单。该程序还为参比电极带来了创新，其中描述了平面液接 Ag/AgCl/KCl 参比电极[132]。使用独特的自动丝网印刷技术同时构建全厚膜条状电极，无需额外的手动、化学或电化学步骤[133]。使用了简单、廉价、市售的材料，例如塑料基材和易于固化的聚合物基糊剂。推荐的构建程序可以大规模和统一生产充分集成的电极系统，该电极系统应用于带有单条和单芯片格式的参比电极和指示电极。丝网印刷技术也被用于制备小型化参比电极，这些参比电极可以与单体中的指示电极集成在一起，其中流动分析微系统是通过低温共烧陶瓷技术生产的[134]。在这项工作中，参比电极通过丝网印刷在包含辅助通道的陶瓷带上制备，0.1mol/L 的 KCl 溶液通过该辅助通道连续流动以提供恒定的参考电位。

3.4 基于电位法的流动分析技术

不同连续流动概念和电位检测之间通过耦合相辅相成并因此衍生出了 200 多篇论文，其中主要采用单通道［图 3.4（1）］或双通道流动进样装置［图 3.4（2）］。在这些基本的优选配置中，快速可靠的浓度测量集中在所用电极的选择性上。流动概念通常确保离子强度和/或 pH 的调整、样品体积的确定及其下游传输。由于设计简单，它们具有环境标准所需的系统便携性[135]，这表明了关于高污染程序的改进替代方案[136]或能够利用一些已知化学反应（间歇处理条件下难以实施）。相关文献报道了一种使用封闭循环流歧管的特殊方法来完成微摩尔范围内的溴酸盐测定[137]。样品中存在的溴酸盐与作为载体的 Fe^{2+}-Fe^{3+} 电位缓冲液中的溴化物反应，产生不稳定的溴素，该溴素通过带有镀金电极和 Ag/ACl 参比电极的流动池检测。虽然流动注射分析系统仍然是电位检测的首选，但近年来越来越多地采用微计算机控制装置的连续流动系统被使用。其中，顺序注射分析（SIA）［图 3.4（3）］被证明非常适合多重测定，因此，可以用于处理在基质中含有不同化学形式物质的样品。表 3.2 总结了最近开发的基于 SIA 的程序，其中只涉及了实际样品的应用程序，关于这个主题的全面和详细的综述已发表在分析期刊[110,162-164]上。最近还通过阀上实验室的概念[165]描述了电位检测，但需要电极和检测池设计的新配置才能减少样品进样体积、达到更高的采样率和在没有接地电极的条件下降低电噪声[111]。关于这些流动概念或其他先前报道的流动概念，例如流动注射分析、单段流动分析、多路换向二元系统，［图 3.4（4）］的最新报告旨在执行附加任务。因此，不同的方法能够实现系统在三个或四个数量级上的自校准，其中观察到 nernstian 检测器响应。在一项建议[166]中，一个精心设计的包括两个蠕动泵、两个计算机接口卡、一个双三通注射阀和一个混合室的装置被用于在单流动进样系统内制备校准溶液。阀门驱动使溶液能够以 22μL~17mL 的体积通过既定编程循环或推送至混合室。尽管如此，由于柔性泵管的使用寿命有限，因此必须执行严格的校准程序。

在第二个建议中，在短开/关循环中驱动的小型化三通电磁阀与顺序进样系统[159]主阀的横向端口相连。在这种方法中，根据电磁阀的开启和关闭次数之间的比率，最多只能将标样稀释100倍。如果保持低样品分散的条件，则使用蠕动泵不会影响校准溶液的准确性。通过改变溶液的化学性质，还可以通过分离和固定的干扰方法[21]实现对干扰的全自动评估，并直接或通过更详细的方案（如添加标样或滴定）进行测定。在同时测定水中的氯化物和硝酸盐时，通过使用管式ISE对这种新方法的通用性进行了评估[149]。在对每个样品的浓度值进行预估后，基于这些预估值对两个电极进行自动在线校准，在干扰离子（硝酸盐）的预测浓度下改变主要离子（氯化物或硝酸盐）的浓度。预估法与标样添加法相比，测量的准确度提高，相对均方根误差从12.1%减小到了2.8%。一种类似的方法通过对氯化物和碘化物（两者的浓度范围均为1~100μmol/L）的滴定实现格氏作图法[167]。该方法的灵活性使得其可以实施简单而稳健的程序，用于实时评估药物控制过程中的准确度[91,92,144,151]。这个概念是基于两种准独立技术对同一物质的同时评估[168]。如果干扰具有不同的特征，则获得的平均结果更准确。还提出了通过顺序注射分析加速电子舌校准程序[169]。电子舌包括一系列化学传感器（对不同物质选择性和交叉敏感性较差）以及用于数据处理的化学计量工具。一个在主流体选择阀横向端口上带有一个耦合混合室的通用顺序进样系统可以对不同物质的储备溶液进行多种采样，并为人工神经网络或其他化学计量算法准备训练集。使用通过几种先前描述的离子提取器和离子载体制备的管状电极组，在对主要离子或干扰物质的响应曲线进行综合分析后，预测了相对标准偏差小于4%的多参数结果。这些电子舌的主要优缺点通过分析水[141,142,170,171]、饮料[172]、肥料[173]和临床样品中的阳离子物质[174]以及水中的阴离子物质[147]得以充分检验。

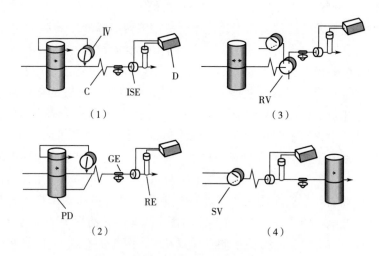

图3.4 用于电位检测的典型连续流歧管
(1) 单通道流动进样系统　(2) 双通道流动进样系统
(3) 具有多任务能力的顺序注射系统　(4) 多路换向流动系统
PD—动力装置　IV—进样阀　SV—三通电磁阀　RV—流选择阀
C—盘管　GE—接地电极　ISE—离子选择电极　RE—参比电极　D—分毫伏表

表 3.2 在实际样品测定中利用电位检测的顺序注射分析（2000—2007年）

分析物	样品	检测方法	线性范围	检测限	R.S.D./%	采样率/h	参考文献
乙酸	醋	用pH玻璃电极滴定	10~90g/L	—	0.4	28	[138]
乙酰水杨酸	阿司匹林药片	酶水解，水杨酸电极	5×10^{-5}~1×10^{-2} mol/L	5×10^{-5} mol/L	0.2	45	[139]
酸度	软饮料	用pH玻璃电极滴定	1~6g/L	—	<0.5	—	[140]
铵	肥料	神经网络，NH_4^+，Na^+，K^+阵列和一个通用电极	5×10^{-3}~4×10^{-2} mol/L	—	1.8~4.2	—	[141]
钡	水（含矿物质）	神经网络，钙阵列电极	$(5$~$35)\times10^{-3}$ mg/L	—	1.8~4.2	—	[142]
碳酸氢盐	水	神经网络加上Legendre分解，两个Cl^-、两个NO_3^-和一个通用电极的阵列	5×10^{-4}~5×10^{-3} mol/L	—	1.8~4.2	—	[143]
钙	水（含矿物质）	神经网络，钙电极阵列	$(5$~$45)\times10^{-3}$ mg/L	—	1.8~4.2	—	[142]
卡托普利	制药	使用Ag/Ag_2S管状电极滴定	$(2.5$~$10)\times10^{-4}$ mol/L	—	<1.3	—	[144]
氯化物	电镀池	管状AgCl/Ag_2S电极	0.1~1.0mol/L	—	—	40	[145]
	牛乳	使用Ag/Ag_2S管状电极滴定	0.01~0.25mol/L	—	<3.4	17	[146]
	药物	使用单段式Ag/Ag_2S管状电极滴定	8×10^{-4}~3×10^{-2} mol/L	—	<1	—	[147]
	水	八乙基（卟啉）钴（III）电极	1×10^{-5}~1×10^{-2} mol/L	—	<0.84	60	[91]
	水	AgCl/Ag_2S电极	10~3500mg/L	1.6mg/L	0.4	30	[148]
	制药	最小二乘多元回归，阀上实验室概念，AgCl/Ag_2S线	10^{-4}~4×10^{-1} mol/L	—	—	—	[149]
	水	神经网络加上Legendre分解，两个Cl^-、两个NO_3^-和一个通用电极的阵列	1×10^{-4}~1.2×10^{-1} mol/L	—	0.7~1.3	50	[150]
	制药	使用单段式Ag/Ag_2S管状电极滴定	5×10^{-3}~4.5×10^{-2} mol/L	—	1.8	—	[143]
	葡萄酒	使用单段式Ag/Ag_2S管状电极滴定	8×10^{-4}~3×10^{-2} mol/L	—	<1	—	[147]
克拉维酸盐	制药	克拉维酸盐选择性电极	2×10^{-3}~1×10^{-1} mol/L	—	0.6	53	[151]
双氯芬酸	制药	环糊精电极	5×10^{-6}~1×10^{-2} mol/L	2×10^{-6} mol/L	<1	33	[92]

续表

分析物	样品	检测方法	线性范围	检测限	R.S.D./%	采样率/h	参考文献
氟化物	自来水	LaF_3 晶体膜电极	$10^{-10} \sim 10^{-6}$ mol/L	1.7×10^{-12} mol/L	0.3	30	[152]
赤霉酸	牙膏	四苯基卟啉锰(Ⅲ)管状电极	$10^{-5} \sim 10^{-1}$ mol/L	2×10^{-6} mol/L	0.3	30	[153]
盐酸	农业增长促进剂	用 pH 玻璃电极滴定	$5 \times 10^{-4} \sim 8 \times 10^{-3}$ mol/L	3×10^{-4} mol/L	<0.4	30	[154]
铁	污水流	稀释、滴定、pH 玻璃电极	$20 \sim 50$ mmol/L	—	<0.22	30	[155]
	制酸厂	神经网络、钙玻璃电极阵列	$5.93 \sim 8.99$ mmol/L	—	<0.4	30	[142]
硝酸盐	水(含矿物质)	神经网络、Legendre 分解、两个 Cl^-、两个 NO_3^- 和一个通用电极铵基电极的阵列	$(5 \sim 40) \times 10^{-3}$ mg/L	—	$1.8 \sim 4.2$	—	[143]
	水	多元回归、叔辛基溴化铵基电极	$5 \times 10^{-4} \sim 5 \times 10^{-3}$ mol/L	—	$1.8 \sim 4.2$	—	[149]
氨	青贮饲料	基于无活菌素的氨电极	$5 \times 10^{-4} \sim 10^{-1}$ mol/L	3mg/L	—	30	[156]
青霉素-G	药物	四苯基卟啉锰(Ⅲ)管状电极	$10 \sim 120$ mg/L	—	<2	25	[157]
钾—游离	药物	K^+ 管状缬氨霉素电极	$2 \times 10^{-4} \sim 10^{-2}$ mol/L	1.5×10^{-4} mol/L	1.8	53	[151]
	佐餐酒和波特酒	在线微波消解后神经网络、NH_4^+、Na^+、K^+ 和两个通用电极的阵列	$2 \times 10^{-3} \sim 10^{-1}$ mol/L	0.8mg/L	0.5	12	[158]
钾—总钾	肥料	神经网络、NH_4^+、Na^+、K^+ 阵列和两个通用电极	$5 \times 10^{-3} \sim 5 \times 10^{-2}$ mol/L	—	$1.8 \sim 4.2$	—	[141]
酸碱度	电镀液	管状 pH、四(对氯苯硼酸)电极	$1 \sim 5$ pH	—	—	40	[145]
吡哆醇	药物	管状吡哆醇电极	$5 \times 10^{-5} \sim 10^{-2}$ mol/L	—	8.6	20	[159]
钠	肥料	神经网络、NH_4^+、Na^+、K^+ 阵列和两个通用电极	$5 \times 10^{-3} \sim 4.5 \times 10^{-2}$ mol/L	—	$1.8 \sim 4.2$	—	[141]
蔗糖	青贮饲料	基于双(三苯基正磷苯胺)铵的高氯酸盐电极	—	0.13%	<2	24	[160]
丙戊酸盐	药物	管状 PVC 四苯基卟啉酸锰(Ⅲ)电极	$5 \times 10^{-4} \sim 10^{-2}$ mol/L	9×10^{-5} mol/L	1.2	30	[93]
尿素	牛乳	带溶胶-凝胶膜的电极	$1 \times 10^{-3} \sim 5 \times 10^{-2}$ mol/L	9×10^{-4} mol/L	2.2	55	[161]
		在气(液)扩散单元之后基于非肌动蛋白的氨电极	$1 \times 10^{-3} \sim 10^{-2}$ mol/L	6×10^{-4} mol/L	1.9	20	

滴定法由于能够响应广泛的活性或浓度变化以及适用于有色或混浊样品，因而特别适用于电位法。流动进样滴定是流动注射分析中最古老的梯度技术，其基于将样品插入用作载体的滴定剂中。在样品分散和反应之后，浓度与瞬时信号宽度（信号上升和下降边沿的当量点之间的距离）的对数关系得以建立[138, 175]。尽管速度很快，但不能认为此过程是绝对的，因为需要系统校准和样品分散重现性来获得准确的结果。同时新的程序被提出以实现作者所说的"真正的滴定"。这些程序通常是通过克服流体动力学波动的影响以及样品在管状路径内物理分散的影响来避免校准。基于反馈的流动比率原理应用于持续时间小于15s的高重现性酸碱滴定，消耗试剂量约为12μL[176]。流动系统需要多达三个蠕动泵和串联微珠单填充反应器来保证良好的混合条件。滴定标准液和标准滴定液之间流速的比率线性增加，直到到达等当量点。然后，通过电信号（控制负责滴定剂流量的泵）在等当量点附近依次减小和增加该比率值。通过反馈控制中使用的最小和最大电信号来进行计算。建议使用快速驱动的电磁三通阀来执行滴定程序，这一次是基于更简单的系统配置[177-179]。该概念基于在管状路径中快速交替注入小体积滴定标准液和标准滴定液并使其充分混合。通过逐步增加开/关循环之间的比率，收集完整的滴定曲线，尽可能地降低溶液的消耗并可以使用传统算法对结果进行处理。一旦证明这种方法在顺序进样中也是可行的，它的多功能性就会增加[147]。最近提出了一种实现抗坏血酸或 Fe^{2+} 滴定的单段式流动系统，其分别产生滴定剂 I_2 和 Ce^{4+}[180]。多达 100μL 的样品插入两个 60μL 体积的滴定剂前体溶液之间，并通过气流携带至检测池。检测池由三个铂电极和一个 Ag/AgCl 参比电极组成，两侧由两个光电开关限制，该开关负责液体单段流动方向的自动反转。增加两种溶液之间的分散和反应可以产生滴定曲线。根据作者的说法，部分渐进的分散符合流动进样原则。在明确标准滴定液的等分试样的情况下不需要使用滴定程序。

通过填充反应器来实现样品自动清洗、预浓缩、形态或生化转化已被证明符合电位检测。在这种情况下，人们提出通过同时测定尿素和肌酐来实时控制血液透析治疗，该设置与血液透析设备中的透析液通道在线耦合[181,182]。流动进样系统需要控制蠕动泵和两个注射阀的计时。第一个阀门位于动力装置之前，并且能够在流过阳离子交换树脂的流体中交替注射 300μL 的无血样品和 2.3mol/L 的 HCl 再生溶液，其中保留了主要的阳离子干扰物质。通过蠕动泵后，将 100μL 清洁过的样品通过第二个阀门注入三羟甲基氨基甲烷盐酸盐（pH8.1）缓冲溶液中，该缓冲溶液在下游分流从而分别通过尿素酶和肌氨酸酐脱亚胺酶对两个检测池的铵选择性电极进行化学锚定修饰。使用填充柱被证明可以有效且准确地测定稀释 10 倍的血清和尿液样品中浓度在 μmol/L 范围内的氟化物[183]。在该歧管中，阴离子交换树脂放置在进样阀之前，以减少水中的阴离子污染物（用作样品载体）。通常观察到的膜晶体中氟化镧的记忆效应、漂移和溶解问题在三通道流动进样歧管中被克服，其中在第一次汇流时通过流动样品引入离子强度为 4.5mol/L 的醋酸盐缓冲液（pH5.3），以尽量减少样品中其他离子引起的氟化物活性波动。通过在缓冲液中加入作为螯合剂的反式-1, 2-二氨基环己烷-N, N, N, N-四乙酸，将阳离子干扰降至最低。26.3μmol/L（或 263μmol/L 尿样）含磷酸盐浓度为 1mmol/L 的氟化物溶液通过第二次汇流加入，以确保稳定的基线电位，并尽可能地减少阴离子对记录信号的干扰。对于水样中的较低浓度测定，可以使用填充有浸渍氧化锆的纤维素的小柱子来预浓缩氟离子[184]。吸附发生在 pH4.8 的条件下，保留的氟化物在 pH13 时解吸。这样可以将检测限提高到 3×10^{-9} mol/L，并提供相对标准偏差优于 1.6% 的结果。对于水样中 As^{3+} 和 As^{5+} 的预浓缩和形态分析，建议使用嵌入硅胶的无定形羟基氧化物填充柱和导电石墨环氧树脂支架上的基于相同材料薄膜的选择性管状电极[185]。色谱柱与进样阀相连，从而插入样品或洗脱液/检测器管线中。为了实现预浓缩，将样品与添加或不添加碘（As^{5+} 或总 As）的磷酸盐缓冲溶液（pH 12）连续合并。在进样后，洗脱和保留砷的 pH 7 的磷酸盐缓冲溶液可以使其转移到检测池中。10mL 水样的预浓缩可使得在 40~500μg/L 浓度范围内，不同批次的结果准确率在 2% 之内。建议使用流动进样系统测定河水样品中非离子型表面活性剂（含有由 6~18 个乙氧基化物单元组成的亲水链）的总含量，在该系统进样阀回路中的 C_{18} 商业树脂上进行固相萃取步骤，以实现样品富集和纯化[186]。管状固态电极用作检测器，其基于聚乙氧基非离子型表面活性剂钡络合物的四苯基硼酸盐[187]。如果 40mL 样品预先流过树脂，则该程序通过 200μL 75% 乙腈水溶液的洗脱可以将检测限降至 3×10^{-6} mol/L。在最近的一篇论文中，作者通过使用十二烷基硫酸钠（作为标样），将该应用拓展到阴离子表面活性剂[188]。

3.5　趋势与展望

尽管并非详尽无遗，但对近期成就的回顾表明，具有电位检测的连续流动技术是未来研究的主要方向。对新型聚合物膜电极（拥有更稳健配置）的响应的理论理解将指导 ISE 的开发并使 ISE 具有更好的选择性特性和更低的检测限。连续流动系统已被用来证明电极阵列和对结果的化学计量处理构成了一种新的低成本多参数工具以通过适度的成本产生良好的结果，而不是将选择性的缺失视为分析质量的限制因素。目前人们可以快速且可重复地制备溶液以充分利用 ISE 响应特性，该过程克服了烦琐的实验室程序（如在宽浓度范围内进行校准、标样添加、滴定和干扰测试），通过使用填充式在位反应器，可以实现生物化学分析并使分离技术能够在接近检测极限的条件下进行高度可重复的测定。此外，缩小样品预处理程序的规模不会导致检测性能和物理稳定性的损失或响应性能的妥协。

参考文献

[1] Nernst, W. and Loeb, K. (1988) Zur Kinetik der in Lösung befindlichen Körper. Zweite Abhanlung Überführungszahlen und Leitvermogen einiger Silbersaze. *Zeitschrift fur Physikalische Chemie - International Journal of Research in Physical Chemistry & Chemical Physics*, 2 (12), 948-963.

[2] Nernst, W. (1897) Die elektrolytische Zersetzung wässriger Lösungen. *Berichte der Deutschen Chemischen Gesellschaft*, 30 (12), 1547-1563.

[3] Cremer, M. (1906) Über die Urasche der elektromotorischen Eigenschaften der Gewebe, zugleich

ein Beitrag zur Lehre von den polyphasischen Elektrolytketten. *Zeitschrift für Biologie*, 347, 562-608.

[4] Haber, F. and Klemensiewicz, Z. (1909) On electrical interfacial potentials. *Zeitschriftfur Physikalische Chemie-Leipzig*, 67, 385-431.

[5] Donnan, F. G. (1911) Theory of membrane equilibrium and membrane potentials in the presence of nondialyzable electrolytes. A contribution to physical-chemical physiology. *Zeitschrift fur Elektrochemie und physikalische Chemie*, 17, 572-581.

[6] Eisenman, G., Rudin, D. O. and Casby, J. U. (1957) Glass electrode for measuring sodium ion. *Science*, 126, 831-834.

[7] Nikolskii, B. P. and Schults, M. M. (1962) Some aspects of glass electrode theory. *Zhurnal Fizicheskoi Khimii*, 36, 1327-1330.

[8] Inzelt, G. (2005) Patent No. 2, 058, 761—or the beginning of electrochemical instrumentation. *Journal of Solid State Electrochemistry*, 9, 181-182.

[9] Radiometer Annual Report, Copenhagen, 1994/1995.

[10] Frant, M. S. and Ross, J. W. (1966) Electrode for sensing fluoride ion activity in solution. *Science*, 154, 1553-1555.

[11] Ross, J. W. (1967) Calcium-selective electrode with liquid ion exchanger. *Science*, 156, 1378-1379.

[12] Stefanac, Z. and Simon, W. (1966) *In-vitro*-verhalten von makrotetroliden in membranen als grundlage fur hochselektive kationenspezifische elektrodensysteme. *Chimia*, 20, 436.

[13] Stefanac, Z. and Simon, W. (1967) Ion specific electrochemical behavior of macrotetrolides in membranes. *Microchemical Journal*, 12, 125-132.

[14] Moody, G. J., Oke, R. B. and Thomas, J. D. R. (1970) Calcium-sensitive electrode based on a liquid ion exchanger in a poly (vinyl-chloride) matrix. *Analyst*, 95, 910-918.

[15] Gunaratna, P. C., Koch, W. F., Paule, R. C., Cormier, A. D., D'Orazio, P., Greenberg, N., O'Connell, K. M., Malenfant, A., Okorodudu, A. O., Miller, R., Kus, D. M. and Bowers, G. N. (1992) Frozen human serum reference material for standardization of sodium and potassium measurements in serum or plasma by ion-selective electrode analyzers. *Clinical Chemistry*, 38, 1459-1465.

[16] Bakker, E., Diamond, D., Lewenstam, A. and Pretsch, E. (1999) Ion sensors: current limits and new trends. *Analytica Chimica Acta*, 393, 11-18.

[17] Guggenheim, E. A. (1929) The conceptions of electrical potential difference between two phases and the individual activities of ions. *Journal of Physical Chemistry*, 33, 842-849.

[18] Guggenheim, E. A. (1930) On the conception of electrical potential difference between two phases. II. *Journal of Physical Chemistry*, 34, 1540-1543.

[19] Bakker, E., Buhlmann, P. and Pretsch, E. (2004) The phase-boundary potential model. *Talanta*, 63, 3-20.

[20] Amemiya, S., Buhlmann, P., Pretsch, E., Rusterholz, B. and Umezawa, Y. (2000) Cationic or anionic sites? Selectivity optimization of ion-selective electrodes based on charged ionophores. *Analytical Chemistry*, 72, 1618-1631.

[21] Buck, R. P. and Lindner, E. (1995) Recommendations for nomenclature of ion-selective electrodes- (IUPAC recommendations 1994). *Pure and Applied Chemistry*, 66, 2527-2536.

[22] Frenzel, W. (1988) Enhanced performance of ion-selective electrodes in flow injection analysis- non-nernstian response, indirect determination, differential detection and modified reverse flow injection

analysis. *Analyst*, 113, 1039-1046.

[23] Chudy, M., Wróblewski, W., Dybko, A. and Brzozka, Z. (2001) Multi-ion analysis based on versatile sensor head. *Sensors and Actuators B*, 78, 320-325.

[24] Trojanowicz, M., Szewczynska, M. and Wcislo, M. (2003) Electroanalytical flow measurements-recent advances. *Electroanalysis*, 15, 347-365.

[25] Krawczyk, T. K. V., Trojanowicz, M. and El-Murr, N. (2000) Enhancement of selectivity of electrochemical detectors by kinetic discrimination in flow injection systems. *Laboratory Robotics and Automation*, 12, 205-215.

[26] Moore, C. and Pressmann, B. C. (1964) Mechanism of action of valinomycin on mitochondria. *Biochemical and Biophysical Research Communications*, 15, 562-567.

[27] Visser, H. C., Reinhoudt, D. N. and de Jong, F. (1994) Carrier-mediated transport through liquid membranes. *Chemical Society Reviews*, 23, 75-81.

[28] Sokalski, T., Zwickl, T., Bakker, E. and Pretsch, E. (1999) Lowering the detection limit of solvent polymeric ion-selective electrodes. 1. Modelling the influence of steady-state ion fluxes. *Analytical Chemistry*, 71, 1204-1209.

[29] Sokalski, T., Ceresa, A., Fibbioli, M., Zwickl, T., Bakker, E. and Pretsch, E. (1999) Lowering the detection limit of solvent polymeric ion-selective membrane electrodes. 2. Influence of composition of sample and internal electrolyte solution. *Analytical Chemistry*, 71, 1210-1214.

[30] Bakker, E., Pretsch, E. and Buhlmann, P. (2000) Selectivity of potentiometric ion sensors. *Analytical Chemistry*, 72, 1127-1133.

[31] Zwickl, T., Sokalski, T. and Pretsch, E. (1999) Steady-state model calculations predicting the influence of key parameters on the lower detection limit and ruggedness of solvent polymeric membrane ion-selective electrodes. *Electroanalysis*, 11, 673-680.

[32] Sokalski, T., Ceresa, A., Zwickl, T. and Pretsch, E. (1997) Large improvement of the lower detection limit of ion-selective polymer membrane electrodes. *Journal of the American Chemical Society*, 119, 11347-11348.

[33] Ceresa, A., Radu, A., Peper, S., Bakker, E. and Pretsch, E. (2002) Rational design of potentiometric trace level ion sensors. A Ag^+-selective electrode with a 100ppt detection limit. *Analytical Chemistry*, 74, 4027-4036.

[34] Qin, W., Zwickl, T. and Pretsch, E. (2000) Improved detection limits and unbiased selectivity coefficients obtained by using ion-exchange resins in the inner reference solution of ion selective polymeric membrane electrodes. *Analytical Chemistry*, 72, 3236-3240.

[35] Ion, A. C., Bakker, E. and Pretsch, E. (2001) Potentiometric Cd^{2+}-selective electrode with a detection limit in the low ppt range. *Analytica Chimica Acta*, 440, 71-79.

[36] Malon, A., Radu, A., Qin, W., Qin, Y, Ceresa, A., Maj-Zurawska, M., Bakker, E. and Pretsch, E. (2003) Improving the detection limit of anion-selective electrodes: An iodide-selective membrane with a nanomolar detection limit. *Analytical Chemistry*, 75, 3865-3871.

[37] Ceresa, A., Bakker, E., Gunther, D., Hattendorf, B. and Pretsch, E. (2001) Potentiometric polymeric membrane electrodes for measurement of environmental samples at trace levels: New requirements for selectivities and measuring protocols, and comparison with ICPMS. *Analytical Chemistry*, 73, 343-351.

[38] Bakker, E. and Pretsch, E. (2005) Potentiometric sensors for trace-level analysis. *Trends Analytical Chemistry*, 24, 199-207.

[39] Morf, W. E., Badertscher, M., Zwickl, T., Rooij, N. F. and Pretsch, E. (2002) Effects of controlled current on the response behavior of polymeric membrane ion-selective electrodes. *Journal of Electroanalytical Chemistry*, 526, 19-28.

[40] Arnold, M. A. and Solsky, R. L. (1986) Ion-selective electrodes. *Analytical Chemistry*, 58, R84-R101.

[41] Yu, R. Q. (1986) Aspects of the development of liquid membrane anion sensitive electrodes. *Ion-Selective Electrode Review*, 3, 153-172.

[42] Wotring, V. J., Johnson, D. M. and Bachas, L. G. (1990) Polymeric membrane anion-selective electrodes based on diquaternary ammonium-salts. *Analytical Chemistry*, 62, 1506-1510.

[43] Simon, W. and Carafoli, E. (1979) Design, properties, and applications of neutral ionophores. *Methods in Enzymology*, 56, 439-448.

[44] Bakker, E., Buhlmann, P. and Pretsch, E. (1997) Carrier-based ion-selective electrodes and bulk optodes. 1. General characteristics. *Chemical Reviews*, 97, 3083-3112.

[45] Beer, P. D. and Gale, P. A. (2001) Anion recognition and sensing: The state of the art and future perspectives. *Angewandte Chemie-International Edition in English*, 40, 486-516.

[46] Siswanta, D., Nagatsuka, K., Yamada, H., Kumakura, K., Hisamoto, H., Shichi, Y., Toshima, K. and Suzuki, K. (1996) Structural ion selectivity of thia crown ether compounds with a bulky block subunit and their application as an ionsensing component for an ion-selective electrode. *Analytical Chemistry*, 68, 4166-4172.

[47] Suzuki, K., Siswanta, D., Otsuka, T., Amano, T., Ikeda, T, Hisamoto, H., Yoshihara, R. and Ohba, S. (2000) Design and synthesis of a more highly selective ammonium ionophore than nonactin and its application as an ionsensing component for an ion-selective electrode. *Analytical Chemistry*, 72, 2200-2205.

[48] Kim, H. S., Park, H. J., Oh, H. J., Koh, Y. K., Choi, J. H., Lee, D. H., Cha, G. S. and Nam, H. (2000) Thiazole containing benzo crown ethers: A new class of ammonium selective ionophores. *Analytical Chemistry*, 72, 4683-4688.

[49] Benco, J. S., Nienaber, H. A. and McGimpsey, W. G. (2003) Synthesis of an ammonium ionophore and its application in a planar ion-selective electrode. *Analytical Chemistry*, 75, 152-156.

[50] Xiao, K. P., Buhlmann, P., Nishisawa, S., Amemiya, S. and Umezawa, Y. (1997) A chloride ion-selective solvent polymeric membrane electrode based on a hydrogen bond forming ionophore. *Analytical Chemistry*, 69, 1038-1044.

[51] Hutchins, R. S., Bansal, P., Molina, P., Alajarin, M., Vidal, A. and Bachas, L. G. (1997) Salicylate selective electrode based on a biomimetic guanidinium ionophore. *Analytical Chemistry*, 69, 1273-1278.

[52] Amemiya, S., Buhlmann, P., Umezawa, Y., Jagessar, R. C. and Burns, D. H. (1999) An ion-selective electrode for acetate based an urea-functionalized porphyrin as a hydrogen-bonding ionophore. *Analytical Chemistry*, 71, 1049-1054.

[53] Fibbioli, M., Berger, M., Schmidtchen, F. P. and Pretsch, E. (2000) Polymeric membrane electrodes for monohydrogen phosphate and sulfate. *Analytical Chemistry*, 72, 156-160.

[54] Berrocal, MJ., Cruz, A., Badr, I. H. A. and Bachas, L. G. (2000) Tripodal ionophore with sulfate recognition properties for anion-selective electrodes. *Analytical Chemistry*, 72, 5295-5299.

[55] Giannetto, M., Mori, G., Notti, A., Pappalardo, S. and Parisi, M. F. (1998) Discrimination

between butylammonium isomers by calix [5] arene-based ISE's. *Analytical Chemistry*, 70, 4631-4635.

[56] Zeng, X., Weng, L., Chen, L., Leng, X., Zhang, Z. and He, X. (2000) Improved silver ion-selective electrodes using novel 1, 3-bis (2-benzothiazolyl) thioalkoxy-p-tert-butylcalix [4] arenes. *Tetrahedron Letters*, 41, 4917-4921.

[57] Mahajan, R. K., Kaur, I. and Kumar, M. (2003) Silver ion-selective electrodes employing Schiff base p-tert-butyl calix [4] arene derivatives as neutral carriers. *Sensors and Actuators B*, 91, 26-31.

[58] Bobacka, J. (2006) Conducting polymer - based solid - state ion - selective electrodes. *Electroanalysis*, 18, 7-18.

[59] Bobacka, J., Ivaska, A. and Lewenstam, A. (2003) Potentiometric ion sensors based on conducting polymers. *Electroanalysis*, 15, 366-374.

[60] Arvand, M., Pourhabib, A., Shemshadi, R. and Giahi, M. (2007) The potentiometric behavior of polymer-supported metallophthalocyanines used as anion-selective electrodes. *Analytical and Bioanalytical Chemistry*, 387, 1033-1039.

[61] Shim, J. H., Jeong, I. S., Lee, M. H., Hong, H. P., On, J. H., Kim, K. S., Kim, H. S., Kim, B. H., Cha, G. S. and Nam, H. (2004) Ion-selective electrodes based on molecular tweezer-type neutral carriers. *Talanta*, 63, 61-71.

[62] Heng, L. Y. and Hall, E. A. H. (2000) Producing self-plasticizing ion-selective membranes. *Analytical Chemistry*, 72, 42-51.

[63] Heng, L. Y. and Hall, E. A. H. (2000) One-step synthesis of K^+-selective methacrylic-acrylic copolymers containing grafted ionophore and requiring no plasticizer. *Electroanalysis*, 12, 178-186.

[64] Heng, L. Y. and Hall, E. A. H. (2000) Taking the plasticizer out of methacrylic - acrylic membranes for K^+-selective electrodes. *Electroanalysis*, 12, 187-193.

[65] Malinowska, E., Gawart, L., Parzuchowski, P., Rokicki, G. and Brzózka, Z. (2000) Novel approach of immobilization of calix [4] arene type ionophore in 'self-plasticized' polymeric membrane. *Analytica Chimica Acta*, 421, 93-101.

[66] Qin, Y, Peper, S. and Bakker, E. (2002) Plasticizer-free polymer membrane ion-selective electrodes containing a methacrylic copolymer matrix. *Electroanalysis*, 14, 1375-1381.

[67] Bereczki, R., Gyurcsányi, R. E., Ágai, B. and Tóth, K. (2005) Synthesis and characterization of covalently immobilized bis-crown ether based potassium ionophore. *Analyst*, 130, 63-70.

[68] Jammal, A. E., Bouklouse, A. A., Patriarche, G. J. and Christian, G. D. (1991) Use of ethylene-vinyl-acetate as a new membrane matrix for calcium ion-selective electrode preparation. *Talanta*, 38, 929-935.

[69] Thomas, J. D. R. (1986) Solvent polymeric membrane ion-selective electrodes. *Analytica Chimica Acta*, 180, 289-297.

[70] Badawy, S. S., Youssef, A. F. and Mutair, A. A. (2004) Construction and performance characterization of ion - selective electrodes for potentiometric determination of phenylpropanolamine hydrochloride applying batch and flow injection analysis techniques. *Analytica Chimica Acta*, 511, 207-214.

[71] Rizk, M. S., Abdel-Ghani, N. T. and El Nashar, R. M. (2001) Construction and performance characteristics of terbutaline plastic membrane electrode in batch and FIA conditions. *Microchemical Journal*, 70, 93-101.

[72] Ibrahim, H. (2005) Chemically modified carbon paste electrode for the potentiometric flow injection analysis of piribedil in pharmaceutical preparation and urine. *Journal of Pharmaceutical and*

Biomedical Analysis, 38, 624-632.

[73] Badawy, S. S., Issa, Y. M. and Mutair, A. A. (2005) PVC membrane ion-selective electrodes for the determination of hyoscyamine in pure solution and in pharmaceutical preparations under batch and flow modes. *Journal of Pharmaceutical and Biomedical Analysis*, 39, 117-124.

[74] Sales, M. G. F., Tomás, J. F. C. and Lavandeira, S. R. (2006) Flow injection potentiometric determination of chlorpromazine. *Journal of Pharmaceutical and Biomedical Analysis*, 41, 1280-1286.

[75] Hassan, S. S. M., Elnemma, E. M., Mahmoud, W. H. and Mohammed, A. H. K. (2006) Continuous potentiometric monitoring of viagra (sildenafil) in pharmaceutical preparations using novel membrane sensors. *Journal of Applied Electrochemistry*, 36, 139-146.

[76] Hassan, S. S. M., Elnemma, E. M. and Mohammed, A. H. K. (2005) Novel biomedical sensors for flow injection potentiometric determination of creatinine in human serum. *Electroanalysis*, 17, 2246-2253.

[77] Ibrahim, H., Issa, Y. M. and Abu-Shawish, H. M. (2005) Potentiometric flow injection analysis of dicyclomine hydrochloride in serum, urine and milk. *Analytica Chimica Acta*, 532, 79-88.

[78] Ibrahim, H., Issa, Y. M. and Abu-Shawish, H. M. (2005) Potentiometric flow injection analysis of mebeverine hydrochloride in serum and urine. *Journal of Pharmaceutical and Biomedical Analysis*, 36, 1053-1061.

[79] El-Nashar, R. M., Abdel-Ghani, N. T. and Bioumy, A. A. (2004) Flow injection potentiometric determination of amitriptyline hydrochloride. *Microchemical Journal*, 78, 107-113.

[80] Sales, M. G. F. and Montenegro, M. C. B. S. M. (2001) Tetracycline-selective electrode for content determination and dissolution studies of pharmaceuticals by flow injection analysis. *Journal of Pharmaceutical Sciences*, 90, 1125-1133.

[81] Abdel-Ghani, N. T., Shoukry, A. F. and El-Nashar, R. M. (2001) Flow injection potentiometric determination of pipazethate hydrochloride. *Analyst*, 126, 79-85.

[82] Martinez-Barrachina, S., del Valle, M., Matia, L., Prats, R. and Alonso, J. (2001) Potentiometric flow injection system for the determination of polyethoxylate nonionic surfactants using tubular ion-selective electrodes. *Analytica Chimica Acta*, 438, 305-313.

[83] Hassan, S. S. M. and Rizk, N. M. H. (2000) Potentiometric dipyridamole sensors based on lipophilic ion pair complexes and native ionic polymer membranes. *Analytical Letters*, 33, 1037-1055.

[84] Abdel-Ghani, N. T. and Hussein, S. H. (2003) Determination of ambroxol hydrochloride in pure solutions and some of its pharmaceutical preparations under batch and FIA conditions. *Il Farmaco*, 58, 581-589.

[85] Sales, M. G. F., Lino, N. F. M. C. and Paiga, P. C. B. (2003) Chlormequat selective electrodes: Construction, evaluation and application at FIA systems. *International Journal of Environmental Analytical Chemistry*, 83, 295-305.

[86] Sales, M. G. F., Pille, A. and Paigu, P. C. B. (2003) Construction and evaluation of cysteine selective electrodes for FIA analysis of pharmaceuticals. *Analytical Letters*, 36, 2925-2940.

[87] Abdel-Ghani, N. T, Rizk, M. S. and El-Nashar, R. M. (2002) Potentiometric flow injection determination of salbutamol. *Analytical Letters*, 35, 39-52.

[88] Issa, Y. M., Badawy, S. S. and Mutair, A. A. (2005) Ion-selective electrodes for potentiometric determination of ranitidine hydrochloride, applying batch and flow injection analysis techniques. *Analytical Sciences*, 21, 1443-1448.

[89] Ibrahim, H., Issa, Y. M. and Abu-Shawish, H. M. (2005) Potentiometric flow injection analysis

of drotaverine hydrochloride in pharmaceutical preparations. *Analytical Letters*, 38, 111-132.

[90] Issa, Y. M., Abdel-Ghani, N. T., Shoukry, A. F. and Ahmed, H. M. (2005) New conventional coated-wire ion-selective electrodes for flow injection potentiometric determination of chlordiazepoxide. *Analytical Sciences*, 21, 1037-1042.

[91] Pimenta, A. M., Araújo, A. N., Montenegro, M. C. B. S. M., Pasquini, C., Rohwedder, J. J. R. and Raimundo, I. M., Jr (2004) Chloride-selective membrane electrodes and optodes based on an indium (III) porphyrin for the determination of chloride in a sequential injection analysis system. *Journal of Pharmaceutical and Biomedical Analysis*, 36, 49-55.

[92] Pimenta, A. M., Araújo, A. N. and Montenegro, M. C. B. S. M. (2002) Simultaneous potentiometric and fluorimetric determination of diclofenac in a sequential injection analysis system. *Analytica Chimica Acta*, 470, 185-194.

[93] Garcia, C. A., Júnior, L. R. and Neto, G. O. (2003) Determination of potassium ions in pharmaceutical samples by FIA using a potentiometric electrode based on ionophore nonactin occluded in EVA membrane. *Journal of Pharmaceutical and Biomedical Analysis*, 31, 11-18.

[94] Santos, E. M. G., Araújo, A. N., Couto, C. M. C. M. and Montenegro, M. C. B. S. M. (2005) Construction and evaluation of PVC and sol-gel sensor membranes based on Mn (III) TPP-Cl. Application to valproate determination in pharmaceutical H. Kahlert, J. R. Porksen, J. Behnert, F. Scholz, FIA acid-base titrations with a new flow-through pH detector. *Analytical and Bioanalytical Chemistry*, 382, 1981-1986.

[95] Farrell, J. R., Iles, P. J. and Dimitrakopoulos, T. (1996) Photocured polymers in ion-selective electrode membranes. Part 5: Photopolymerised sodium sensitive ion-selective electrodes for flow injection potentiometry. *Analytica Chimica Acta*, 334, 133-137.

[96] Farrell, J. R., Iles, P. J. and Dimitrakopoulos, T. (1996) Photocured polymers in ion-selective electrode membranes Part 6; Photopolymerized lithium sensitive ion-selective electrodes for flow injection potentiometry. *Analytica Chimica Acta*, 335, 111-116.

[97] Alexander, P. W., Dimitrakopoulos, T. and Hibbert, D. B. (1997) A photo-cured coated-wire calcium ion selective electrode for use in flow injection potentiometry. *Talanta*, 44, 1397-1405.

[98] Edmonds, T. E. and Coutts, G. (1983) Flow injection analysis system for determining soil pH. *Analyst*, 108, 1013-1017.

[99] Hongbo, C., Ruzicka, J. and Hansen, E. H. (1985) Evaluation of critical parameters for measurement of pH by flow injection analysis determination of pH in soil extracts. *Analytica Chimica Acta*, 169, 209-220.

[100] del Mundo, F. R., Cardwell, T. J., Catrail, R. W., Iles, P. J. and Hamilton, I. C. (1989) An ultraviolet-cured, pH-sensitive membrane electrode for use in flow injection analysis. *Electroanalysis*, 1, 353-356.

[101] Marzouk, S. A. M. (2003) Improved Electrodeposited Iridium Oxide pH Sensor Fabricated on Etched Titanium Substrates. *Analytical Chemistry*, 75, 1258-1266.

[102] Hassan, S. S. M., Marzouk, S. A. M. and Badawy, N. M. (2002) Solid State Iridium Oxide-Titanium Based Sensor for Flow Injection pH Measurements. *Analytical Letters*, 35, 1301-1311.

[103] de Roiij, N. F., and Vlekkert, H. H. (1991) Microstructured ISFETs. *Chemical Sensor Technology*, 3, 213-231.

[104] Zampronio, C. G., Rohwedder, J. J. R. and Poppi, R. J. (2000) Development of a potentiometric flow cell with a stainless steel electrode for pH measurements. Determination of acid mixtures

using flow injection analysis. *Talanta*, 51, 1163-1169.

[105] Kahlert, H., Porksen, J. R., Behnert, J. and Scholz, F. (2005) FIA acid-base titrations with a new flow-through pH detector. *Analytical and Bioanalytical Chemistry*, 382, 1981-1986.

[106] Kahlert, H., Porksen, J. R., Isildak, I., Ardac, M., Yolau, M., Behnert, J. and Scholz, F. (2005) Application of a new pH-sensitive electrode as a detector in flow injection potentiometry. *Electroanalysis*, 17, 1085-1090.

[107] Teixeira, M. F. S., Ramos, L. A., Cassiano, N. M., Fatibello-Filho, O. and Bocchi, N. (2000) Evaluation of a Fe_2O_3-based graphite-epoxy tubular electrode as pH sensor in flow injection potentiometry. *Journal of the Brazilian Chemical Society*, 11, 27-31.

[108] Nagels, L. J. (2004) Potentiometric detection for high-performance liquid chromatography is a reality Which classes of organic substances are the targets? *Pure and Applied Chemistry*, 76, 839-845.

[109] Bohets, H., Vanhoutte, K., de Maesschalck, R., Cockaerts, P., Vissers, B. and Nagels, L. G. (2007) Development of *in situ* ion selective sensors for dissolution. *Analytica Chimica Acta*, 581, 181-191.

[110] Prieto-Simon, B., Campas, M., Andreescu, S. and Marty, J. L. (2006) Trends in flow-based biosensing systems for pesticide assessment. *Sensors*, 6, 1161-1186.

[111] Amorim, C. G., Araújo, A. N. and Montenegro, M. C. B. S. M. (2007) Exploiting sequential injection analysis with lab-on-valve and miniaturized potentiometric detection: epinephrine determination in pharmaceutical products. *Talanta*, 72, 1255-1260.

[112] Nagels, L. G., Bazylak, G. and Zielinska, D. (2003) Designing potentiometric sensor materials for the determination of organic ionizable substances in HPLC. *Electroanalysis*, 15, 533-538.

[113] Nagels, L. G. and Poels, I. (2000) Solid state potentiometric detection systems for LC, CE and MTAS methods. *Trends in Analytical Chemistry*, 19, 410-417.

[114] Bazylak, G. and Nagels, L. J. (2002) Potentiometric detection of N, N′-diethylaminoethanol and lysosomotropic amino alcohols in cation exchange high-performance liquid chromatography systems. *Analytica Chimica Acta*, 472, 11-26.

[115] Poels, L. and Nagels, L. J. (2001) Potentiometric detection of amines in ion chromatography using macrocycle-based liquid membrane electrodes. *Analytica Chimica Acta*, 440, 89-98.

[116] Bazylak, G. and Nagels, L. J. (2002) Potentiometric detection of exogenic beta-adrenergic substances in liquid chromatography. *Journal of Chromatography*. A, 973, 85-96.

[117] Bazylak, G. and Nagels, L. J. (2003) A novel potentiometric approach for detection of beta-adrenergics and beta-adrenolytics in high-performance liquid chromatography. *Il Famtaco*, 58, 591-603.

[118] Picioreanu, S., Poels, I., Frank, J., van Dam, J. C., van Dedem, G. W. K. and Nagels, L. G. (2000) Potentiometric detection of carboxylic acids, phosphate esters, and nucleotides in liquid chromatography using anion-selective coated-wire electrodes. *Analytical Chemistry*, 72, 2029-2034.

[119] Poels, I., Nagels, L. G., Verreyt, G. and Geise, H. J. (1998) Conducting polymer based potentiometric detection applied to the determination of organic acids with narrow-bore LC systems. *Biomedical Chromatography*, 12, 124-125.

[120] Poels, I., Schasfoort, R. B. M., Picioreanu, P., Frank, J., van Dedem, G. W. K., van den Berg, A. and Nagels, L. G. (2000) An ISFET-based anion sensor for the potentiometric detection of organic acids in liquid chromatography. *Sensors and Actuators B*, 67, 294-299.

[121] Zielinska, D., Poels, I., Pietraszkiewicz, M., Radecki, J., Geise, H. J. and Nagels, L. J.

(2001) Potentiometric detection of organic acids in liquid chromatography using polymeric liquid membrane electrodes incorporating macrocyclic hexaamines. *Journal of Chromatography. A*, 915, 25-33.

[122] Bao, Y, Everaert, J., Pietraszkiewicz, M., Pietraszkiewicz, O., Bohets, H., Geise, H. J., Peng, B. X. and Nagels, L. J. (2003) Behaviour of nucleotides and oligonucleotides in potentiometric HPLC detection. *Analytica Chimica Acta*, 550, 130-136.

[123] Toth, K., Stulik, K., Ktner, W., Feher, Z. and Lindner, E. (2004) Electrochemical detection in liquid flow analytical techniques: Characterization and classification (IUPAC Technical Report). *Pure and Applied Chemistry*, 76, 1119-1138.

[124] Pungor, E. (1998) The theory of ion-selective electrodes. *Analytical Sciences*, 14, 249-256.

[125] Gyurcsányi, R., Rangisetty, N., Clifton, S., Pendley, B. D. and Lindner, E. (2004) Microfabricated ISEs: critical comparison of inherently conducting polymer and hydrogel based inner contacts. *Talanta*, 63, 89-99.

[126] Gyurcsányi, R., Nyback, A. S., Toth, K., Nagy, G. and Ivaska, A. (1998) Novel polypyrrole based all-solid-state potassium-selective microelectrodes. *Analyst*, 123, 1339-1344.

[127] Tymecki, L., Glab, S. and Koncki, R. (2006) Miniaturized, planar ion-selective electrodes fabricated by means of thickfilm technology. *Sensors*, 6, 390-396.

[128] Bohm, S., Olthius, W. and Bergveld, P. (2000) A generic design of a flow-through potentiometric sensor array. *Mikrochimica Acta*, 134, 237-243.

[129] Hassan, S. S. M., Sayour, H. E. M. and al-Mehrezi, S. S. (2007) A novel planar miniaturized potentiometric sensor for flow injection analysis of nitrates in wastewaters, fertilizers and pharmaceuticals. *Analytica Chimica Acta*, 581, 13-18.

[130] Toczylowska, R., Pokrop, R., Dybko, A. and Wróblewski, W. (2005) Planar potentiometric sensors based on Au and Ag microelectrodes and conducting polymers for flow-cell analysis. *Analytica Chimica Acta*, 540, 167-172.

[131] Ciosek, P., Brzózka, Z. and Wróblewski, W. (2006) Electronic tongue for flow-through analysis of beverages. *Sensors and Actuators B*, 118, 454-460.

[132] Tymecki, L., Zwierkowska, E. and Koncki, R. (2004) Screen-printed reference electrodes for potentiometric measurements. *Analytica Chimica Acta*, 526, 3-11.

[133] Yoon, H. Y., Shin, J. H., Lee, S. D., Nam, H., Cha, G. S., Strong, T. D. and Brown, R. B. (2000) Solid-state ion sensors with a liquid junction-free polymer membrane-based reference electrode for blood analysis. *Sensors and Actuators B*, 64, 8-14.

[134] Ibanez-Garcia, N., Mercader, M. B., Rocha, Z. M., Seabra, C. A., Gongora-Rubio, M. R. and Chamarro, J. A. (2006) Continuous flow analytical microsystems based on low-temperature co-fired ceramic technology. Integrated potentiometric detection based on solvent polymeric ion-selective electrodes. *Analytical Chemistry*, 78, 2985-2992.

[135] Watanabe, A., Kayanne, H., Nozaki, K., Kato, K., Negishi, A., Kudo, S., Kimoto, H., Tsuda, M. and Dickson, A. G. (2004) A rapid, precise potentiometric determination of total alkalinity in seawater by a newly developed flow-through analyzer designed for coastal regions. *Marine Chemistry*, 85, 75-87.

[136] Dantan, N., Kroning, S., Frenzel, W. and Kuppers, S. (2000) Comparison of spectrophotometric and potentiometric detection for the determination of water using Karl Fischer method under flow injection analysis conditions. *Analytica Chimica Acta*, 420, 133-142.

[137] Ohura, H., Imato, T., Kameda, K. and Yamasaki, S. (2004) Potentiometric determination of bromate using an Fe(III)-Fe(II) potential buffer by circulatory flow injection analysis. *Analytical Sciences*, 20, 513-518.

[138] van Staden, J. F., Mashamba, M. G. and Stefan, R.-I. (2002) An on-line potentiometric sequential injection titration process analyser for the determination of acetic acid. *Analytical and Bioanalytical Chemistry*, 374, 141-144.

[139] Paseková, H., Sales, M. G. F., Montenegro, M. C. B. S. M., Araújo, A. N. and Polásek, M. (2001) Potentiometric determination of acetylsalicylic acid by sequential injection analysis (SIA) using a tubular salicylate-selective electrode. *Journal of Pharmaceutical and Biomedical Analysis*, 24, 1027-1036.

[140] van Staden, J. F., Mashamba, M. G. and Stefan, R.-I. (2005) Determination of the total acidity in soft drinks using potentiometric sequential injection titration. *Talanta*, 58, 1109-1114.

[141] Cortina, M., Gutés, A., Alegret, S. and del Valle, M. (2005) Sequential injection system with higher dimensional electrochemical sensor signals. Part 2. Potentiometric e-tongue for the determination of alkaline ions. *Talanta*, 66, 1197-1206.

[142] Calvo, D., Grossi, M., Cortina, M. and Del Valle, M. (2007) Automated SIA system using an array of potentiometric sensors for determining alkaline-earth ions in water. *Electroanalysis*, 19, 644-651.

[143] Cortina, M., Duran, A., Alegret, S. and del Valle, M. (2006) A sequential injection electronic tongue employing the transient response from potentiometric sensors for anion multidetermination. *Analytical and Bioanalytical Chemistry*, 385, 1186-1194.

[144] Pimenta, A. M., Araújo, A. N. and Montenegro, M. C. B. S. M. (2001) Sequential injection analysis of captopril based on colorimetric and potentiometric detection. *Analytica Chimica Acta*, 438, 31-38.

[145] Silva, J. E., Pimentel, M. F., Silva, V. L., Montenegro, M. C. B. S. M. and Araújo, A. N. (2004) Simultaneous determination of pH, chloride and nickel in electroplating baths using sequential injection analysis. *Analytica Chimica Acta*, 506, 197-202.

[146] Reis-Lima, M. J., Fernandes, S. M. V. and Rangel, A. O. S. S. (2004) Sequential injection titration of chloride in milk with potentiometric detection. *Food Control*, 15, 609.

[147] Vieira, J. A., Raimundo, I. M., Jr, Reis, B. F., Montenegro, M. C. B. S. M. and Araújo, A. N. (2003) Monosegmented flow potentiometric titration for the determination of chloride in milk and wine. *Journal of the Brazilian Chemical Society*, 14, 259-264.

[148] Andrade-Eiroa, A., Erustes, J. A., Forteza, R., Cerda, V. and Lima, J. L. F. C. (2002) Determination of chloride by multisyringe flow injection analysis and sequential injection analysis with potentiometric detection. *Analytica Chimica Acta*, 467, 25-33.

[149] Santos, E., Montenegro, M. C. B. S. M., Couto, C., Araújo, A. N., Pimentel, M. F. and Silva, V. L. (2004) Sequential injection analysis of chloride and nitrate in waters with improved accuracy using potentiometric detection. *Talanta*, 63, 721-727.

[150] Jakmunee, J., Patimapornlert, L., Suteerapataranon, S., Lenghor, N. and Grudpan, K. (2005) Sequential injection with lab-at-valve (LAV) approach for potentiometric determination of chloride. *Talanta*, 65, 789-793.

[151] Pimenta, A. M., Araújo, A. N. and Montenegro, M. C. B. S. M. (2002) A sequential injection analysis system for potassium clavulanate determination using two potentiometric detectors. *Journal of Pharmaceutical and Biomedical Analysis*, 30, 931-937.

[152] van Staden, J. F., Stefan, R.-I. and Birghila, S. (2000) Evaluation of different SIA sample-

buffer configurations using a fluoride-selective membrane electrode as detector. *Talanta*, 52, 3-11.

[153] Santos, E. M. G., Couto, C. M. C. M., Araújo, A. N., Montenegro, M. C. B. S. and Reis, B. F. (2004) Determination of Gibberellic acid by sequential injection analysis using a potentiometric detector based on Mn (III) -Porphyrin with improved characteristics. *Journal of the Brazilian Chemical Society*, 15, 701-707.

[154] van Staden, J. F., Mashamba, M. G. and Stefan, R. -I. (2002) On-line determination of hydrochloric acid in process effluent streams by potentiometric sequential injection acid-base titration. *South African Journal of Chemistry*, 55, 43-55.

[155] van Staden, J. F., Mashamba, M. G. and Stefan, R. -I. (2002) On-line dilution and determination of the amount of concentrated hydrochloric acid in the final products from a hydrochloric acid production plant using a sequential injection titration system. *Talanta*, 58, 1089-1094.

[156] Silva, F. V., Nogueira, A. R. A., Souza, G. B. and Cruz, G. M. (2000) Determination of total, volatile and acid detergent insoluble nitrogen in silage by sequential injection. *Analytical Sciences*, 16, 361-364.

[157] Santos, E. M. G., Araújo, A. N., Couto, C. M. C. M., Montenegro, M. C. B. S. M., Kejzarova, A. and Solich, P. (2004) Ion selective electrodes for penicillin-G based on Mn (III) TPP-Cl and their application in pharmaceutical formulations control by sequential injection analysis. *Journal of Pharmaceutical and Biomedical Analysis*, 36, 701-709.

[158] Zárate, N., Araújo, A. N., Montenegro, M. C. B. S. M. and Pérez-Olmos, R. (2003) Sequential injection analysis of free and total potassium in wines using potentiometric detection and microwave digestion. *American Journal of Enology and Viticulture*, 54, 46-49.

[159] Fernandes, R. N., Sales, M. G. F., Reis, B. F., Zagatto, E. A. G., Araújo, A. N. and Montenegro, M. C. B. S. M. (2001) Multitask flow system for potentiometric analysis: its application to tire determination of vitamin B-6 in pharmaceuticals. *Journal of Pharma-ceutical and Biomedical Analysis*, 25, 713-720.

[160] Silva, F. V., Souza, G. B. and Nogueira, A. R. A. (2001) Use of yeast crude extract for sequential injection determination of carbohydrates. *Analytical Letters*, 34, 1377-1388.

[161] Silva, F. V., Nogueira, A. R. A., Souza, G. B., Reis, B. F., Araujo, A. N., Montenegro, M. C. B. S. M. and Lima, J. L. F. C. (2000) Potentiometric determination of urea by sequential injection using Jack bean meal crude extract as a source of urease. *Talanta*, 53, 331.

[162] Perez-Olmos, R., Soto, J. C., Zarate, N., Araújo, A. N., Lima, J. L. F. C. and Saraiva, M. L. M. F. S. (2005) Application of sequential injection analysis (SIA) to food analysis. *Food Chemistry*, 90, 471-490.

[163] Perez-Olmos, R., Soto, J. C., Zarate, N., Araújo, A. N., Lima, J. L. F. C. and Montenegro, M. C. B. S. M. (2005) Sequential injection analysis using electrochemical detection: A review. *Analytica Chimica Acta*, 554, 1-16.

[164] Pimenta, A. M., Araújo, A. N., Montenegro, M. C. B. S. M. and Calatayud, J. M. (2006) Application of sequential injection analysis to pharmaceutical analysis. *J. Pharmaceutical and Biomedical Analysis*, 40, 16-34.

[165] Kikas, T. and Ivaska, A. (2007) Potentiometric measurements in sequential injection analysis lab-on-valve (SIA-LOV) flow-system. *Talanta*, 71, 160-164.

[166] Dorneanu, S. A. Coman, V., Popescu, I. C. and Fabry, P. (2005) Computer-controlled system

for ISEs automatic calibration. *Sensors and Actuators B*, 105, 521-531.

[167] Araújo, A. N., Montenegro, M. C. B. S. M., Kousalova, L., Skleranova, H., Solich, P. and Perez-Olmos, R. (2004) Sequential injection system for simultaneous determination of chloride and iodide by a Gran's plot method. *Analytica Chimica Acta*, 505, 161-166.

[168] Oliveira, C. C., Sartini, R. P., Zagatto, E. A. G. and Lima, J. L. F. C. (1997) Flow analysis with accuracy assessment. *Analytica Chimica Acta*, 350, 31-36.

[169] Duran, A., Cortina, M., Velasco, L., Rodriguez, J. A., Alegret, S. and del Valle, M. (2006) Virtual instrument for an automated potentiometric e-tongue employing the SIA technique. *Sensors*, 6, 19-24.

[170] Gallardo, J., Alegret, S., de-Roman, M. A., Munoz, R., Hernandez, P. R., Leija, L. and del Valle, M. (2003) Determination of ammonium ion employing an electronic tongue based on potentiometric sensors. *Analytical Letters*, 36, 2893-2908.

[171] Gallardo, J., Alegret, S., Munoz, R., de-Roman, M., Leija, L., Hernandez, P. R. and del Valle, M. (2003) An electronic tongue using potentiometric all-solid-state PVC-membrane sensors for the simultaneous quantification of ammonium and potassium ions in water. *Analytical and Bioanalytical Chemistry*, 377, 248-256.

[172] Gallardo, J., Alegret, S. and del Valle, M. (2005) Application of a potentiometric electronic tongue as a classification tool in food analysis. *Talanta*, 66, 1303-1309.

[173] Gallardo, J., Alegret, S., Munoz, R., Leija, L., Hernandez, P. R. and del Valle, M. (2005) Use of an electronic tongue based on all-solid-state potentiometric sensors for the quantitation of alkaline ions. *Electroanalysis*, 17, 348-355.

[174] Gutierrez, M., Alegret, S. and del Valle, M. (2007) Potentiometric bioelectronic tongue for the analysis of urea and alkaline ions in clinical samples. *Biosensors & Bioelectronics*, 22, 2171-2178.

[175] Kahlert, H., Porksen, J. R., Behnert, J. and Scholz, F. (2005) FIA acid-base titrations with a new flow-through pH detector. *Analytical and Bioanalytical Chemistry*, 382, 1981-1986.

[176] Dasgupta, P. K., Tanaka, H. and Jo, K. D. (2001) Continuous on-line true titrations by feedback based flow ratiometry application to potentiometric acid-base titrations. *Analytica Chimica Acta*, 435, 289-297.

[177] Borges, E. P., Martelli, P. B. and Reis, B. F. (2000) Automatic stepwise potentiometric titration in a monosegmented flow system. *Mikrochimica Acta*, 135, 179-184.

[178] Almeida, C. M. V., Lapa, R. A. S. and Lima, J. L. F. C. (2001) Automatic flow titrator based on a multicommutated unsegmented flow system for alkalinity monitoring in wastewaters. *Analytica Chimica Acta*, 438, 291-298.

[179] Paim, A. P. S., Almeida, C. M. N. V., Reis, B. F., Lapa, R. A. S., Zagatto, E. A. G. and Lima, J. L. F. C. (2002) Automatic potentiometric flow titration procedure for ascorbic acid determination in pharmaceutical formulations. *Journal of Pharmaceutical and Biomedical Analysis*, 28, 1221-1225.

[180] Aquino, E. V., Rohwedder, J. J. R. and Pasquini, C. (2006) A new approach to flow-batch titration. A monosegmented flow titrator with coulometric reagent generation and potentiometric or biamperometric detection. *Analytical and Bioanalytical Chemistry*, 386, 1921-1930.

[181] Koncki, R., Radomska, A. and Glab, S. (2000) Bioanalytical flow injection system for control of hemodialysis adequacy. *Analytica Chimica Acta*, 418, 213-224.

[182] Radomska, A., Koncki, R., Pyrzynska, K. and Glab, S. (2004) Bioanalytical system for

control of hemodialysis treatment based on potentiometric biosensors for urea and creatinine. *Analytica Chimica Acta*, 523, 193-200.

[183] Itai, K. and Tsunoda, H. (2001) Highly sensitive and rapid method for determination of fluoride ion concentrations in serum and urine using flow injection analysis with a fluoride ion-selective electrode. *Clinica Chimica Acta*, 308, 163-171.

[184] Hosseini, M. S. and Rahiminegad, H. (2006) Potentiometric determination of ultratrace amounts of fluoride enriched by zirconia in a flow system. *Journal of Analytical Chemistry*, 6, 166-171.

[185] Rodriguez, J. A., Barrado, E., Vega, M. and Lima, J. L. F. C. (2005) Speciation of inorganic arsenic in waters by potentio-metric flow analysis with on-line precon-centration. *Electroanalysis*, 17, 504-511.

[186] Martinez-Barrachina, S., del Valle, M., Matia, L., Pratz, R. and Alonso, J. (2002) Determination of polyethoxylated nonionic surfactants using potentiometric flow injection systems. Improvement of the detection limits employing an on-line pre-concentration stage. *Analytica Chimica Acta*, 454, 217-227.

[187] Jones, D. L., Moody, G. J. and Thomas, J. D. R. (1981) Potentiometry of alkoxylates. *Analyst*, 106, 439-447.

[188] Martinez-Barrachina, S. and del Valle, M. (2006) Use of a solid-phase extraction disk module in a FI system for the automated preconcentration and determination of surfactants using potentiometric detection. *Microchemical Journal*, 83, 48-54.

4 流动伏安法

Ivano G. R. Gutz、Lúcio Angnes 和 Andrea Cavicchioli

4.1 引言

电化学为流动分析（FA）提供了各种各样的多功能检测器，从通用传感器到选择性很高的检测器，这些检测器能够测量浓度或活性、评估金属的不稳定性、提供物质的氧化还原形态、在检测器上预浓缩分析物、研究吸附甚至进行分子识别。目前，许多可用于流动分析的电化学检测器（ECDs）可以从市场上获得，或者可以由为一般用途或 HPLC 生产的流动池、电极和仪器轻松改装。ECDs 也可以在实验室中构建，如本章所示，作者开发的检测池包括微流控检测池。

先进的 ECDs 完美契合了常规流动分析和微流动分析以及 HPLC 和 CE 中涉及的时间-大小-流量尺度。它们具有死体积小和响应快速的特点，需要极少量的样品并涵盖不同数量级的分析物浓度（最多12个），检测限范围 $10^{-14} \sim 10^{-4}$ mol/L 或低至毫微微克级。电极插入低功耗固态仪器并基于现代模拟和数字集成电路，这使得它们简单、可靠、坚固、小型化、便携、便宜，并可直接连接到计算机。与光谱法流动分析检测器相比，ECDs 没有光学元件、火焰、炬管、烘箱、泵、真空系统或任何移动的机械部件。大多数 ECDs 具有中等选择性并且不受有色、混浊或盐水基质的干扰。然而，电极与样品基质、试剂和产品的直接物理接触使得几乎所有 ECDs（包括伏安法电化学检测）都容易受到分析物逐渐失活或对分析物的响应发生改变的影响。该缺陷在流动分析中得到了缓解，因为流动分析中的暴露时间比间歇操作中的短，并且通过重新校准进行补偿，直到灵敏度的损失严重而不得不重新激活电极。因此，最近的研究工作所取得的一些进展主要与提高传感器响应（例如，新的或改进的电极材料、通过电化学或其他方式进行原位再激活、膜屏蔽、样品预处理、一次性电极等）的持久性有关。使用改进的电极，尤其是电流生物传感器，保质期和浸出是需要解决的额外问题。目前在选择性、灵敏度、应用谱图方面不断取得进步，对于伏安检测器，扩大电位域并且将间歇伏安溶出法转换为流动分析，扩大了例如方波伏安法和多脉冲安培法等技术的使用、同步测试的开发、ECDs 与其他技术的耦合、仅用流动分析才能实现的对电极上某些吸附过程的研究，以及微流体系统的开发。

但是，电化学可以为流动分析提供的远不止检测器。流入式电化学操作包括：分析物的预浓缩（通过电镀、吸附、迁移、汞齐化）、载体或样品的纯化（去除电活性干扰物）、相转移的促进、有机材料的破坏（氧化、矿化或在金刚石电极处"燃烧"）、表面极性调谐、离子迁移传输、通过电渗泵置换流体（详见本书第一卷第5章）、通过电溶解金属和合金采集样品、通过库仑法产生试剂（包括不稳定试剂）、电聚合、有机电合成、电化学发光、光电催化、光谱电化学等。微流体（见本书第一卷第6章）开辟了另外一种途径将功能丰富的电化学传质和化学转化手段与 ECDs 和非电化学检测器联系起来。

为了说明 ECD-FA 的趋势和进展，本章汇编了过去十年中发布的大量但并不详尽的代表性示例，其专为有经验的用户和该领域的新手而设计。为简洁起见，本章省略了大量的理论讨论和计算，但这些都可以在引用的参考文献中找到。

4.2 伏安/安培流动分析

4.2.1 原理与技术

近两个世纪以来,科学技术的进步为电化学领域的发展奠定了坚实的基础,电分析化学的理论、仪器、实验实践和应用在超过20多本权威系列专著[1]、近年手册[2]和众多书籍[3-8]中有所提及。关于流动分析与ECDs,最全面的论文仍然是1987年出版的一本书——流动液体中的电分析测量[9],其涵盖了原理、电池设计、特性和应用。同时,针对一种或另一种形式的流动分析,关于ECDs的更具体的工作已在相关书籍章节[10-12]和综述[13-15]中出现。用于色谱[16-18]和毛细管电泳[19]的ECDs的大量文献对流动分析来说也是有价值的。互联网(例如[20,21])提供有关带流动池的伏安法和电流法ECDs的可靠教程,这些教程对新手和从业者是有帮助的。

在整个流动分析中,包含电解的检测器,如安培和伏安系统,比不含电解的检测器(主要是电位和电导检测器)应用更广泛。电位检测器(在第3章中详述)对活性的感测依赖于在电流基本为零的条件下对指示电极(例如,离子选择性电极)平衡电位的测量。在电导检测(广泛用作离子色谱中的流量检测器)中,通过交替激发信号可以规避离子置换过程(在电场的作用下)中的电解。在大约1MHz的频率下,由于电容耦合,不再需要电极与样品直接接触。基于该特性并通过在毛细管周围安装双金属环,毛细管电泳中引入和发展了非接触式电导检测[22-24]。非接触式电导检测对流动分析而言是不错的选择[25],只要通过化学手段或其他方式可以确保其选择性。真正的非接触式检测不能扩展到安培(或电位)传感器,但有时可以通过其他方式避免直接暴露于整个样品基质中,例如,对于挥发性或气态物质,可以通过克拉克电池(一种适用于流动分析的非常流行的安培氧传感器)中插入疏水膜实现。

在对工作电极/溶液界面处的电位差进行线性(或阶梯)扫描期间(固定条件下)可以实施连续(或采样)电流测量,并且在定义的电位(三角波)处可以反转扫描方向的循环伏安法,普遍用于氧化还原表征离子和分子。以各种扫描速率($v = dE/dt$)所得的电流-电压曲线可以提供以下相关信息,例如异质电荷转移过程的数量、它们的条件标准电位、$E^{0'}$、自身电子转移和氧化还原过程之前或之后的耦合化学反应的动力学参数,以及反应物和/或产物在电极表面的吸附[5]。金属配合物的稳定性和不稳定性以及样品中电活性物质的主要氧化状态(形态)也可以通过伏安法进行评估。

除了流入或流出电极以促进阳极(I_{an})或阴极(I_{cat})过程的法拉第电流(I_f)之外,伏安法期间施加的电位的任何变化都伴随着电容电流(I_c),该电流弥补了用于在电极双电层的扩散和致密区域重新排列中性和带电化学物质的电荷,包括吸附/解吸。I_c是背景或剩余电流(I_r)的重要贡献者,其通常还包括法拉第组分,源自溶液、电解质或电极中的微量电活性杂质,或电极材料本身的表面氧化/还原。

可用于正负方向伏安法的电位范围受到一些区域的限制,在这些区域中,I_r由于以下原因呈指数增长，即电极材料的大量氧化、溶剂的电解或为了处理迁移电流和调整介质而在样品基质或电解质中添加了一种组分。必须选择条件以使分析物（或由分析物产生或消耗的物质）在此"清洁"范围内的电位下具有电活性。

循环伏安法（CV）作为热力学、机械、动力学和传质研究工具，其丰富性源于许多因素的相互作用，这些因素可能会阻碍人们为新 ECD-FA 方法选择最佳的操作条件，电化学方面的背景知识很有价值。对于流动注射分析来说，循环伏安法扫描必须与进样同步，并且必须在分析物塞通过流动池期间快速执行。然而，V 越高，I_r 就越重要，即使是对检测限损害的采样电流模式也是如此。在固定（或脉冲）电位下伏安法优于电流法，因此在流动分析中应用更广泛，包括对多分析物的定量和更好的干扰诊断/免疫（如带有多 λ 的流动分析与单 λ 分光光度检测相比）。方波伏安法（SWV）[6]是一种快速扫描技术，它在间歇伏安法中获得认可，但在流动分析中使用不多。它类似于微分脉冲伏安法（DPV），因为在分析物的 $E^{0'}$ 处记录的是峰形信号而不是 S 形波。但与微分脉冲伏安法不同的是，它扫描速率要高得多，并且记录的 I 值是指在每个脉冲期间（I_c 的短暂衰减之后，在顶部和底部电位）通过双电流采样测量的 $I_{an}-I_{cat}$。另一方面，方波伏安法受到电极过程可逆性的强烈影响。

所有阶梯和脉冲伏安技术，包括微分脉冲伏安法和方波伏安法，探索如何通过质量传输使不需要的 I_c（I_r 的一个组成部分）的衰减速度比 I_f 的更快[11]。因此，通过在每个电位的增量或脉冲之后以及电流采样之前留出若干毫秒来减少干扰。尽管如此，通过固体和糊状电极将 I_r 最小化所需的时间比液体汞更长，因此，只有在恒定电位操作条件下才能达到流动分析中信号值与背景电流的最佳比率。这种行为和简易性解释了为什么每当单个电活性成分（带有定义明确且不包含干扰的伏安波）需要通过流动分析进行量化时，选择的技术通常是直流电流检测（或脉冲安培法）。在伏安图中谨慎地选择一个值，然后在该值处电位（或脉冲宽度）被"冻结"，同时电流被监控。打开系统时观察到的高 I_r 值衰减到低且稳定的基线水平，就可以开始测定；所获的峰形 I_f 类似于在单波长检测中观察到的吸光度峰，高度与浓度成正比（可以使用面积积分代替，但这种做法不常见）。因此，就检测限而言，（脉冲）安培操作模式非常适用于流动注射分析、间歇注入分析、顺序注射分析、连续流动分析和 HPLC[16]。

通过安培法同时测定多种电活性分析物需要在越来越正（或负）电位下串联的各种检测器。在安培测量中，只有一小部分电活性分析物被电解。因此，最容易氧化（或还原）的物质不仅会在第一个工作电极而且会在所有工作电极上产生电流，并且贡献的代数减法可能会有损准确性。具有高电极面积与样品流量比率的库仑电池（多孔或管状工作电极或某些低流量薄层几何形状），可以通过在每个检测阶段消耗一种物质来克服这个问题。

虽然在流动分析中不经常使用，但极谱检测至少对于存在严重"电极污染"的系统而言是一种有趣的替代方法。极谱法是唯一一种使用汞滴作为电极的技术，在每次线性、阶梯或脉冲电位扫描期间，汞滴经常更换（每 0.3~3s）。使用现代带阀的自动静态汞滴电极，扫描的（可回收）汞量约为 50 滴（每滴 1μL）。液滴的更换会引起足够的对流使溶液均质化，以使正向和反向扫描为电活性物质定义相同的 sigmoid 曲线。

文献中有数以千计的适用于流动分析的极谱（间歇）方法，其借助于适合商用汞滴电极的流动适配器，正如下文所示。

如前所述，流动伏安法和电流分析法受益于分析物到电极的对流传输（以及可溶产物的去除），降低扩散边界层厚度，从而增加 I_f。这就是在连续进样或再循环的流动池中使用恒定面积电极也可以获得 S 形曲线（如极谱曲线）的原因；在扩散性/对流传质控制和低速情况下，正向和反向扫描变得无法区分，至少对于可溶和不可吸附的电极反应产物而言是如此。受电极测量的电流和流动池几何形状影响的方程式可用于最常见的设计[9,10]。在带有通道电极（穿过流道的矩形条带，在微流体电池中也常见）的流动池中对伏安响应的理论和数值模拟方面取得了进展，同时 Cooper 和 Compton[26] 对此进行了综述。工作曲线可在线获得，此研究包含了从 10 mm 到微米级的各种宽度的电极，以及从 10^{-4} 到超过 10^{-1} cm^3/s 的流速，以及不同的电极工艺机制。

通过使用快速旋转的圆盘电极[27]，进一步增强了间歇注入分析中的质量传输和灵敏度。然而，该流动池也适用于流动注射分析或顺序注射分析，但其构造和操作不那么简单。通过将电极的暴露面积减小到微米级，可以获得电流密度（I_f/a）的可比增益[28]，进而也可以促进径向扩散。尽管在提升 I_f/I_c 比率方面有优势，但这种微小电极的 I_f 绝对值非常低，需要更高的电子放大率，这会导致不太有利的信噪比。微电极矩阵或阵列的互联或使用带状或线状电极似乎是一个很好的折中方案[29]。

在流动条件下在电极界面处原位积累分析物然后进行伏安测定的能力，通常称为阳极或阴极溶出分析（ASV 或 CSV），对于达到非常低的检测限具有决定性意义，可与复杂的光谱技术相媲美[30]。必须仔细选择溶出分析的实验条件，因为当各种金属同时积聚在固体电极或汞齐上时，就有可能形成金属间化合物，并在与纯元素不同的电位处显示 ASV 波。CSV 主要是在电活性有机分子或金属配合物与配体诱导的吸附积累之后应用。预浓缩过程遵循吸附等温线，涉及分子相互作用和上限的表面饱和，这意味着校准曲线将在有限范围内呈线性。由于分子比分析物更易吸附而产生的竞争性表面饱和也给分析应用带来了问题。这就是在金属痕量分析之前对带有有机配体的样品进行光解、消化或矿化的原因之一，不包括对形态和不稳定性的研究。

用于电流检测的最小仪器由恒电位仪、用于数据采集的记录器或计算机，以及带有工作电极、参比电极和辅助电极的流动池组成。电池和电极将在接下来的部分中讨论。简单的恒电位仪可以在实验室中使用低成本运算放大器[3,5,11]构建，并使用数据采集卡与计算机连接。多种恒电位仪和电化学系统可在市场上买到，它们中的大多数为 FA-ECD 指定了一个非常广泛的技术库。代表厂商品牌有：Ecochemie、BAS、PAR、Metrohm、Pine、Radiometer、PamlSens、Solartron、Uniscan、Gamry、EDAQ 和 Bank（Wenking）。这些公司还生产各种电池和电极。相关技术数据可在公司网站上获得。为流动伏安法/电流法设计的仪器由一些提供流动池的公司出售，将在接下来的部分中引用。

无论流动分析中使用何种恒电位仪，在扩散/对流质量传输控制下用正常尺寸的电极记录的电流信号呈现波动，这些波动是由蠕动泵和电磁泵操作自带的流动脉动引起的。径向扩散方案下的微电极对这种脉动相当不敏感。其他减弱这种干扰的方法是：

气动阻尼（带有气泡的腔室，然后是流动限制）；模拟或数字信号滤波，电流采样与脉动同步；由电机驱动的注射泵提供"无脉冲"流动（在顺序注射分析中常见，但在流动注射分析中不常见）、电动移液器（典型用于间歇注入分析）或通过重力或压缩气体作用对载体（和试剂）加压（与流动注射分析兼容）提供的"无脉冲"流动。一种用于在水族箱中吹入空气的廉价膜泵非常适合此目的[31]。

4.2.2 电极材料

伏安工作电极最常见的材料是汞（可再生液滴、薄膜或汞齐）、贵金属（主要是Pt和Au）和各种导电碳同素异形体（如碳糊、石墨、碳纤维、玻璃碳、网状碳、高度规整的热解石墨、C_{60}-富勒烯、碳纳米管和掺杂金刚石）[30]。由于具有比玻璃碳更有利的电位窗口、更低的背景电流以及更高的化学和机械阻力，制备的合成金刚石薄膜（掺杂硼以获得导电性）在最近的文献中受到了极大的关注[30,32,33]。

面积 $1 \sim 20 mm^2$ 的扁平电极（通常是圆形），紧密嵌入绝缘聚合物块中，这是大多数流动池采用的典型的几何形状，因为块状聚合物可以很容易进行拆卸和抛光。其他形状、尺寸和安装方式并不少见，包括印刷电极或溅射电极、管状电极、多孔柱状电极和微电极阵列。

特定的吸附过程和固体电极形态或表面氧化的变化并不总是可逆的，因此，使用可再生汞滴或汞膜以外的电极复制的伏安图可能会出现波的某些位移或扭曲以及电流强度的降低，新的或重新抛光的固体（或糊状）电极需要在工作电位下进行调节，直到背景电流衰减到一个低而稳定的值；偏离进入氢和/或氧开始析出的区域，或其他脉冲序列，在许多情况下被发现对激活电极有效[16]。最终，伏安检测器在无人看管的情况下工作数月[11]。对电极表面进行机械抛光是在其他活化程序失败时的惯用方法，替代方法包括超声处理、用腐蚀性溶液（例如食人鱼溶液）进行化学处理、加热或激光处理。

使用可再生汞滴电极是获得完美重现的伏安图的最佳方法，当涉及负电位以及NHE下的氧化还原反应时该结论仍然成立。商用汞滴电极发生器是自动的，通过计算机、手动或由伏安分析仪本身触发液滴的变化。可以在每次进样之前（典型）或在一系列进样之后进行更新。对于快速阻塞电极[34]的分析物，可能需要更快的液滴变化（例如每秒）。

除了固体汞合金，铋电极或镀铋电极已被提议作为汞电极的替代品，因为它们的无毒性质和有利的氢过电压，虽然其电压不如汞高[35]。双电极已成功应用于金属的ASV-SIA[36,37]中。为了提高选择性和耐久性，改性电极也被广泛应用于流动分析，这将在本章末有所描述。

4.2.3 商业流动池

设计用作液相色谱检测器的电流式流动池自1974年就已上市[11]。如今，由于HPLC广泛采用ECDs来测定易氧化和可还原的化合物，因此一般的电化学仪器制造商和特殊的HPLC设备制造商提供了各种流动池。例如，BAS（美国拉斐特）提供了各种薄层流动池,具有串联或并联的一个至四个电极[38];沃特世生产 2465 ECD[39];Antec Leyden 销售 VT-03 电化学流动池和用于电化学衍生化的反应池[40];岛津制造的 L-ECD-6A EQD[41];Dionex 将 ICS-3000 ECD[42]商业化;Cypress Systems 的流动池 66-CL010 是一种通用型流动池,适用于流动注射分析[43];带有生物传感器芯片的流动池由 TRACE[44]生产。

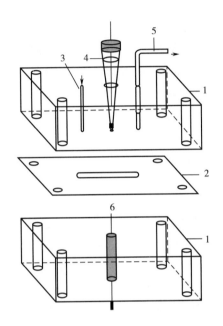

图 4.1 作者实验室制作的通用伏安法/电流法流动池

1—上下有机玻璃(或聚四氟乙烯)块 2—间隔垫片 3—流动溶液的入口 4—参比电极,微型 Ag/AgCl[48]或其他带适配器的类型 5—不锈钢(皮下注射针)出口管也用作对电极 6—盘状工作电极,例如,紧嵌在下块中的铂、金或玻璃碳圆柱体,或嵌入顶部并填充碳糊的黄铜机螺钉。这些块用(未显示)固定螺钉或夹子压在一起。

大多数所提到的装置都是错流薄层型,流线平行于带有嵌入式铂或玻璃碳盘工作电极的扁平块的表面。图 4.1 中描绘的在作者实验室中构建的电池说明了这种基本设计。一些制造商更喜欢壁式喷射设计(流入口垂直于电极盘的中心)。用于 HPLC 的检测器由不锈钢和聚合物(如 Teflon 和 Kel-F)构成,对几乎所有溶剂都呈惰性,可耐受相当高的压力,并且响应体积为 $1\mu L$ 或更小,适用于微孔 HPLC。虽然适用于流动注射分析和顺序注射分析,但对于在接近大气压的水溶液中工作,它们的规格和价格都过高。

带有多孔工作电极(玻璃碳颗粒)的库仑流动池由 ESA(美国切姆斯福德)提供[45]。它们提供最终的灵敏度,但不一定是最佳的 S/N 比[11],因为增加了背景电流。利用法拉第定律足以获得在该流动池条件下的分析物浓度,从而省却了标准溶液的校准程序。在具有两个或多个库仑测定池的流动系统中,可测定多种分析物;第一种分析物也可用于从电解液中去除痕量的电活性污染物。

具有多个(最多 65 个)叉指微电极($2\sim20\mu m$ 宽,$2\sim3mm$ 长条带,Au 或 Pt)的

流动池也可从至少两家供应商处获得。

4.2.4 用于商业间歇流动池的适配器

虽然流动系统与伏安或电流检测器的接口可以通过流动池实现，就像上一段中描述的那样，但也可以使用间歇池的适配器。两者的区别在于通过两种不同的策略将流体引至工作电极表面，在该表面上发生检测步骤中涉及的氧化还原过程。严格来说流动池具有较低的死体积，并将三电极系统紧密地结合在同一空间内，而流动池适配器是使流动溶液撞击浸入常规电解槽中的工作电极的装置，该电解槽通常具有几毫升的体积，其中还浸入了参比电极和辅助电极。实际上，两种方法各有优缺点。流动池更复杂，需要将微型电极精确地安装到微小而紧凑的外壳中，这需要一些技巧，包括组装参比电极也是如此[48-51]。而适配器可以有效地安装在任何商业流动池上，该过程比较简单。然而，它们通常具有较大的死体积，并且根据实际设计，它们会受到来自电解槽的已经处理过的分析物的再污染，并且在移除和重新安装时会造成它们的位置发生偏移。

流动适配器特别方便的一个背景是使用汞滴作为工作电极（MDEs）并将流动歧管和检测器进行耦合。MDEs仍然是一种有趣的电极类型，因为与它们在使用过程中具有一些重要的优势，如高氢过压、光滑和均匀的表面，以及在现代设备中，电极表面很容易通过自动敲击器实现原位更新。显然，在这种情况下，流动池和适配器的设计必须考虑到每次测定时流出的一滴脱落的汞滴（1μL）。文献［52］中可以找到对截至2003年相关文献提到的用于MDEs的几个流动注射分析接口的综述以及对为流动系统中的电流法和溶出伏安法检测而设计的四种模型性能的比较。在那篇论文中，在评估每个模型时考虑了灵敏度、流速相关性、响应时间、液滴稳定性、对气泡通过的容忍度、未补偿阻力的程度和再污染等方面，如图4.2所示。

尽管由于使用汞而受到限制，但人们仍然对使用流量适配器和用于MDEs的流动池抱有很大的兴趣，这反映在过去几年中提出的各种新模型中。由圆柱形丙烯酸块组成的流动池，轴向通道直径为0.7mm，纵向中心有一个孔用于安装垂直于流口的汞毛细管，该流动池已在多项研究中得到利用，其中汞电极和流动注射分析的使用分析模式非常关键[53-57]。在环境分析中，类似的装置被开发以测定实际水样品中的U^{3+}[58]而圆柱形丙烯酸适配器（其中流入的流体在进入电极室之前分离）用于流动注射分析测定土壤提取物中的莠去津[59]和HMDE上水和废水中的Cd^{2+}、Pb^{2+}和Cu^{2+}[60,61]。在一个非常简单的解决方案中，一个L形PTFE管通过硅胶环连接到用于输送水银滴的玻璃毛细管上，并以将液体撞击到水滴上的方式使用[62]。当然，在流动条件下进料的底部有磁力搅拌器的小型化间歇池和顶部的玻璃毛细管可用于此目的[63,64]。也有人提出使用汞半球超微电极作为微射流电极，即溶液的细射流在高传质速率下从喷嘴上射出并撞击实际电极表面[65]。在为MDEs设计接口时必须牢记许多问题，例如，通过原始螺线管或气动敲击器有效分离汞滴的必要性、液滴自由流出到达池底部、系统的流体动力学以及其如何影响积累/溶出伏安法中的传质、对气泡通过的豁免、适配器和氧气流入管的正确密封（负电势以及ECS的电活性）等。

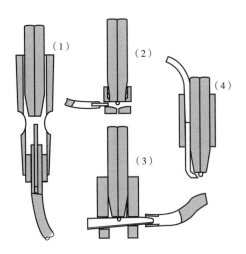

图 4.2 经许可带有自动汞滴电极（HMDE 和 SMDE）[52]的为商业间歇流动池设计的四种简单流动适配器
(1) 由移液管尖端制成的上流式适配器，压力安装在电极的玻璃毛细管上；(2) 侧流适配器，由丙烯酸制成，压力配合；(3) 侧流适配器，由一段带有上部 V 形切口的移液管尖端制成，以容纳液滴，由多孔硅管固定；(4) L 形聚四氟乙烯管，由硅胶环固定。版权所有 (2004) Wiley-VCH。

对于固体电极，不太经常使用适配器，因为在这种情况下，流动池代表了更实用的解决方案，最常见的是专门设计的用于间歇注射分析的流动池，这是本书第一卷第 4 章详述的一类流动池。尽管如此，还是有一些使用适配器的例子，如 Walcarius 等给出的适配器，其中对铵离子、碱金属和碱土金属离子的电流检测是在沸石改性的碳糊电极[66,67]或 Bergelin 和 Wasberg[68]的电极上进行的。本文提出了一种用于工作电极和液体出口的适配器配置，其基于界面区域的部分浸没，这导致在固体电极和流动出口之间形成弯液面，从而在 WE 和流动池电解质之间形成屏蔽。

4.2.5 专门设计的流动池和系统

Macpherson, J. V. 和 Unwin, P. R 在 1997 年对伏安微电池的构造进行了综述[69]。从那时起，文献中出现了许多新颖的设计。Gun 等将以前由 Hua 等开发的由镀汞金微丝制成的适配器横向安装到一段硅胶管中 [3cm×5mm（o. d.）×2mm（i. d.）][70,71]。

图 4.3 中再现了一个通用电池模型，其适用于任何平面固体电极，包括从 Angnes 和同事[72,73]提出的 CD-Rs（计算机用光盘）中获得的廉价（一次性）金电极。其性能几乎与昂贵的商业产品一样好，并且如果组装在固体、平坦的介电材料（陶瓷、聚合物或玻璃质）中/上，还可以适用于夹层微电极或多个带状电极或丝网印刷电极。

丝网印刷（一次性）电极由于其生产技术的不断发展[74-76]，越来越频繁地应用于流动伏安检测。一个有趣的例子是制备电化学过滤器，即用于在串联电流生物传感器之前从生物流体中去除电化学活性成分的电解微反应器[77]。用于流动分析的一次性检测器包括最近提出的一种由激光蚀刻技术制成的聚酯-碳纤维通道结构[78]。相关文献还描述了通过将乙酰胆碱酯酶固定在丝网印刷电极上的替代解决方案,用于在流动注射分析条件下监测水样中的有机磷农药[79-81]。Gao 等开发的基于丝网印刷碳电极的免疫电极条用于带有电化学检测的流动进样酶联免疫测定系统[82]。在相同的工作中,更多酶免疫测定与流动进样系统和电流检测相结合的例子被引用。

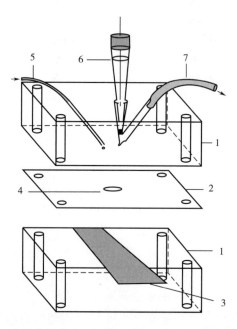

图 4.3 实验室制造的用于扁平电极的伏安法/电流法流动池

1—上下有机玻璃块,用固定螺钉(未显示)压在一起　2—垫片　3—工作电极,例如,一片金溅射型 CD-R(三井金)[72,73]、Pt 或玻璃碳板、带有微电极矩阵的芯片[97]或叉指电极[98]　4—垫片上的孔(定义电极面积和流道几何形状)　5—流动溶液的入口管　6—微型 Ag/AgCl 参比电极[48]　7—不锈钢(皮下注射针)出口管,用作对电极

在电化学系统的微设计和微制造进步的刺激下,薄层流动池(流道厚度低于 100μm)的重要性正在增加[83,84],并有可能实现大规模生产。这一趋势导致总流动池体积和溶液通量的减少,从而使样品和试剂的消耗量降至最低。Bratten 等[85]为了分析仪器中单个流动池的响应,使用光刻程序生产了一个微机械加工的 200μm(直径)× 20μm(深度)的电化学流动池。虽然它们的工作不是为了流动分析,但它为后续薄层流动池的发展构建了基础[86]。薄层流动池在壁面喷射圆盘电极中作用很大,用于分析测定或动力学研究。Toda 等[87]通过微加工为壁面喷射圆盘电极设计了深度低至 5μm 的流动池。在通过流动注射分析安培法检测氯霉素的应用中,这种薄层流动池是为了优化环工作电极对圆盘电极壁上所形成的衍生产物的收集[88]。一种测微电极被开发并通过毛细管流动注射分析系统用于半乳糖的测定[89]。相关文献还实现了具有高圆盘电极生成率/环收集效率的双薄层流动池的想法[90,91]。Gutz 和 Daniel[92]使用激光打印墨粉掩模获得了一个 19μm 深的电池。图 4.4 中描绘的电池与 TiO_2 介导的光降解反应器同时工作(用来自发光二极管的紫外线辐射照射,具有光催化剂固定在金电极上,作为电子清除剂界面)并且作为检测 Cu^{2+} 的伏安检测器。薄层可以快速促进对流过电极表

面的液体体积中有机物的光催化破坏。

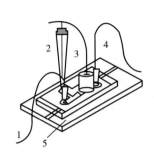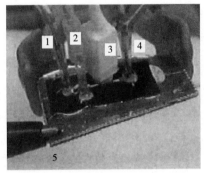

图 4.4　带有芯片上光电催化消化有机物的微流控伏安电池

1—流动池入口　2—小型化的 Ag/AgCl 参比电极[48]　3—UV-LED 辐射源　4—用作辅助电极和电池出口的不锈钢管　5—带有金电极的电池,如参考文献所述制备[98]

(经 Elsevier 许可,来自文献[92]的示意图,版权(2007)。)

虽然没有使用物理封闭的薄层电池,但相关文献中描述了一种流体动力学系统,其中液体从距离带有环形电极的平面基板上 $20\mu m$ 处的喷嘴中排出,以便径向流动到池内,即实际上如同在 $20\mu m$ 的通道内流动[93]。电化学检测器特别适合小型化,近年来,在微流体伏安电池方面取得了显著进展,这些进展导致以芯片实验室的形式组装整个流动系统的趋势增加(参见本书第一卷第 6 章)。

至少一个尺寸小于扩散层厚度的微电极(微线或微带或微盘电极)通过径向扩散呈增加了传质(而线性扩散主要用于较大的电极),如在文献中详细论证的那样[94]。对于任何尺寸的电极,对流传输都会造成的电流增加。较高的电流密度(通过径向扩散和/或对流)是有利的,因为对于给定的基于面积的双层容量,i_f/i_r 比率增加。对于微电极,由于需要放大弱信号,这种优势被部分抵消,并由此导致更多的噪声。然而,彼此相距较远的微电极阵列可以通过相互连接以仅保留优点。在流动系统中,微电极对由蠕动泵产生的脉动而引起的电流调制不太敏感。此外,对于可逆分析物,流路中的叉指电极可以通过氧化还原循环产生放大信号[95]。相反,每个电极阵列可以在不同的电位下运行和/或进行不同的表面改性,从而有利于同时测定各种分析物[96]。我们已经提出了通过简单的方法生产可单独处理的金电极阵列[97]以及多频和叉指电极[98]。

4.3　提高选择性、灵敏度和耐久性的策略

4.3.1　预浓缩

随着现代仪器和技术的改进,对各种样品中浓度越来越低的众多物质进行检测和定量的必要性越来越高,这刺激了预浓缩策略的发展。在文献中,有大量研究涉及流动技术与预浓缩过程的关联,不仅旨在获得更好的检测限,而且还旨在消除不同类型

样品研究中的干扰。大多数预浓缩方案基于分析物的可逆化学键合（例如，在离子交换树脂上）、吸附过程（例如，在固体材料或电极上）或沉积在导电表面之后进行的氧化还原过程（即电化学电极）。

Miro 等最近评检了涉及在线分离和/或预浓缩的流动注射分析，重点是监测环境中的金属离子[99]。这项工作的重点是流动注射分析与光谱技术的关联。在同一方向上，Camel[100]对固相萃取（SPE）进行了第二次评检，也涵盖了许多流动进样应用。还回顾了流动注射分析与原子吸收相结合的方法学和实际应用的进展[101]。最近，相关文献提出将微型色谱柱和选择性程序（称为阀上实验室与电热原子吸收光谱联用）相结合，用于量化超痕量金属[102,103]。

FA-ECD 与电极预浓缩的结合具有许多优点。分析物积聚在传感器表面，这一特征显著提高了检测限。通过选择最有利的电位，可以优化预浓缩，同时将干扰降至最低。甚至在流动注射分析出现之前，Zirino 和 Liberma 就发表了关于连续流动和阳极溶出伏安法关联性的第一项研究[104]。涉及流动注射分析加微分脉冲[105]或金属剥离电位[106]的过程被开发用于测定土壤中的铊以及海洋和自来水中的汞。相关文献已经描述了在实验室[107]，甚至在海船[108]上在线执行金属预浓缩的自动化系统。汞是 FIA-ASV 中首选的电极材料；玻璃碳上的相同预镀 Hg 膜可以多次重复使用，用于离线 UV 消解后测定废水中的 Cd^{2+}、Cu^{2+}、Pb^{2+} 和 Zn^{2+} [109]。Masini 的团队在离线酸消解后，通过 SI-ASV 测定河流沉积物[110]和涂料工业废水中的相同元素[111]。

在许多应用中也广泛探索了有机物质的吸附预浓缩，例如，分析葡萄酒、茶或番茄汁中的类黄酮[112]或测定药物中的帕拉西汀[113]和核黄素[114]。对于高纯铁样品中低浓度钴的定量，电化学预浓缩与化学发光的耦合是一个有用的过程，因为钴可以被选择性地剥离回溶液[115]。

4.3.2 介质交换

迄今为止，流动分析没有在溶出伏安法中的介质交换方面探索其管理溶液的潜力。电极在沉积阶段暴露于酸化载体和含金属离子的样品，而更合适的介质（缓冲络合溶液）被选择用于汽提步骤。通过预期一种金属相对于另一种金属的剥离，配体可以提供一些额外的选择性。介质交换被用于对复杂样品中的金属进行痕量分析，例如盐水[116]或土壤[117]或量化复杂样品中的有机化合物[118-120]。相关文献还探索了改良电极和培养基交换的关联，以提高灵敏度[117,119,121,122]。Ivaska 和 Kubiak[123]也证明了顺序注射分析在介质交换方面的固有优势，用于定量测定自来水中的铜。

Sawamoto 在 20 世纪 90 年代率先将流动分析和介质交换用于基础研究，其研究涉及在其他方式无法达成的电极上的吸附/解吸[124,125]，随着这些研究的发展，人们甚至可以测量在先前形成的吸附单层之上的第二种吸附物质[126]，在另一项研究中，喹啉在汞电极上的重新排列得到了证实[127]。相关文献还使用流动分析和微分容量-时间曲线研究了表面活性剂在汞上的可逆或不可逆吸附[128]。

4.3.3 除氧

在静态和流动操作模式下，在负电位伏安测量中，应特别注意溶解氧的存在。由于其在低于大约 NHE 的电位下电还原为 H_2O_2 或 H_2O，它通常会干扰测定过程并需要去除（对于主要响应可逆电极过程的技术，如 SWV，受影响较小）。在间歇分析中，通常通过预先在溶液中用惰性气体鼓泡（过程比较耗时）来除氧。这对于流动分析显然是不切实际的，迄今为止已经提出了各种替代解决方案。一种可能性是通过可渗透的窄孔硅胶管的壁从溶液中提取溶解氧，该管盘绕在低压或充满惰性气体（例如 N_2）的小室中[129]，可以将硅胶管取而代之的是微孔毛细管，如 Oxyphan 毛细管，一种聚丙烯中空纤维，在室温下通以 N_2。

已经比较了其他膜材料以及脱气室中的除氧剂[130]，甚至考虑使用酶促交联除氧凝胶[131]。此外，Reinke 和 Simon 建议使用基于色谱膜的流动池，该流动池由聚四氟乙烯微孔和大孔组成。在这种方法中，利用毛细管张力的特性（液体溶液无法透过），气相通过微孔被选择性地消除[132]。

4.3.4 催化电极工艺

在伏安法和安培法中增强法拉第电流以获得灵敏度的一种有效方法是探索催化电极工艺，其中分析物被重复循环，最简单的例子是金属离子的氧化还原反应，即在电极上金属离子被还原，由较高的氧化态转变为较低的氧化态。其他机制在综述[133]中介绍，其中包含 100 多种催化伏安法；其中许多方法还利用吸附原理在电极界面处积累分析物从而达到额外的灵敏度[134]，大多数方法依赖于汞电极（液滴或薄膜）并显示出极好的检测限，检测限为 $10^{-14} \sim 10^{-7}$ mol/L。这些方法的选择性通常很好，但必须首先消解含有有机物质的样品，因为表面活性剂可能会干扰电极的吸附。一些催化伏安法已经在流动系统中得以应用。铀的催化流动注射分析体现了这一概念。在汞滴电极处，U^{4+} 被还原为 U^{3+}。在含有大量过量硝酸盐的电解质中，该阴离子将 U^{3+} 氧化回 U^{4+}，使其可用于新的电还原步骤。这种催化循环很快，并且在正常流动注射分析条件下在汞滴处重复发生，大大提高了测量电流和检测限[58]。

4.3.5 光谱电化学

通过光谱技术对电极过程进行原位观察，可以扩大从样品中提取的信息量。人们为流动分析光谱电化学提出了不同的电池设计，并将其应用于基础研究和分析方法，主要是为了提高选择性。在光谱的紫外-可见光区域测量电极处/附近的光吸收和/或发射是最简单的方法。例如，有人提出了一种长光路（10mm）流动池，它带有双金工作电极，可装入标准尺寸的石英、玻璃或丙烯酸比色皿中［图 4.5（1）］[135]。将电池插入二极管阵列分光光度计的未经修改的比色皿支架中，并应用于异丙嗪和氯丙嗪的光谱电化学测定[136,137]。可以推导出吸收光谱与电位的 3-D 表面的关系，如图 4.5（2）所示，以进行机制分析研究以及为使用电流分析和单波长或多波长检测的分析方法选择最佳条件。多年前，在循环荧光伏安法[138]技术发明期间，标准石英比色皿被用作固定电化学电池。通过使用图 4.5（1）中所示的相同概念，该池可以转换为用于荧光伏安法和荧光安培法的流动池。最近一本关于电化学发光的书[139]介绍了它与流动分析的关联，激发了许多分析应用[140]，例如，氯霉素对 RuO_2^{2+} 电生化学发光的抑制是该药物使用流动注射分析方法的基础[141],将电化学发光检测缩小到微芯片电泳的趋势(在 Du 和 Wang[19]的综述中提到并涉及 200 多篇参考文献)也可扩展到微流动注射分析[142]。

表面增强共振拉曼光谱(SERRS)与 ECD-FA[143]耦合已应用于铁的测定。在注入的样品中,金属离子络合为 Fe(phen)$_2^{2+}$,该络合物通过吸附在镀银玻碳电极上进行预浓缩,并通过激光束照射以获得散射光谱,然后与银层一起剥离。该光谱可以选择性测定低至 fmol 水平的复合物。对于常规分析应用,目前缺乏价格合理、随时可用的拉曼光谱仪和光谱电化学池是一个限制。

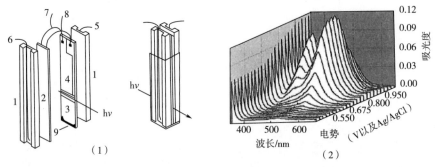

图 4.5 (1)用于光谱电化学的长光程流动池(经许可转载自文献[135],版权所有(2001 年)WileyVCH。) (2)在电池(1)中获得的 0.1mol/L H_2SO_4 中异丙嗪电氧化的 3-D 光谱伏安图 (经许可转自文献[136],版权所有(2003 年)Elsevier。)

1—丙烯酸基板 2—Au-CDtrode 的聚碳酸酯背板 3—Au-CDtrode 工作电极区[72] 4—辅助电极区 5—溶液入口管 6—到参比电极[48]的盐桥管(未显示) 7—双辅助电极的电互联 8—双工作电极的电互联 9—硅橡胶垫片,200μm 厚

4.3.6 改良电极

改良电极的开发为创造用于大量应用的传感器提供了可能性。大多数这些改良电极可以很好地用作流动分析中的检测器。通过吸附、浸涂、刷涂、喷漆、旋涂、Langmuir-Blodgett 薄膜沉积、电聚合、共电沉积和分子自组装,将薄膜或涂层,有时是多层的,施加到裸电极上。改性剂在碳糊或丝网印刷油墨中的分散也被广泛采用。应用一种或多种改性剂的主要目的:①通过介导或催化目标物质的电荷转移来促进目标物质的电解;②有利于目标分析物在传感器上的预浓缩,提高检测限;③保护电极表面免受干扰物质和/或钝化剂的影响;④固定或避免其他改性剂的浸出。Brainina 及其同事[30]对溶出伏安法的综述涵盖了通过原位和非原位过程修改电极,区分表面涂层和大量插入,并提供了数百个关于此类传感器及其在环境和食品分析中应用的参考文献,其中大部分传感器仍以间歇方式进行。

近年来,改良电极与流动注射分析的关联显著增加,用于表面改性的材料范围也在扩大。用生物材料(主要是酶)修饰的电极仍然占主导地位[144-147]。有关酶在流动分析中使用的更多信息,请参见本书第 2 章。修饰电极的化合物如普鲁士蓝和其他类

似盐类化合物[148,149]、大环化合物（酞菁、卟啉、金属和卟啉的络合物）[150,151]、SAMs[152,153]和许多其他材料正在迅速增加。复合材料的制备[154,155]，新电极的使用，例如掺杂金刚石[30,156]、碳纳米管[157]、导电聚合物[158]或在同一传感器中组合不同材料[159-162]变得越来越重要。

电极改性的另一种方法是在其表面使用聚合物薄膜。这种策略可以提高选择性和稳定性，而对复杂基质中成分快速分析和直接分析的需求日益增加，提高这两种特性是满足该需求的必然要求。这些需求激励了研究人员探索聚合物膜，它对于分离非常有效。选择性可以基于膜的静电效应（离子交换剂，例如全氟磺酸膜）、分子大小（例如透析膜）和/或疏水/亲水特性（例如 PTFE）。电极表面上的膜通常通过滴注和电聚合的方式应用[158,163,164]。关于这一点，应将膜放置在两个流动流体之间作为修饰电极的替代方法；电极改性已广泛用于基于离子渗析[165-167]或气态物质扩散[168-170]的分离。

4.4 趋势与展望

对过去 15 年关于流动分析文献的检查表明，涉及安培检测的出版物每年增加约 5%，与分光光度检测的增长率大致相同，分光光度检测的使用更为广泛，而伏安法的使用，尤其是 ASV 和 CSV，增长速度是原来的两倍（每年 10%）。

新电极材料、酶活性中心的合成类似物、用于表面修饰的纳米材料和改进的电极修饰技术的出现对流动分析产生了有利影响，因为它们为电流分析传感器和生物传感器提供了更大的选择性和耐用性。基于人们对这些技术众多优势的广泛认可，例如，多分析物进行快速伏安扫描（例如 SWV）的能力以及高灵敏度、合理选择性伏安溶出分析与流动注射分析、间歇注入分析或顺序注射分析的完美兼容性，关于此类检测技术的研究将会在未来不断增加。

中等或低成本的商业仪器（电极、流动池、恒电位仪和多恒电位仪）的可获得性促成了这一趋势，许多这样的仪器可以由计算机控制。用于同时控制流动分析系统（阀门、泵、样品转换器等）和伏安检测器（无限的用户可编程脉冲和扫描序列、电流采样方案、电极激活程序和同步样本注入）的更灵活和用户友好的软件将受到拥有更复杂需求并且对自己编写程序不感兴趣的从业者的欢迎。

用于个人医疗保健的便携式 FA-ECD 系统的开发、在现场或无人值守操作的偏远地方使用越来越引起人们的兴趣。由于伏安检测器很容易满足低成本和尺寸、电池可操作性、耐用性和自动化等要求，因此挑战性主要在于电极耐用性（此类低成本设备可以考虑系统冗余）、自动校准和试剂/电解质供应以实现长期无人监督的操作等方面。

FA-ECD 小型化的趋势是明显且可行的，更多步骤的逐步整合，如取样、样品消解处理、萃取、反应，然后是预浓缩、介质交换、剥离等，最终与微流动注射分析的整合即将实现。

参考文献

[1] Bard, A. J. and Rubinstein, I. (eds) (1966 – 2004) *Electroanalytical Chemistry: a Series of Advances*, Vols. 1–22, Marcel Dekker, New York.

[2] Zoski, C. G. (2007) *Handbook of Electrochemistry*, Elsevier, Boston.

[3] Sawyer, D. T., Sobkowiak, A. and Roberts, J. L. (1995) *Electrochemistry for Chemists*, John Wiley & Sons, New York.

[4] Brett, C. M. A. and Brett, A. M. O. (1998) *Electroanalysis*, Oxford University Press, New York.

[5] Bard, A. J. and Faulkner, L. R. (2001) *Electrochemical Methods: Fundamentals and Applications*, Wiley, New York.

[6] Scholz, F. (2002) *Electroanalytical Methods: Guide to Experiments and Applications*, Springer, Berlin.

[7] Bard, A. J., Stratmann, M. and Unwin, P. R. (eds) (2003) *Instrumentation and Electro-analytical Chemistry*, Vol. 3, Wiley-VCH, Weinheim.

[8] Wang, J. (2006) *Analytical Electrochemistry*, Wiley-VCH, Weinheim.

[9] Stulik, K. and Pacakova, V. (1987) *Electroanalytical Measurements in Flowing Liquids*, Ellis Horwood, Chichester.

[10] Gunasingham, H. and Fleet, B. (1990) Hydrodynamic Voltammetry in Continuous – Flow Analysis, in *Electroanalytical Chemistry: a Series of Advances*, 16, Marcel Dekker, New York, pp. 89–180.

[11] Kissinger, P. T. and Heineman, H. (1996) *Laboratory Techniques in Electroanalytical Chemistry*, Marcel Dekker, New York.

[12] Trojanowicz, M. (2000) *Flow Injection Analysis: Instrumentation and Applications*, World Scientific, Singapore.

[13] Trojanowicz, M., Szewczynska, M. and Wcislo, M. (2003) Electroanalytical flow measurements-recent advances. *Electroanalysis*, 15, 347–365.

[14] Pérez-Olmos, R., Soto, J. C., Zarate, N., Araújo, A. N. and Montenegro, M. C. B. S. M. (2005) Sequential injection analysis using electrochemical detection: a review. *Analytica Chimica Acta*, 554, 1–16.

[15] Quintino, M. S. M. and Angnes, L. (2004) Batch injection analysis: an almost unexplored powerful tool. *Electroanalysis*, 16, 513–523.

[16] LaCourse, W. R. (1997) *Pulsed Electro – chemical Detection in High – Performance Liquid Chromatography*, Wiley-Interscience, New York.

[17] Flanagan, R. J., Perrett, D. and Whelpton, R. (2005) *Electrochemical Detection in HPLC*, Royal Society of Chemistry, Cambridge.

[18] LaCourse, W. R. and Modi, S. J. (2005) Microelectrode applications of pulsed electrochemical detection. *Electroanalysis*, 17, 1141–1152.

[19] Du, Y. and Wang, E. K. (2007) Capillary electrophoresis and microchip capillary electrophoresis with electrochemical and electrochemiluminescence detection. *Journal of Separation Science*, 30, 875–890.

[20] http://epsilon – web.com/Lc/manuals / Principles/principles.html, last accessed on 10/Aug/2007.

[21] http://www.chem.agilent.com/scripts/LiteraturePDF.asp?iWHID=20644, last accessed on 10/Aug/2007.

[22] Zemann, A. J., Schnell, E., Volgger, D. and Bonn, G. K. (1998) Contactless conductivity detection for capillary electrophoresis. *Analytical Chemistry*, 70, 563-567.

[23] Fracassi da Silva, J. A. and do Lago, C. L. (1998) An oscillometric detector for capillary electrophoresis. *Analytical Chemistry*, 70, 4339-4343.

[24] Kubáň, P. and Hauser, P. C. (2004) Contactless conductivity detection in capillary electrophoresis: a review. *Electroanalysis*, 16, 2009-2021.

[25] Hoherčáková, Z. and Opekar, F. (2005) A contactless conductivity detection cell for flow injection analysis: determination of total inorganic carbon. *Analytica Chimica Acta*, 551, 132-136.

[26] Cooper, J. A. and Compton, R. G. (1998) Channel electrodes - a review. *Electroanalysis*, 10, 141-155.

[27] Tur'yan, Y. I., Strochkova, E. M., Kuselman, I. and Shenhar, A. (1996) Microcell for anodic stripping voltammetry of trace metals. *Fresenius' Journal of Analytical Chemistry*, 354, 410-413.

[28] Macpherson, J. V., Simjee, N. and Unwin, P. R. (2001) Hydrodynamic ultramicroelectrodes: kinetic and analytical applications. *Electrochimica Acta*, 47, 29-45.

[29] Nascimento, V. B., Augelli, M. A., Pedrotti, J. J., Gutz, I. G. R. and Angnes, L. (1997) Arrays of gold microelectrodes made from split integrated circuit chips. *Electroanalysis*, 9, 335-339.

[30] Brainina, K. Z., Malakhova, N. A. and Stojko, N. Y. (2000) Stripping voltammetry in environmental and food analysis. *Fresenius' Journal of Analytical Chemistry*, 368, 307-325.

[31] Matos, R. C., Gutz, I. G. R., Angnes, L., Fontenele, R. S. and Pedrotti, J. J. (2001) A versatile and pulsation-free pneumatic impeller for flow analysis systems. *Quimica Nova*, 24, 795-798.

[32] Fujishima, A., Einaga, Y., Rao, T. N. and Tryk, D. A. (2005) *Diamond Electrochemistry*, Elsevier, Amsterdam.

[33] Fujishima, A. (2007) *Electrochemistry of Diamond*, BKS, Tokyo.

[34] Hidalgo, P. and Gutz, I. G. R. (2001) Determination of low concentrations of the flotation reagent ethyl xanthate by sampled DC polarography and flow injection with amperometric detection. *Talanta*, 54, 403-409.

[35] Wang, J., Lu, J. M., Kirgöz, U. A., Hocevar, S. B. and Ogorevc, B. (2001) Insights into the anodic stripping voltammetric behavior of bismuth film electrodes. *Analytica Chimica Acta*, 434, 29-34.

[36] Kefala, G. and Economou, A. (2006) Polymer-coated bismuth film electrodes for the determination of trace metals by sequential-injection analysis/anodic stripping voltammetry. *Analytica Chimica Acta*, 576, 283-289.

[37] Economou, A. and Voulgaropoulos, A. (2007) On-line stripping voltammetry of trace metals at a flow-through bismuth-film electrode by means of a hybrid flow injection/sequential-injection system. *Talanta*, 71, 758-765.

[38] http://www.bioanalytical.com/products/lc/flowcells.html, last accessed on 10/Aug/2007.

[39] http://www.waters.com/WatersDivision/ContentD.asp?watersit=JDRS-5LWJX8, last accessed on 10/Aug/2007.

[40] http://www.antecleyden.com/products/vt03cell.shtml, last accessed on 10/Aug/2007.

[41] http://www.ssi.shimadzu.com/products/product.cfm?product=l-ecd-6a, last accessed on 10/Aug/2007.

[42] www.dionex.com/en-us/ ins_ dxprn29100.html, last accessed on 10/Aug/2007.

[43] http://www.cypresssystems.com/Cells/ FIA.html, last accessed on 10/Aug/2007.

[44] http://www.trace.de/en/onlinemessung/ fliesszelle-bild.htm, last accessed on 10/ Aug/2007.

[45] http://www.esainc.com/products/ type/instruments/H PLC/ specialty_ detectors/coularray, last accessed on 10/Aug/2007.

[46] http://www.ijcambria.com/ ALS_ printed_ electrodes.htm, last accessed on 10/Aug/2007.

[47] http://www.abtechsci.com/labproducts.html, last accessed on 10/Aug/2007.

[48] Pedrotti, J. J., Angnes, L. and Gutz, I. G. R. (1996) Miniaturized reference electrodes with microporous polymer junctions. *Electroanalysis*, 8, 673-675.

[49] Keller, O. C. and Buffle, J. (2000) Voltammetric and reference microelectrodes with integrated microchannels for flow through microvoltammetry. 1. The microcell. *Analytical Chemistry*, 72, 936-942.

[50] Hashimoto, M., Upadhyay, S., Kojima, S., Suzuki, H., Hayashi, K. and Sunagawa, K. (2006) Needle-type Ag/AgI reference electrode with a stagnant electrolyte layer and an active liquid junction. *Journal of the Electrochemical Society*, 153, 155-160.

[51] Polk, B. J., Stelzenmuller, A., Mijares, G., MacCrehan, W. and Gaitan, M. (2006) Ag/ AgCl microelectrodes with improved stability for microfluidics. *Sensors and Actuators B-Chemical*, 114, 239-247.

[52] Cavicchioli, A., Daniel, D. and Gutz, I. G. R. (2004) Critical comparison of four simple adaptors for flow amperometric and stripping voltammetry with mercury drop electrodes in batch cells. *Electroanalysis*, 16, 391-398.

[53] Colombo, C., van den Berg, C. M. G. and Daniel, A. (1997) A flow cell for on-line monitoring of metals in natural waters by voltammetry with a mercury drop electrode. *Analytica Chimica Acta*, 346, 101-111.

[54] Colombo, C. and van den Berg, C. M. G. (1998) Determination of trace metals (Cu, Pb, Zn and Ni) in soil extracts by flow analysis with voltammetric detection. *International Journal of Environmental Analytical, Chemistry*, 71, 1-17.

[55] Al-Farawati, R. and van den Berg, C. M. G. (1999) Metal-sulfide complexation in seawater. *Marine Chemistry*, 63, 331-352.

[56] Al-Farawati, R. and van den Berg, C. M. G. (2001) Thiols in coastal waters of the western North Sea and English Channel. *Environmental Science & Technology*, 35, 1902-1911.

[57] Achterberg, E. P., van den Berg, C. M. G. and Colombo, C. (2003) High resolution monitoring of dissolved Cu and Co in coastal surface waters of the Western North Sea. *Continental Shelf Research*, 23, 611-623.

[58] Aguiar, M. A. S., Marquez, K. S. G. and Gutz, I. G. R. (2000) Determination of traces of uranium in environmental samples using a flow injection system with amperometric catalytic detection. *Electroanalysis*, 12, 742-746.

[59] Abate, G., Lichtig, J. and Masini, J. C. (2002) Construction and evaluation of a flow-through cell adapted to a commercial static mercury drop electrode (SMDE) to study the adsorption of Cd (II) and Pb (II) on vermiculite. *Talanta*, 58, 433-443.

[60] dos Santos, L. B. O., Abate, G. and Masini, J. C. (2005) Application of sequential injection-square wave voltammetry (SI-SWV) to study the adsorption of atrazine onto a tropical soil sample. *Talanta*, 68, 165-170.

[61] dos Santos, L. B. O., Abate, G. and Masini, J. C. (2006) Developing a continuous flow-square

wave voltammetry method for determination of atrazine in soil solutions using the hanging mercury drop electrode. *Journal of the Brazilian Chemical Society*, 17, 36-42.

[62] Pedrotti, J. J. and Gutz, I. G. R. (2003) Ultra-simple adaptor to convert batch cells with mercury drop electrodes in voltammetric detectors for flow analysis. *Talanta*, 60, 695-705.

[63] Korolczuk, M. and Grabarczyk, M. (1999) Voltammetric determination of Cr (VI) in a flow system in the presence of diethylenetriaminepentaacetic acid (DTPA) following its deposition in the metallic state. *Analytica Chimica Acta*, 387, 97-102.

[64] Korolczuk, M. (1999) Voltammetric determination of Cr (VI) in natural water in the presence of bipyridine following its deposition to the metallic state. *Electroanalysis*, 11, 1218-1221.

[65] Macpherson, J. V. and Unwin, P. R. (1997) Characterization and application of a mercury hemisphere microjet electrode. *Analytical Chemistry*, 69, 5045-5051.

[66] Walcarius, A., Mariaulle, P., Louis, C. and Lamberts, L. (1999) Amperometric detection of nonelectroactive cations in electrolyte-free flow systems at zeolite modified electrodes. *Electroanalysis*, 11, 393-400.

[67] Walcarius, A., Vromman, V. and Bessiere, J. (1999) Flow injection indirect amperometric detection of ammonium ions using a clinoptilolite-modified electrode. *Sensors and Actuators B-Chemical*, 56, 136-143.

[68] Bergelin, M. and Wasberg, M. (1998) The impinging jet flow method in interfacial electrochemistry: an application to beadtype electrodes. *Journal of Electroanalytical Chemistry*, 449, 181-191.

[69] Tur'yan, Y. I. (1997) Microcells for voltammetry and stripping voltammetry. *Talanta*, 44, 1-13.

[70] Hua, C., Jagner, D. and Renman, L. (1987) Automated-determination of total arsenic in seawater by flow constant-current stripping analysis with gold fiber electrodes. *Analytica Chimica Acta*, 201, 263-268.

[71] Gun, J., Salaun, P. and van den Berg, C. M. G. (2006) Advantages of using a mercury coated, micro-wire, electrode in adsorptive cathodic stripping voltammetry. *Analytica Chimica Acta*, 571, 86-92.

[72] Angnes, L., Richter, E. M., Augelli, M. A. and Kume, G. H. (2000) Gold electrodes from recordable CDs. *Analytical Chemistry*, 72, 5503-5506.

[73] Munoz, R. A. A., Matos, R. C. and Angnes, L. (2001) Amperometric determination of dipyrone in pharmaceutical formulations with a flow cell containing gold electrodes from recordable compact discs. *Journal of Pharmaceutical Sciences*, 90, 1972-1977.

[74] Nascimento, V. B. and Angnes, L. (1998) Screen-printed electrodes. *Quimica Nova*, 21, 614-629.

[75] Neufeld, T., Eshkenazi, I., Cohen, E. and Rishpon, J. (2000) A micro flow injection electrochemical biosensor for organophosphorus pesticides. *Biosensors & Bioelectronics*, 15, 323-329.

[76] Hsu, C. T., Chung, H. H., Lyuu, H. J., Tsai, D. M., Kumar, A. S. and Zen, J. M. (2006) An electrochemical cell coupled with disposable screen-printed electrodes for use in flow injection analysis. *Analytical Sciences*, 22, 35-38.

[77] Okawa, Y, Kobayashi, H. and Ohno, T. (1995) Direct and simultaneous determination of uric-acid and glucose in serum with electrochemical filter/ biosensor flow injection analysis system. *Analytica Chimica Acta*, 315, 137-143.

[78] Kilbey, G., Karousos, N. G., Eglin, D. and Davis, J. (2006) Laser etched carbon fibre

composites: disposable detectors for flow analysis applications. *Electrochemistry Communication*, 8, 1315-1320.

[79] Rippeth, J. J., Gibson, T. D., Hart, J. P., Hartley, I. C. and Nelson, G. (1997) Flowinjection detector incorporating a screen-printed disposable amperometric biosensor for monitoring organophosphate pesticides. *The Analyst*, 122, 1425-1429.

[80] Shi, M. H., Xu, J. J., Zhang, S., Liu, B. H. and Kong, J. L. (2006) A mediator-free screen-printed amperometric biosensor for screening of organophosphorus pesticides with flow injection analysis (FIA) system. *Talanta*, 68, 1089-1095.

[81] Law, K. A. and Higson, S. P. J. (2005) Sonochemically fabricated acetylcho-linesterase micro-electrode arrays within a flow injection analyser for the determination of organophosphate pesticides. *Biosensors & Bioelectronics*, 20, 1914-1924.

[82] Gao, Q., Ma, Y., Cheng, Z. L., Wang, W. D. and Yang, X. R. (2003) Flow injection electrochemical enzyme immunoassay based on the use of an immunoelectrode strip integrate immunosorbent layer and a screen-printed carbon electrode. *Analytica Chimica Acta*, 488, 61-70.

[83] Pernaut, J. M. and Matencio, T. (1999) Thin layer electrochemical cells, principle and application. *Quimica Nova*, 22, 899-902.

[84] Toda, K. (2004) Development of miniature key devices for flow analysis and their applications. *Bunseki Kagaku*, 53, 207-219.

[85] Bratten, C. D. T., Cobbold, P. H. and Cooper, J. M. (1997) Micromachining sensors for electrochemical measurement in subnanoliter volumes. *Analytical Chemistry*, 69, 253-258.

[86] Muck, A., Jr, Wang, J. and Barek, J. (2003) Microfluidic platform for FIA with electrochemical detection. *Chemicke Listy*, 97, 957-960.

[87] Toda, K., Oguni, S., Takamatsu, Y. and Sanemasa, I. (1999) A wall-jet ring disk electrode fabricated within a thin-layered micromachined cell. *Journal of Electroanalytical Chemistry*, 479, 57-63.

[88] Liao, C. Y., Chang, C. C., Ay, C. and Zena, J. M. (2007) Flow injection analysis of chloramphenicol by using a disposable wall-jet ring disk carbon electrode. *Electroanalysis*, 19, 65-70.

[89] Kovalcik, K. D., Kirchhoff, J. R., Giolando, D. M. and Bozon, J. P. (2004) Copper ringdisk microelectrodes: fabrication, characterization, and application as an amperometric detector for capillary columns. *Analytica Chimica Acta*, 507, 237-245.

[90] Jusys, Z., Kaiser, J. and Behm, R. J. (2004) A novel dual thin-layer flow cell doubledisk electrode design for kinetic studies on supported catalysts under controlled mass-transport conditions. *Electrochimica Acta*, 49, 1297-1305.

[91] Paixão, T. R. L. C., Matos, R. C. and Bertotti, M. (2003) Design and characterisation of a thin-layered dualband electrochemical cell. *Electrochimica Acta*, 48, 691-698.

[92] Daniel, D. and Gutz, I. G. R. (2007) Microfluidic cell with a TiO_2-modified gold electrode irradiated by an UV-LED for *in situ* photocatalytic decomposition of organic matter and its potentiality for voltammetric analysis of metal ions. *Electrochemistry Communication*, 9, 522-528.

[93] Macpherson, J. V. and Unwin, P. R. (1998) Radial flow microring electrode: development and characterization. *Analytical Chemistry*, 70, 2914-2921.

[94] Wittstock, G., Gründig, B., Strehlitz, B. and Zimmer, K. (1998) Evaluation of microelectrode arrays for amperometric detection by scanning electrochemical microscopy. *Electroanalysis*, 10, 526-531.

[95] Wittstock, G. (2002) Sensor arrays and array sensors. *Analytical and Bioanalytical Chemistry*,

372, 16-17.

[96] Matos, R. C., Augelli, M. A., Lago, C. L. and Angnes, L. (2000) Flow injection analysis-amperometric determination of ascorbic and uric acids in urine using arrays of gold microelectrodes modified by electrodeposition of palladium. *Analytica Chimica Acta*, 404, 151-157.

[97] Augelli, M. A., Nascimento, V. B., Pedrotti, J. J., Gutz, I. G. R. and Angnes, L. (1997) Flow-through cell based on an array of gold microelectrodes obtained from modified integrated circuit chips. *The Analyst*, 122, 843-847.

[98] Daniel, D. and Gutz, I. G. R. (2005) Microfluidic cells with interdigitated array gold electrodes: fabrication and electrochemical characterization. *Talanta*, 68, 429-436.

[99] Miro, M., Estela, J. M. and Cerda, V. (2004) Application of flowing stream techniques to water analysis Part III. Metal ions: alkaline and alkaline-earth metals, elemental and harmful transition metals, and multielemental analysis. *Talanta*, 63, 201-223.

[100] Camel, V. (2003) Solid phase extraction of trace elements. *Spectrochimica Acta Part B-Atomic Spectroscopy*, 58, 1177-1233.

[101] Begak, O. Y. and Borodin, A. V. (1998) Flow-injection analysis in atomic absorption spectrometry (review). *Industrial Laboratory*, 64, 371-381.

[102] Long, X. B., Miro, M. and Hansen, E. H. (2005) An automatic micro-sequential injection bead injection Lab-on-Valve (μSI-BI-LOV) assembly for speciation analysis of ultra trace levels of Cr (III) and Cr (VI) incorporating on-line chemical reduction and employing detection by electrothermal atomic absorption spectrometry (ETAAS). *Journal of Analytical Atomic Spectrometry*, 20, 1203-1211.

[103] Long, X. B., Miro, M., Jensen, R. and Hansen, E. H. (2006) Highly selective micro-sequential injection lab-on-valve (μSI-LOV) method for the deter-mination of ultra-trace concentrations of nickel in saline matrices using detection by electrothermal atomic absorption spectrometry. *Analytical and Bioanalytical Chemistry*, 386, 739-748.

[104] Zirino, A. and Lieberma, S. (1973) Continuous-flow anodic stripping voltammetry of Zn, Cd, and Pb in San Diego bay water. *Journal of the Electrochemical Society*, 120, C254-C254.

[105] Lukaszewski, Z. and Zembrzuski, W. (1992) Determination of thallium in soils by flow injection differential pulse anodic stripping voltammetry. *Talanta*, 39, 221-227.

[106] Richter, E. M., Augelli, M. A., Kume, G. H., Mioshi, R. N. and Angnes, L. (2000) Gold electrodes from recordable CDs for mercury quantification by flow injection analysis. *Fresenius' Journal of Analytical Chemistry*, 366, 444-448.

[107] Wang, J., Setiadji, R., Chen, L., Lu, J. M. and Morton, S. G. (1992) Automated-system for online adsorptive stripping voltammetric monitoring of trace levels of uranium. *Electroanalysis*, 4, 161-165.

[108] Daniel, A., Baker, A. R. and van den Berg, C. M. G. (1997) Sequential flow analysis coupled with ACSV for on-line monitoring of cobalt in the marine environment. *Fresenius' Journal of Analytical Chemistry*, 358, 703-710.

[109] Suteerapataranon, S., Jakmunee, J., Vaneesorn, Y. and Grudpan, K. (2002) Exploiting flow injection and sequential injection anodic stripping voltammetric systems for simultaneous determination of some metals. *Talanta*, 58, 1235-1242.

[110] da Silva, C. L. and Masini, J. C. (2000) Determination of Cu, Pb, Cd, and Zn in river sediment extracts by sequential injection anodic stripping voltammetry with thin mercury film electrode. *Fresenius' Journal of Analytical Chemistry*, 367, 284-290.

[111] Santos, A. C. V. and Masini, J. C. (2006) Development of a sequential injection anodic stripping voltammetry (SI-ASV) method for determination of Cd (II), Pb (II) and Cu (II) in wastewater samples from coatings industry. *Analytical and Bioanalytical Chemistry*, 385, 1538-1544.

[112] Volikakis, G. J. and Efstathiou, C. E. (2005) Fast screening of total flavonols in wines, tea-infusions and tomato juice by flow injection/adsorptive stripping voltammetry. *Analytica Chimica Acta*, 551, 124-131.

[113] Nouws, H. P. A., Delerue - Matos, C., Barros, A. A. and Rodrigues, J. A. (2006) Electroanalytical determination of paroxetine in pharmaceuticals. *Journal of Pharmaceutical and Biomedical Analysis*, 42, 341-346.

[114] Kubiak, W. W., Latonen, R. M. and Ivaska, A. (2001) The sequential injection system with adsorptive stripping voltammetric detection. *Talanta*, 53, 1211-1219.

[115] Economou, A., Clark, A. K. and Fielden, P. R. (2001) Determination of Co (II) by chemiluminescence after *in situ* electrochemical pre-separation an a flow-through mercuty film electrode. *The Analyst*, 126, 109-113.

[116] Romanus, A., Muller, H. and Kirsch, D. (1991) Application of adsorptive stripping voltammetry (AdSV) for the analysis of trace-metals in brine. 2. Development and evaluation of a flow injection system. *Fresenius' Journal of Analytical Chemistry*, 340, 371-376.

[117] Vazquez, M. D., Tascon, M. L. and Debran, L. (2006) Determination of Pb (II) with a dithizone-modified carbon paste electrode. *Journal of Environmental Science and Health Part A*, -Toxic/ Hazardous Substances & Environmental Engineering, 41, 2735-2746.

[118] Han, J., Chen, H. and Gao, H. (1991) Alternating-current adsorptive stripping voltammetry in a flow system for the determination of ultratrace amounts of folic-acid. *Analytica Chimica Acta*, 252, 47-52.

[119] Khodari, M. (1993) Voltammetric determination of the antidepressant trimipramine at a lipid-modified carbon-paste electrode. *Electroanalysis*, 5, 521-523.

[120] Villar, J. C. C., Garcia, A. C. and Blanco, P. T. (1992) Determination of mitoxantrone using phase-selective ac adsorptive stripping voltammetry in a flow system with selectivity enhancement. *Analytica Chimica Acta*, 256, 231-236.

[121] Guo, S. X. and Khoo, S. B. (1999) Highly selective and sensitive determination of silver (I) at a poly (8-mercaptoquinoline) film modified glassy carbon electrode. *Electroanalysis*, 11, 891-898.

[122] Thomsen, K. N., Kryger, L. and Baldwin, R. P. (1998) Voltammetric determination of traces of nickel (II) with a medium exchange flow system and a chemically modified carbon paste electrode containing dimethylglyoxime. *Analytical Chemistry*, 60, 151-155.

[123] Ivaska, A. and Kubiak, W. W. (1997) Application of sequential injection analysis to anodic stripping voltammetry. *Talanta*, 44, 713-723.

[124] Sawamoto, H. and Gamoh, K. (1990) Desorption studies at a hanging mercury drop electrode by a flow injection method. *Journal of Electroanalytical Chemistry*, 283, 421-424.

[125] Sawamoto, H. (1993) Adsorptiondesorption studies of alcohols at a hanging mercury drop electrode by a flow injection method. *Journal of Electroanalytical Chemistry*, 361, 215-220.

[126] Sawamoto, H. (1994) Adsorption of methanol, 1-propanol and 1-octanol on a mercury-electrode and adsorption of 1 - octanol on the adsorbed layer of methanol at a mercury - electrode. *Journal of Electroanalytical Chemistry*, 375, 391-394.

[127] Sawamoto, H. (1997) The study of adsorption and reorientation of quinoline at mercury electrodes

by measuring differential capacity-potential and differential capacity-time curves. *Journal of Electroanalytical Chemistry*, 432, 153-157.

[128] Sawamoto, H. (2003) Reversible and irreversible adsorption of Surfactants at a hanging mercury drop electrode. *Analytical Sciences*, 19, 1381-1386.

[129] Pedrotti, J. J., Angnes, L. and Gutz, I. G. R. (1994) A fast, highly efficient, continuous degassing device and its application to oxygen removal in flow injection analysis with amperometric detection. *Analytica Chimica Acta*, 298, 393-399.

[130] Colombo, C. and van den Berg, C. M. G. (1998) In-line deoxygenation for flow analysis with voltammetric detection. *Analytica Chimica Acta*, 377, 229-240.

[131] Tercier-Waeber, M. L. and Buffle, J. (2000) Submersible online oxygen removal system coupled to an *in situ* voltammetric probe for trace element monitoring in freshwater. *Environmental Science & Technology*, 34, 4018-4024.

[132] Reinke, R. and Simon, J. (2002) The online removal of dissolved oxygen from aqueous solutions used in voltammetric techniques by the chromatomembrane method. *Analytical and Bioanalytical Chemistry*, 374, 1256-1260.

[133] Bobrowski, A. and Zarebski, J. (2000) Catalytic systems in adsorptive stripping voltammetry. *Electroanalysis*, 12, 1177-1186.

[134] Czae, M. Z. and Wang, J. (1999) Pushing the detectability of voltammetry how low can we go? *Talanta*, 50, 921-928.

[135] Daniel, D. and Gutz, I. G. R. (2001) Long-optical-path thin-layer spectroelectrochemical flow cell with inexpensive gold electrodes. *Electroanalysis*, 13, 681-685.

[136] Daniel, D. and Gutz, I. G. R. (2003) Flow injection spectroelectroanalytical method for the determination of promethazine hydrochloride in pharmaceutical preparations. *Analytica Chimica Acta*, 494, 215-224.

[137] Daniel, D. and Gutz, I. G. R. (2005) Spectroelectrochemical determination of chlorpromazine hydrochloride by flow injection analysis. *Journal of Pharmaceutical and Biomedical Analysis*, 37, 281-286.

[138] Rubim, J. C., Gutz, I. G. R. and Sala, O. (1985) Cyclic-fluorovoltammetry as a technical tool in tire study of passivating films generated on electrode surfaces. *Journal of Electroanalytical Chemistry*, 190, 55-63.

[139] Bard, A. J. (2004) *Electrogenerated Chemiluminescence*, CRC Press, New York.

[140] Fähnrich, K. A., Pravda, M. and Guilbault, G. G. (2001) Recent applications of electrogenerated chemiluminescence in chemical analysis. *Talanta*, 54, 531-559.

[141] Lindino, C. A. and Bulhões, L. O. S. (2004) Determination of chloramphenicol in tablets by electrogenerated chemiluminescence. *Journal of the Brazilian Chemical Society*, 15, 178-182.

[142] Al-Gailani, B. R. M., Greenway, G. and McCreedy, T. (2007) A miniaturized flow injection analysis (μFIA) system with online chemiluminescence detection for the determination of iron in estuarine water. *International Journal of Environmental Analytical*, *Chemistry*, 87, 637-646.

[143] Gouveia, V. J. P., Gutz, I. G. R. and Rubim, J. C. (1994) A new spectroelectrochemical cell for flow injection analysis and its application to the determination of Fe (II) down to the femtomol level by surface-enhanced resonance Raman-scattering (SERRS). *Journal of Electroanalytical Chemistry*, 371, 37-42.

[144] Gorton, L. (1995) Carbon-paste electrodes modified with enzymes, tissues, and cells.

Electroanalysis, 7, 23-45.

[145] Mello, L. D. and Kubota, L. T. (2002) Review of the use of biosensors as analytical tools in the food and drink industries. *Food Chemistry*, 77, 237-256.

[146] Baeumner, A. J. (2003) Biosensors for environmental pollutants and food contaminants. *Analytical and Bioanalytical Chemistry*, 377, 434-445.

[147] Prieto-Simon, B., Campas, M., Andreescu, S. and Marty, J. L. (2006) Trends in flow-based biosensing systems for pesticide assessment. *Sensors*, 6, 1161-1186.

[148] Koncki, R. (2002) Chemical sensors and biosensors based on Prussian blues. *Critical Reviews in Analytical Chemistry*, 32, 79-96.

[149] Jayasri, D. and Narayanan, S. S. (2007) Manganese (II) hexacyanoferrate based renewable amperometric sensor for the determination of butylated hydroxyanisole in food products. *Food Chemistry*, 101, 607-614.

[150] Ozoemena, K. I. and Nyokong, T. (2006) Novel amperometric glucose biosensor based on an ether-linked cobalt (II) phthalocyanine-cobalt (II) tetraphenylporphyrin pentamer as a redox mediator. *Electrochimica Acta*, 51, 5131-5136.

[151] da Rocha, J. R. C., Angnes, L., Bertotti, M., Araki, K. and Toma, H. E. (2002) Amperometric detection of nitrite and nitrate at tetr aruthenated porphyrin-modified electrodes in a continuous-flow assembly. *Analytica Chimica Acta*, 452, 23-28.

[152] Wang, J., Wu, H. and Angnes, L. (1993) Online monitoring of hydrophobic compounds at self-assembled monolayer modified amperometric flow detectors. *Analytical Chemistry*, 65, 1893-1896.

[153] Pedrosa, V. A., Lowinsohn, D. and Bertotti, M. (2006) FIA determination of paracetamol in pharmaceutical drugs by using gold electrodes modified with a 3 - mercaptopropionic acid monolayer. *Electroanalysis*, 18, 931-934.

[154] Zacco, E., Pividori, M., Llopis, X., Del Valle, M. and Alegret, S. (2004) Renewable protein a modified graphite-epoxy composite for electrochemical immunosensing. *Journal of Immunological Methods*, 286, 35-46.

[155] Brahim, S., Narinesingh, D. and Guiseppi-Elie, A. (2002) Polypyrrolehydrogel composites for the construction of clinically important biosensors. *Biosensors & Bioelectronics*, 17, 53-59.

[156] Chailapakul, O., Siangproh, W. and Tryk, D. A. (2006) Boron-doped diamondbased sensors: a review. *Sensor Letters*, 4, 99-119.

[157] Arribas, A. S., Bermejo, E., Chicharro, M., Zapardiel, A., Luque, G. L., Ferreyra, N. F. and Rivas, G. A. (2006) Analytical applications of a carbon nanotubes composite modified with copper microparticles as detector in flow systems. *Analytica Chimica Acta*, 577, 183-189.

[158] Nagels, L. J. and Staes, E. (2001) Polymer (bio) materials design for amperometric detection in LC and FIA. *Trends in Analytical Chemistry*, 20, 178-185.

[159] Ricci, F. and Palleschi, G. (2005) Sensor and biosensor preparation, optimisation and applications of Prussian Blue modified electrodes. *Biosensors & Bioelectronics*, 21, 389-407.

[160] Trojanowicz, M. (2005) Electroanalytical flow measurements. *Annali di Chimica*, 95, 421-435.

[161] Hart, J. P., Crew, A., Crouch, E., Honeychurch, K. C. and Pemberton, R. M. (2004) Some recent designs and developments of screen-printed carbon electrochemical sensors/biosensors for biomedical, environmental, and industrial analyses. *Analytical Letters*, 37, 789-830.

[162] Vidal, J. C., Espuelas, J., Garcia - Ruiz, E. and Castillo, J. R. (2004) Amperometric

cholesterol biosensors based on the electropolymerization of pyrrole and the electr ocatalytic effect of Prussian-Blue layers helped with self-assembled monolayers. *Talanta*, 64, 655-664.

[163] Wang, J. (1992) in *Biosensors and Chemical Sensors* (eds P. Edelman and J. Wang), America Chemical Society, Washington, DC.

[164] Guerrieri, A., Lattanzio, V., Palmisano, F. and Zambonin, P. G. (2006) Electrosynthesized poly (pyrrole) /poly (2-naphthol) bilayer membrane as an effective anti-interference layer for simultaneous determination of acetylcholine and choline by a dual electrode amperometric biosensor. *Biosensors & Bioelectronics*, 21, 1710-1718.

[165] Ortuno, J. A., Sanchez-Pedreno, C. and Gil, A. (2005) Flow-injection pulse amperometric detection based on ion transfer across a water-plasticized polymeric membrane interface for the determination of verapamil. *Analytica Chimica Acta*, 554, 172-176.

[166] Ortuno, J. A., Sanchez-Pedreno, C., Hernandez, J. and Oliva, D. J. (2005) Flow-injection potentiometric determination of triiodide by plasticized poly (vinyl chloride) membrane electrodes and its application to the determination of chlorine-containing disinfectants. *Talanta*, 65, 1190-1195.

[167] Silva, I. S., Richter, E. M., do Lago, C. L, Gutz, I. G. R., Tanaka, A. A. and Angnes, L. (2005) FIA-potentiometry in the sub-Nernstian response region for rapid and direct chloride assays in milk and in coconut water. *Talanta*, 67, 651-657.

[168] Amini, N., Cardwell, T. J., Cattrail, R. W. and Kolev, S. (2005) Determination of mercury (II) at trace levels by gas-diffusion flow injection analysis with amperometric detection. *Analytica Chimica Acta*, 539, 203-207.

[169] Oshima, M., Wei, Y. L., Yamamoto, M., Tanaka, H., Takayanagi, T. and Motomizu, S. (2001) Highly sensitive determination method for total carbonate in water samples by flow injection analysis coupled with gas-diffusion separation. *Analytical Sciences*, 17, 1285-1290.

[170] Santos, J. C. C. and Korn, M. (2006) Exploiting sulphide generation and gas diffusion separation in a flow system for indirect sulphite determination in wines and fruit juices. *Mikrochimica Acta*, 153, 87-94.

5 使用流动分析对蛋白质-蛋白质和蛋白质-配体的相互作用进行亲和相互作用分析

Jeroen. Kool, N. P. E. Vermeulen, Herk. Lingeman, R. J. E. Derks 和 Hubertus Irth

5.1 引言

在过去十年中，制药行业的进步推动了分析化学的发展。分离科学和光谱方法的显著进步促进了药物发现关键领域的研究。近年来，生化分析（BCA）作为分析化学中的一种检测方法已变得非常流行。生化分析主要有两种策略，即基于微孔板格式的高通量筛选（HTS）技术，以及结合流动分析（FA）和高效液相色谱（HPLC）分析的连续流动生化分析。在本章中，我们概述了连续流生化分析方法及其与关键分析技术［如质谱（MS）］的集成。

质谱已成为最流行的分析技术之一，主要是由于分辨率和灵敏度方面的性能提高、与生物来源基质的兼容性增加以及新的 LC-MS 接口的出现。质谱与生化分析结合使用是相当新的，因为大多数生化分析基于荧光或辐射检测。虽然荧光和放射性生化分析通常提供出色的灵敏度和高重现性，但需要合成对受体或酶保持显著亲和力的荧光探针或底物，这通常会阻碍生化分析的发展速度。新的基于质谱的方法的另一个显著特点是技术设置。虽然传统的 HTS—BCA 通常在微量滴定板中进行，但基于质谱的新型生化分析是在连续流动环境中进行的。此外，这些格式能够将生化分析与分离技术（如 HPLC）直接耦合，从而能够分析来自生物或合成来源的高度复杂的混合物。FA-MS 的另一个有趣应用是分析非共价蛋白质复合物、蛋白质-配体相互作用和其他生物相关的蛋白质-蛋白质相互作用，随着高性能质谱仪的开发和商业化，这些应用迅速发展。这个领域现在被称为天然质谱。

主要是由于流动注射分析（FIA）模式下的生化分析可以与强大的分离技术相结合，最近开发了许多基于荧光的不同流动注射分析技术。除了基于荧光和基于质谱的技术外，文献中还描述了其他检测方法，这些基于质谱或荧光检测的生化分析方法难以开发。当生化分析法依赖质谱读数时，必须在流动（注射）分析中开发这些生化分析，以便将示踪分子或蛋白质引入质谱。然而，与微量滴定板格式（每小时 50~2000 个样本）相比，FIA—BCA 模式在处理量（每小时 30~60 个样本）方面受到了影响。此外，由于这些线圈中的谱带变宽，在注射后反应盘管中具有真正在线生化孵育的 FIA—BCA 仅限于较短的反应时间（<5min）。除了化合物混合物中的生化分析读数仅代表整个样品中存在的生物活性之外，没有获得关于样品中引起这种生物活性的化合物身份的信息，也没有获得有关这些生物活性化合物浓度的信息，由此，会提出问题：生物活性是由低浓度的强效化合物还是高浓度的弱化合物引起的？分析（非常）复杂的样品时可能出现的另一个问题是样品成分与生化分析的生物相互作用的干扰，导致读数不准确。可以使用分析分离技术来克服这些问题。几种不同的方法已被用于将生化相互作用与分析分离技术相结合。这些生化相互作用在分析分离之前、分离中和分离之后结合起来。在分析分离之前使用生化相互作用的一个例子是使用固定亲和蛋白柱来预浓缩目标化合物并在分析分离步骤之前"净化"样品[1,2]。亲和色谱是一种基于生物特异性相互作用的分离技术，是在分析分离步骤中使用生化相互作用的一个例子[1,3]。这两种方法都增加了分析的选择性，但没有观察到灵敏度的真正增强。第三种

方法是在分析分离后结合生化相互作用。与其他两种方法相比，这种方法具有多种优势。首先，生化分析的检测原理还是和分批生化分析一样，因此，应该预期有类似的敏感性。其次，通过分离（复杂的）样品和去除干扰的基质成分，该方法的选择性也得到了提高。最简单的方法是在分析分离后收集样品的馏分，去除溶剂，将馏分重新溶解到生化分析兼容的缓冲液中，并对每个馏分进行生化分析。这种方法已广泛应用于许多不同的领域，例如生物分析、环境分析、天然产物筛选和代谢分析[4-7]。然而，这种方法有几个缺点。色谱分馏和所有分馏物的离线测量很费力，但更严重的缺点是在分馏步骤和再溶解步骤后去除溶剂的过程中可能会丢失活性化合物。通过将生化检测（BCD）和在线分析分离步骤结合起来，这些缺点就可以避免。

本章介绍了各种基于 FA-MS 的生化分析模式，可用于蛋白质-蛋白质和蛋白质-配体相互作用的亲和相互作用分析以及生物活性化合物的筛选。它还侧重于其他主要基于荧光的流动（注射）分析模式的生化分析模式，这些模式是为在线耦合分析分离技术（梯度 HPLC）而开发的。

5.2 基于质谱流动分析的非共价蛋白质-蛋白质和蛋白质-配体相互作用的分析

对于基于 FA-MS 的生化分析，可以选择几种根本不同的方法。生物活性化合物与蛋白质的相互作用可以通过测量来监测。

（1）非共价蛋白-配体复合物的形成；

（2）报告分子的浓度变化，即目标蛋白与分析物相互作用时已知的活性化合物；

（3）在蛋白质-分析物复合物与未结合的化合物分离后，从蛋白质中解离的分析物。

图 5.1 总结了所涉及的不同交互；质谱检测到的核素以灰色框突出显示。在酶抑制剂筛选的情况下，抑制剂与酶的相互作用也可以通过测量底物浓度或酶产物浓度的变化来检测。

5.2.1 非共价配合物的流动注射和连续灌注电喷雾电离质谱分析

5.2.1.1 引言

电喷雾电离（ESI）和基质辅助激光解吸/电离（MALDI）是生成大生物分子气相离子的最合适方法，这两种电离方法都是所谓的"软电离"技术。

ESI 和 MALDI 已被用于研究各种形式的复合物。非共价键能比共价键能低大约一个数量级。这使得非共价复合物难以用质谱检测。

5.2.1.2 非共价配合物的电喷雾电离质谱

在过去的几十年里，ESI 已经被证明是一种非常好的电离技术，可以用质谱检测非共价复合物。Katta 等人和 Ganem 等是第一个报告使用 ESI-MS 检测非共价配合物的小组，他们在研究肌红蛋白的球蛋白-血红蛋白相互作用[8]和受体-配体配合物[9]时，研究的非共价复合物的例子有：受体-配体、酶-底物、主体-客体、完整的多聚体蛋白、双链和四链 DNA、寡核苷酸与药物和蛋白的复合物以及蛋白-药物复合物[8,10-17]。几年前，van den Heuvel 和 Heck[18]全面综述了快速扩展的天然质谱领域，此后该领域的研

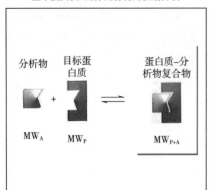
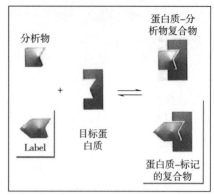
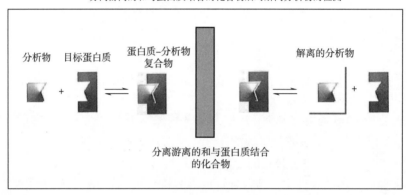

图 5.1　与质谱结合使用的生化分析法

究仍在增加。

5.2.1.3　非共价配合物研究中电喷雾电离质谱的局限性

用质谱检测非共价复合物的一个非常重要的要求是所使用的溶液要与质谱兼容。然而，在 ESI-MS 中用于实现最大灵敏度的典型溶剂条件并不总是维持完整的非共价复合物的最佳条件。一般来说，磷酸盐缓冲液和盐类（如氯化钠）被用来模拟保持非共价配合物完整的生理条件。此外，经常添加阻断剂（如吐温 20）以防止非共价配合物（和蛋白质）与表面的非特异性结合。然而，洗脱液中的非挥发性添加剂，如磷酸盐缓冲液和阻断剂，会造成离子抑制，与质谱检测不兼容。有机改性剂（即甲醇、乙腈或异丙醇）的百分比必须尽可能低，以防止非共价配合物解离甚至变性。然而，ESI 更适用于酸性（对于阳极 ESI）环境和一定量的有机改性剂。考虑到上述所有情况，需要在提供足够的加热/活化以使离子脱溶和保持界面条件足够温和，以在 ESI 过程中保持非共价复合物之间取得（微妙的）平衡。

5.2.2　使用报告分子的流动进样质谱分析

5.2.2.1　引言

将质谱耦合到生化分析有多种可能性。一种可能性是将质谱与（基于荧光的）生化分析并行耦合。近年来，已经为此技术开发了多种应用[19-23]。这种方法与直接基于质谱的方法有很大不同。不使用质谱检测生物活性，而是使用基于荧光的生化读数（分析方案，见图5.2）。通过这种方式同时获得生化和化学信息（这将在后面详细讨论）。传统的荧光生化分析方法通常具有出色的灵敏度和高重现性。然而，尽管有这些优点，但需要合成对受体或酶保持显著亲和力的荧光探针或底物或配体，这通常会阻碍生化分析的发展。除了开发荧光标记的底物或配体，在线生化分析中使用的任何底物、形成的产物或配体也可以通过质谱在直接基于质谱的方法中进行监测。

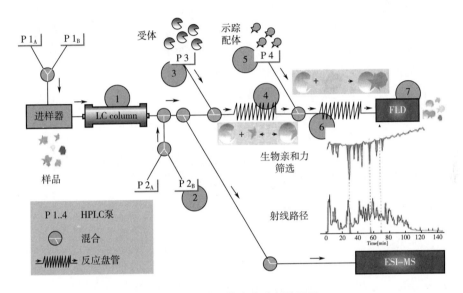

图5.2 在线受体生化分析的配置

1—进样后梯度HPLC　2—反向梯度，然后分流到质谱和生化分析　3—受体溶液的试剂泵
4—反应盘管　5—示踪配体溶液的试剂泵　6—反应盘管　7—荧光检测器（FLD）

5.2.2.2 基于质谱的生化分析的要求

与连续流生化分析的要求相反，质谱更适用于高有机改性剂含量。因此，应仔细优化有机改性剂的含量，以便在生化分析中获得良好的反应动力学和在质谱中获得良好的可检测性/灵敏度。第二个重要的要求是连续流动的生化分析与质谱中使用的缓冲液和缓冲液添加剂的兼容性。通常在不与质谱耦合的批量生化分析和连续流动生化分析中，可以使用生理缓冲液和添加剂/盐。通常，磷酸盐缓冲液和盐（例如氯化钠）用于模拟生化分析所需的生理条件。此外，通常还向洗脱液中加入封闭剂以防止蛋白质或蛋白质复合物黏附在反应盘管或管道的表面上。所有这些缓冲盐、添加剂和封闭剂都会导致离子抑制，因此不适用于与质谱耦合的连续流动生化分析。通常，磷酸盐缓冲液可以用挥发性缓冲液代替，如甲酸铵或醋酸铵溶液。pH约为6~7.5的甲酸铵或醋酸铵溶液是生化分析的良好折中方案，并且与质谱兼容。

5.2.2.3 使用质谱作为检测器的流动进样配体结合测定

Hogenboom等人首先描述了一个基于质谱的竞争性生化分析的例子[24]。它们开发了两种不同的竞争性绑定生化分析来获得原理证明。分别选择链霉亲和素和荧光素-生物素作为生化分析的模型分子靶标和报告配体。由于生物素和链霉亲和素之间的亲和相互作用是目前自然界已知的最强的相互作用之一（$K_a=0.6\times10^{15}\text{L/mol}$），因此亲和复合物很可能在电离过程中得以保留。荧光报告配体用于证明基于质谱的生化分析格式产生与其基于荧光的对应物相似的结果。为了证明弱亲和力相互作用的原理，选择抗地高辛和地高辛作为第二个基于质谱的生化分析（$K_a=\sim10^9\text{L/mol}$）的目标蛋白和报告配体。

5.2.2.4 使用质谱作为检测器的流动进样酶测定

DeBoer 等最近报道了一种基于质谱的酶促生化分析模式[25]。在这种情况下，组织蛋白酶 B 用作模型分子靶标。在生化反应过程中，组织蛋白酶 B 将底物转化为两种产物，分别为 Z-FR 和 AMC，然后由质谱连续监测。如果不注入抑制剂，则产物 Z-FR 和 AMC 的离子电流达到最大水平。然而，在注射抑制剂后，酶转化率降低，导致 Z-FR 和 AMC 的浓度降低。筛选酶促转化产物的一种略有不同但重要的方法是脉冲超滤质谱[26]。该技术也可应用于受体-配体相互作用筛选，可用于酶促产品的中等通量筛选。在早期应用中，质谱物质文库混合物中包含的配体与大分子受体结合，然后通过超滤纯化配体-受体复合物，最后用甲醇将配体洗脱到电喷雾质谱仪中进行检测[27]。后来，同一小组开发了类似的方法来评估药物的代谢稳定性[28]。

5.2.3 蛋白质-配体复合物解离后的无报告基因检测

除了连续流生化分析法之外，最近还开发了几种基于质谱的技术，这些技术无需报告分子即可检测配体[29-32]，这些所谓的无报告分子基于质谱的方法已被用于执行合成化合物混合物或生物来源样品中的"配体捕捞"。Hsieh 等、Kaur 等和 Lenz 等已经报道了全自动 LC-UV/MS 系统[33-35]。通常，蛋白质靶标与化合物库一起孵育，然后分离亲和结合和未结合的配体。最近报道了一种类似的方法，其中使用两个按顺序配置的受限访问列来分离复杂基质中存在的配体[29]。图 5.3 显示了所使用的流程方案。这种

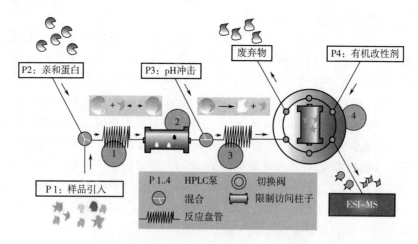

图 5.3 基于质谱的生化分析方案

1—联用反应盘管　2—限制访问柱捕获未结合的配体　3—用于解离的反应盘管　4—限制进入柱捕获生物活性分子

方法与第5.2.2.4节中描述的基于超滤的方法有许多相似之处。另一种用于测量蛋白质-蛋白质和蛋白质-配体相互作用的无报告系统是表面等离子共振（例如Biacore系统）。尽管这不是基于质谱的流动分析方法，但它是测量生物分子相互作用的真实亲和力以及关联和解离率的最重要技术之一。由于该系统或多或少具有质量选择性，因此它最适合大型配体相互作用，但是随着技术的最新改进，小型配体结合研究也变得不再遥远[36]。

5.3 流动分析和高效液相色谱的集成用于混合物的生物亲和性筛选

5.3.1 引言

化合物活性筛选是药物研发中的常见做法，通常以高通量方式进行，以促进大量化合物的筛选[37]。化合物筛选通常在各种不同的生化分析中进行。虽然基于放射性配体的亲和生化分析是过去的主要生化分析方法，但与基于荧光的生化分析相比，使用放射性标记示踪分子和生化分析的异质性使其不太适合高通量自动化。这导致在药物研发的所有领域都开发了基于荧光的生化分析。在这方面可以列举诸如时间分辨荧光共振能量转移、荧光各向异性、荧光相关光谱、荧光波动光谱和基于正常荧光的生化分析等实例[38]。化合物筛选的主要瓶颈是鉴定在复杂混合物中负责生物活性的一个或多个化合物，例如组合和天然化合物库。这也适用于代谢混合物的筛选[39]。代谢物的活性和/或亲和力，通常与母体化合物的结构非常相似，是药物最终作用的决定因素。为了分离在混合物中的化合物（通常在96~1536孔板中），通常必须执行烦琐的劳动密集型和昂贵的去重复化过程，然后才能将分离的化合物重新提交给原始生化分析[4,40,41]。图5.4展现了一个示例。

受体筛选技术[42,43]确实可以对单个成分进行亲和性筛选，但它们通常不能检测混合物中的单个化合物。一些受体亲和力筛选策略涉及活性代谢物与受体的结合，然后进行离线离心超滤[27,44]或限制进入的柱子分离[29]，从而导致收集结合的配体，并允许通过LC-MS进行表征。然而，这些技术缺乏在高亲和力配体存在的情况下或在代谢混合物的高浓度母体化合物存在的情况下有效捕集低亲和力配体的能力。这个主要的瓶颈可以通过直接在线柱后连接所需的流动分析生化分析到HPLC[45]来规避，表示为HPLC模式下的流动分析FA-BCD-HPLC。当与质谱测量同时进行时，单个化合物的化学信息可以直接与它们的生物活性联系起来[21,22]。本节回顾了FA-BCD-HPLC的发展和现状，讨论了原理和应用，并评估了优缺点。

5.3.2 流动生化分析与HPLC的在线耦合

FA-BCD-HPLC可追溯到20世纪90年代，随着第一个真正的在线柱后生化分析[45]的开发而被引入。随着FA-BCD-HPLC的出现，新的筛选方法出现了，从而为快速有效地筛选复杂混合物开辟了道路。并行化的简易性允许同时筛选多个目标[23]。图5.5所示为采用生化分析并行化的FA-BCD-HPLC设置的一般方案。以下部分详细描述了不同的生化分析法。

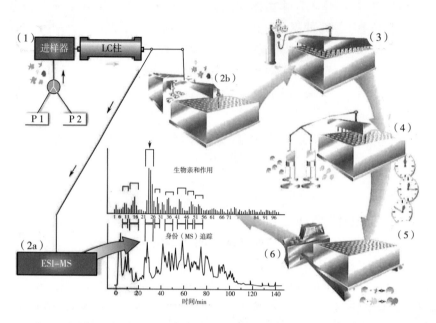

图 5.4 使用基于 HPLC 的去重复化模式和 96 孔板进行活性代谢物分析

生化分析引导的活性代谢物检测和质谱表征允许鉴定代谢混合物中的活性代谢物。代谢物混合物通过 HPLC(1) 分离,然后分流到质谱(2a)和馏分收集器,用于分馏到 96 孔板(2b)。然后将级分干燥(3),与生化分析组分混合(4),孵育(5)并用酶标仪(6)筛选活性。最后,在分级生物亲和力和质谱迹线之间建立相关性。

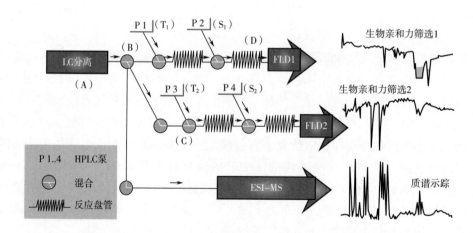

图 5.5 具有两个平行生化分析的 FA-BCD-HPLC 设置的一般方案

使用柱后补充梯度(A)进行梯度 HPLC 后,总流量被分流(B)到两个平行的生化分析。通过随后的 BCAT 结合(C),将目标蛋白(T_1 和 T_2)添加到反应盘管(D)中 HPLC 流出物的分离部分。然后将底物或配体(S_1 和 S_2)添加到以下反应盘管中。在线生物反应之后,通常采用荧光读数(FLD)。并行质谱读数允许识别活性化合物。当使用两个相似的目标(例如,ER-α 和 ER-β)时,并行生化分析模式提供选择性数据。

5.3.2.1 一般原则

基于FA-BCD-HPLC的生化分析的主要特征是目标蛋白质和示踪分子的连续混合，通常来自不同的储液器，分离技术的流出物，如HPLC[46]。在通常由盘管或非盘管组成的反应室中混合后，反应产物直接进入检测单元（均相形式）或首先通过某种分离步骤以分离结合或反应的来自未结合或未反应的示踪分子（异质形式）的示踪分子。为了随着时间的推移保持相对稳定的目标和示踪剂溶液，它们通常储存在适当的温度（0~4℃）下，并通过超级回路引入反应单元[47]，实际上是大容量注射器。当洗脱化合物与筛选的目标相互作用时，会在洗脱时报告检测器信号的临时变化。

基本上，FA-BCD-HPLC检测模式可分为两大类，即异质和均质生化分析。在异质分析形式中，未与生化分析靶标结合的示踪配体在检测之前被捕获（例如，使用在线限制访问或亲和柱）。异质生化分析法的主要优点是由于低背景荧光和潜在干扰化合物的捕获而具有固有的高灵敏度。Oosterkamp等给出了一个例子，它同时开发了一种用于雌激素受体（ER）的异质和同质生化分析法[48]。主要依赖于均质形式的生化分析是酶促生化分析，其中测试化合物对非荧光底物转化为荧光产物的抑制作用是衡量酶亲和力的标准[49]。虽然由于使用的受限访问柱的堵塞风险，异质生化分析通常被证明不太稳定，就像雌激素受体的情况一样，背景荧光的去除通常会使生化分析更加敏感。异质形式的基于FA-BCD-HPLC的受体生化分析几乎与图5.2中讨论的均相FA-BCD-HPLC方案相同。在异质情况下，以下部分是有区别的：①HPLC流出物；②逆梯度，然后分流到质谱和生化分析；③添加雌激素受体；④反应盘管；⑤添加荧光配体香豆素拟雌内酯；⑥反应盘管，然后第二个反应盘管后面是一个限制访问的柱（图中未显示）；⑦荧光检测器。当将梯度HPLC在线耦合到生化分析时，众所周知的大多数生化靶标与有机改性剂的不相容性可通过相对较短的柱后生化分析反应时间稍微弥补。在这问题中的另一个重要方面是当应用梯度HPLC分离时有机改性剂的浓度增加。这个问题可以通过添加反作用柱后梯度来有效解决（图5.2；第2点），允许含有恒定浓度有机改性剂的流出物进入生化分析[22]。

5.3.2.2 与HPLC耦合的在线配体结合流动测定

第一个柱后在线BCD系统是由Przyjazny等开发的[50]。这种原理设置的证明利用了亲和素与生物素和生物胞素的高亲和力相互作用。具有亲和染料的亲和素与HPLC流出物连续混合。在生物素或生物胞素洗脱后，染料从抗生物素蛋白的活性位点被置换，导致通过紫外线测量的光学变化。虽然这种在线生化分析显示了快速筛选混合物中生化相互作用的新可能性，但基于紫外线检测方法的生化分析法对未知混合物的筛选不太有用，因为混合物中的许多化合物在所用波长下显示出吸光度。随后，Przyjazny等开发了一种类似的柱后均质荧光团连接的生化分析[51]。虽然这种设置在很大程度上克服了与紫外线相关的问题，但这种生化分析格式给类生物素化合物带来了问题，这些化合物在结合在活性位点时不会导致荧光增强。随着另一种在线生物检测策略的发展[45]，这个普遍问题被规避了。该策略利用固定的地高辛和荧光标记的抗地高辛片段来检测地高辛样分子。洗脱的地高辛可以与荧光抗体反应，从而占据它们的活性位点并使它们失去活性以进一步结合固定在上游的地高辛亲和柱。随后可以用荧光检测器检测未保留的荧光抗体。

表 5.1 列出了大多数当前开发的 FA-BCD-HPLC 检测格式及其生化分析特性。本章讨论了所有这些 FA-BCD-HPLC 分析。从开发的 FA-BCD-HPLC 分析来看，大多数生化分析都基于荧光检测。然而，在过去几年中，已经开发了使用质谱仪检测 BCD 的 FA-BCD-HPLC 分析。少数例外之一是 Neungchamnong 等开发的 FA-BCD-HPLC 分析。为了检测混合物中的抗氧化剂，它们开发了一种 FA-BCD-HPLC 策略，能够基于紫外-可见检测对混合物中的抗氧化剂进行在线柱后检测[52]。模型化合物的在线柱后化学氧化导致连续形成有色产物，可以通过分光光度法对其进行监测。

表 5.1　目前开发的在线流动生化分析法（按检测格式排序）

年份	作者	靶标	分析物	参考文献
与目标结合或通过底物转化时配体的光学变化				
1990	Przyjazny	亲和素	类生物素	[50]
2000、2005	Ingkanman，Fabel	乙酰胆碱酯酶	乙酰胆碱酯酶抑制剂	[68, 69, 86]
2005	Nuengchamnong	自由基	抗氧化剂	[52]
与靶标结合时配体的荧光增强				
1993	Przyjazny	亲和素	类生物素	[51]
1996、2001、2004	Oosterkamp、Schobel、van Elswijk	雌激素受体	异雌激素	[22, 23, 48, 63, 64]
2006	Kool、van Liempd			
2003	Schenk	荧光团标记的磷酸结合蛋白	磷酸盐消耗或释放酶	[20]
荧光产品的形成				
2003	Rhee	乙酰胆碱酯酶（AChE）	乙酰胆碱酯酶抑制剂	[49]
2003	Schenk	磷酸二酯酶（PDEs）	磷酸二酯酶抑制剂	[21]
2005、2006	Kool	细胞色素 P450（CYPs）	细胞色素抑制剂	[77, 78]
2007	Kool	谷胱甘肽 S-转移酶	谷胱甘肽 S-转移酶抑制剂	[79]
2007	Kool	ROS 靶向生物分子	产生 ROS 的化合物和抗氧化剂	[85]
基于时间分辨 FRET 的生物亲和筛选				
2004	Hirata	酪氨酸激酶	酪氨酸激酶抑制剂	[70]
基于固定化抗原/配体柱的异质分析形式				
1993、1994	Irth、Oosterkamp	荧光素标记的抗体	地高辛样物质	[45, 46]
1996	Miller	荧光素标记的抗体	粒细胞集落刺激因子	[55]
1999、2000	Van Bommel	酶标链霉亲和素	类生物素	[58-60]
2001	Schenk	荧光素标记的抗体	细胞因子	[56]

续表

年份	作者	靶标	分析物	参考文献
基于固定化抗体/目标蛋白柱的异质分析方法				
1996	Oosterkamp	抗体	白三烯	[57]
1998	Oosterkamp	荧光素标记的尿激酶受体	尿激酶受体配体	[67]
基于流式细胞术的异质分析方法				
2003	Schenk	荧光素标记的抗体	地高辛样物质	[47]
基于中空膜过滤器的异构化验方法				
1996	Lutz	抗体	类生物素	[61]
基于自由流动电泳的异质分析方法				
2000	Mazereeuw	链霉亲和素	类生物素	[62]
在线筛选的理论概念				
1997	Oosterkamp	雌激素受体	雌二醇	[53]
1997	Oosterkamp	基于生物素的亲和树脂	生物素	[54]
与光学生物传感器结合				
2000	Haake	抗体	异丙隆杀虫剂	[32]
基于MS的亲和交互				
2001、2003	Hogenboom、Derks	抗体	生物素和地高辛	[24, 87]
2003、2005、2006	Krabbe	铁（Ⅲ）甲基钙黄绿素蓝	对金属离子具有高亲和力的分析物（例如磷酸化肽）	[88-90]
2007	Krabbe	铁（Ⅲ）甲基钙黄绿素蓝	金属配体配合物	[91]
基于MS的酶预测形成				
2003、2006	Van Eswijk、de Jong	血管紧张素转换酶（ACE）	血管紧张素转换酶（ACE）抑制剂	[19, 92]
2004、2005	De Boer	组织蛋白酶B	组织蛋白酶B抑制剂	[25, 72, 80]
2007	Bruyneel	蛋白质鉴定	柱后消化肽	[93]

当将类似间歇式的生化分析模式转移到在线系统中时，重要的问题之一是间歇式模式中目标蛋白质的生化特征是否与在线模式中的特征相似。由于目标分子上的多个结合位点可能会使相互作用复杂化，Oosterkamp等用理论模型表明，在采用标记配体的在线系统中[53]，或在使用标记的蛋白质时，如果存在多个结合位点，也会保留生物化学的相互作用[54]。理论模型最终分别通过（基于荧光增强的）受体亲和力检测生化分析[48]和荧光素标记的链霉亲和素和生物素进行了验证。

由于抗体固有的稳定性和高亲和力，基于抗体的FA-BCD-HPLC测定形式通常非常稳定和灵敏。因此，这些生化分析格式经常用于证明新 FA-BCD-HPLC 方法的原理研究。基于高亲和力抗体相互作用的在线生化分析法是在线柱后甲硫氨酰粒细胞集落刺激因子分析[55]。Schenk 等开发的对细胞因子敏感的在线柱后生化分析并且基于免疫化学创造了快速确定细胞因子相对浓度的可能性，从而允许测量生物基质中重要的细胞因子谱[56]。另一个类似的在线生化分析依赖于荧光标记的白三烯通过固定化抗体柱与溶液中的抗体结合。洗脱的白三烯与荧光标记的白三烯竞争溶液中存在的有限数量的抗体，从而改变检测到的荧光标记白三烯的数量[57]。van Bommel 等[58,59]检查了酶促信号放大。与 Oosterkamp 等描述的策略相比[48]，使用了荧光素标记的链霉亲和素和生物素，这里链霉亲和素上的荧光素标记被过氧化物酶取代。这种基于酶的信号放大确实提供了比基于荧光的在线生化分析更灵敏的生化分析。van Bommel 等[60]也描述了一个类似的系统，它是为测定蛋白质而开发的，随着基于中空纤维的在线生化分析的开发，连续分离游离和结合示踪剂的新方法成为可能，流动免疫化学检测出现了[61]。此外，这种方法的使用也用小的酶标记来证明。另一种基于免疫亲和的 BCA 是基于对农药异丙隆的高度交叉反应性抗体的在线耦合无标记光学生物传感器[32]。另一种技术，是使用自由流动电泳，利用电场产生层流垂直于在线生化分析流[62]。由于其不同的电泳特性，所有生化试剂都以不同的迁移率经历这种层流。这允许分离结合和非结合示踪剂，之后通过透明电泳池进行光学检测。其他新颖的和最近的基于 FA-BCD-HPLC 的方法是使用流式细胞仪进行在线柱后检测[47]。使用该技术，不会受到洗脱化合物和未结合的荧光免疫试剂的荧光干扰。该原理基于细胞仪对结合抗原标记珠的荧光免疫试剂的唯一测量。在这方面，生化分析原理类似于 Oosterkamp 等[46]所描述的原理，也可以指定为异构生化分析，其中固定的支持物连续泵送通过。这导致生化分析法具有同质生化分析的稳定性特征，又具有异质生化分析的优点，例如荧光背景消除和增强的灵敏度[48]。本节中描述的基于抗体或亲和素的生化分析法在大多数情况下用于演示新的 FA-BCD-HPLC 方法。然而，大多数提到的方法仅限于与高亲和力相互作用一起使用。由于与基于抗体或亲和素的系统相比，酶系统通常表现出低得多的亲和力，因此它们通常不适合这些策略。这也适用于受体系统，尽管受体系统通常比酶具有更高的亲和力。对于受体系统，通常较低的受体浓度和复杂的生物学行为使其不太适合这些生化分析法。此外，许多酶和受体是膜结合的，因此可能会堵塞在线异质生化分析法，其中游离和结合的配体在检测前要分离。

人体 ERα 的柱后生化分析是第一个在线受体亲和生化分析[48]。这个在线生化分析已经在第 1 节中进行了更详细的讨论。生化分析在分析含有多种雌激素的混合物中显示了它的价值，例如含有污染雌激素的环境混合物。Schobel 等[23]和 van Elswijk 等[22]报告了这种方法的应用。它们使用 ERα 和 ERβ 与质谱并行使用的技术快速筛选天然提取物中的 ERα 和 ERβ 选择性化合物。2006 年，该方法被用于演示模型药物（如他莫昔芬和雷洛昔芬）代谢混合物中单个代谢物的在线 ERα 亲和筛选和质谱鉴定[63,64]。图 5.6 显示了在微粒体孵育期间形成的他莫昔芬（TM）代谢物（质谱迹线）和相应的 ERα-与柱后生化分析的亲和力迹线。然而，必须说明的是，由于受体-配体动力学速度快，可以非常迅速地测量 ERα-亲和力反应，从而使 FA-BCD-HPLC 方法取得成功[48]。此外，研究中使用的 ERα 对有机改性剂的高稳定性使生化分析法非常适合 FA-BCD-HPLC 的目的。在这方面，人们可以想到不太合适的受体系统，如 GPCR，它需要非常精细的生化分析模式，以防止由于受体破坏或膜溶解而导致的活性损失。

尿激酶受体在细胞黏附、迁移、增殖和关键的基质降解中起重要作用，并参与细胞内降解介导的疾病和影响纤溶酶原激活的蛋白水解级联系统[66]，该受体用荧光素标记，用于异质在线生化分析法类似于大多数异质免疫亲和生化分析[67]。对于基于受体的 FA-BCD-HPLC 检测，生化分析中的受体浓度必须相对较高，以便在较短的反应时间内获得良好的信噪比。此外，许多基于受体的间歇式生化分析的读数基于第二信使的形成。由于在这些（通常基于贴壁细胞的）生化分析中形成第二信使的时间跨度通常不适合 FA-BCD-HPLC 分析中相对较短的柱后生化分析时间，不能以这种方式监测受体相互作用。因此，FA-BCD-HPLC 分析只能用于有限数量的受体系统，并且具有当前可用的分析形式和生化检测方法。

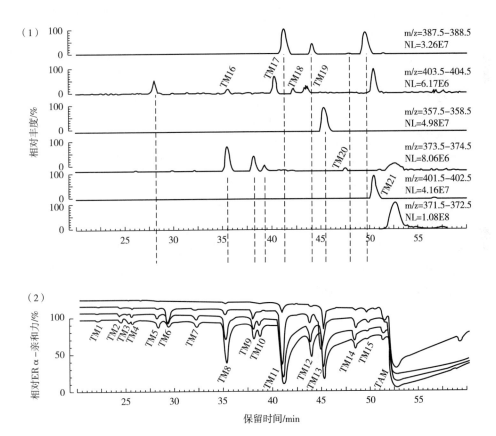

图 5.6 (1)对应于具有或不具有 ERα 亲和力的它莫昔芬代谢物的质量范围的 TIC 色谱图。随着孵育时间的增加，ERα-亲和力的痕迹。(2)从上到下，亲和力痕迹对应于 0min、7min、17min 和 26min 微粒体孵育。TIC 色谱图与在线 ERα-亲和力生化分析的 ERα-亲和力迹线相关

5.3.2.3 与 HPLC 耦合的在线酶流动分析

Ingkaninan 等[68]描述了第一个在线柱后酶亲和生化分析。这种乙酰胆碱酯酶

(AChE) 活性的比色生化分析使用了常见的策略,其中将底物转化为有色产物。由于产物的形成暂时减少,洗脱配体引起的酶抑制导致紫外线基线的变化。Ingkaninan 等后来将此方法与平行质谱结合使用,以快速检测植物提取物中的单个活性化合物并同时进行鉴定[69]。2003 年,Rhee 等改进了在线 AChE 抑制剂生化分析,将比色探针替换为底物 7-乙酰氧基-1-甲基碘化喹啉[49]。为实现在线柱后开发的通用磷酸盐消耗生化分析允许筛选混合物中的磷酸盐消耗或释放酶[20]。磷酸二酯酶在细胞信号传导中发挥重要作用,例如,通过水解 cAMP 和 cGMP 等第二信使,是 Schenk 等[21]描述的在线柱后检测方法的目标。磷酸二酯酶将荧光底物 Mant-cGMP 转化为高度荧光的 Mant-GMP 是这种在线生化分析的关键。在血管紧张素转化酶(ACE)的 FA-BCD-HPLC 系统中采用了类似的策略[19]。使用这种方法,水解牛乳样品中发现的生物活性肽可以同时用质谱进行鉴定。虽然大多数基于荧光的酶促生化分析依赖于产品形成时直接的荧光变化,但 Hirata 等[70]使用了一种更精细的方法,即时间分辨荧光共振能量转移(TR-FRET)。这种在线 FIA BCA 使用来自荧光标记底物的 TR-FRET 依赖性光发射,该底物在被酪氨酸激酶磷酸化时与荧光标记抗体结合。在 Hirata 等基于 FRET 的后续研究中,应用了另一种不同于通常使用的 HPLC 分离的分离策略,即尺寸排阻色谱法[71]。Hirata 等描述了分离含有蛋白酶抑制剂的化合物混合物,然后针对枯草杆菌蛋白酶 Carlsberg 对抑制剂进行在线柱后筛选。该生化分析基于用两种荧光化合物标记的底物肽。蛋白酶消化后,大量的分子内 FRET 淬灭减少,导致荧光增加。对于昂贵目标的筛选,小型化是非常可取的。为了减少目标消耗,已经开发了一种能够执行在线生化分析的连续流微流体系统[72]。总流速仅为 4μL/min,与传统的在线生化分析格式相比,成本显著降低。然而,使用半胱氨酸蛋白酶组织蛋白酶 B 作为模型酶,与传统的在线生化分析相比,获得了较低的敏感性,但非特异性结合降到了最低。

生物转化酶在内源性化合物和外源性化合物的代谢中起着至关重要的作用。细胞色素 P450(CYPs)是最重要的 I 期代谢酶,在药物和其他外源性物质的氧化代谢反应中发挥核心作用[73,74]。药物和代谢物对 CYPs 的亲和力会导致药物代谢水平上不必要的药物相互作用。II 期代谢也可导致药物不良反应。II 期代谢酶,包括葡萄糖醛酸转移酶、磺基转移酶和谷胱甘肽 S-转移酶(GSTs),在化合物的官能团上添加取代基,通常会导致亲水性强烈增加并促进排泄。由于药物的代谢会导致形成具有药理活性、药物间相互作用和/或反应性的代谢物,因此,在药物发现和开发计划中对它们进行早期考虑是必不可少的[75]。化合物的筛选过程通常采用 HTS 方法[37]。然而,大多数使用的 HTS 方法不允许识别代谢混合物中的单个配体。在对单个化合物进行亲和筛选之前,这些混合物必须进行色谱分离[39,76]。通过将新型 CYPs 酶亲和检测(EAD)系统在线耦合到梯度 HPLC,首次对 CYPs 亲和筛选的 Kool 等启用了混合物中的单个化合物[77]。通过不同 CYP—EAD 系统的并行配置,设计了一种 FA-BCD-HPLC 方法,该方法能够同时对混合物中的单个化合物进行亲和筛选,以针对多达三个相关 CYPs 的面板[78]。除了目前用于筛选 CYPs 相互作用的 FA-BCD-HPLC 方法外,使用类似方法进行的 GST 相互作用筛选也可以帮助药物发现和开发过程。由于 GST 清除反应性中间体和亲电化合物,因此 GST 的 FA-BCD-HPLC 系统可用于鉴定混合物中的谷胱甘肽结合物和亲电化合物。最近，一种用于筛选混合物中同工酶特异性 GSTs 抑制剂的平行 FA-BCD-HPLC 方法[79]。例如，图 5.7 显示了从测试混合物中获得的平行 EAD 色谱图（使用平行 FA-BCD-HPLC 配置测量，柱后分离为细胞溶质 GSTs BCA 和人 GST Pi BCA）。这种 FA-BCD-HPLC 方法的问题之一是非极性化合物的严重拖尾，这些化合物在较高浓度的有机改性剂下洗脱（图 5.7）。这可能是由于柱后反作用梯度混合后非极性化合物的沉淀造成的。然而，具有亲和力的化合物在彼此洗脱后不久就会出现叠加，仍然可以被区分出来。

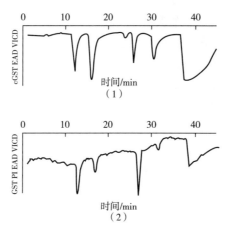

图 5.7 GST EAD 的痕迹，在总 FA-BCD-HPLC 系统中注入了 6 种化合物的混合物，用于 GSTs 检测。注入的化合物是：丙烯醛（3600μmol/L；12.5min）、巴豆醛（2900μmol/L；16.5min）、2-氯-5-硝基吡啶（2400μmol/L；24.0min）、肉桂醛（1500μmol/L；26.5min）、4-氯-3-硝基苯甲酸丁酯（300μmol/L；33.0min）和乙丙烯酸（300μmol/L；38.5min）。（1）cGSTs EAD 示踪。（2）GST Pi EAD 示踪

其他分离方法，如高温液相色谱法（HTLC）或尺寸排除色谱法（SEC）可用于降低在线生化分析中的有机改性剂浓度，但获得的分辨率仍不如常规 HPLC[71,80]。

产生 ROS 的化合物，有时在生物激活时产生，通常被认为是有害的化合物。由于产生 ROS 的化合物的产物可以与同一混合物中的抗氧化剂发生反应，传统的间歇式生化分析[81,82]通常对混合物进行检测，可能会产生假阴性。原则上，基于 HPLC 的在线生化分析已经存在[83,84]，可以测量混合物中的单个抗氧化剂，但是，这些生化分析不适合同时筛选混合物中的促氧化剂和抗氧化剂。随着对产生 ROS 的化合物和抗氧化剂的 FA-BCD-HPLC 方法的开发，可以在一次运行中快速筛选混合物中的此类化合物[85]。

当 EAD 系统中荧光产物的形成速率较低时，检测配体的分辨率也会下降，因为需要更长的反应时间（因此需要更长的反应盘管）来创造足够的信噪比。因此，基于酶或受体的 FA-BCD-HPLC 系统给出快速和敏感的反应[19,22]，对于 FA-BCD-HPLC 系统与 HTS 系统竞争是至关重要的。然而，Fabel 等[86]已经报告了在最小的反应盘管带宽的情况下允许更长的在线柱后反应时间的尝试。这是通过在这些反应盘管中应用气体分段流来实现的。由于需要荧光靶标或配体，如果不容易得到的话，可能会使生化分

析的开发严重复杂化,因此,开发了基于质谱的在线生化分析来克服这个问题[24,25,87]。在这些生化分析中,示踪剂不是通过其荧光特性来监测,而是通过其特征分子质量来监测[24]。Derks等[87]证明了质谱的主要优势之一,即在一个生化分析中容易测量不同的配体-靶体相互作用。DeBoer等[25]将ESI-MS作为一种检测方法,通过测量特定组织蛋白酶B底物的酶学产物(AMC和Z-FR)的形成变化来鉴定组织蛋白酶B抑制剂。质谱在线柱后BCD的另一个用途涉及检测对金属离子具有高亲和力的分析物[88],对一个示踪配体(甲基钙蓝)的选择性离子监测被用作这种类型分析的基础,该配体在金属配体复合物(Fe^{3+}甲基钙蓝)的配体交换中被释放。该系统能够检测到作为竞争亲和化合物的磷酸化肽。非磷酸化和磷酸化的肽在在线BCD之前通过HPLC分离。这种生化分析的改进形式允许在在线配体交换反应后筛选出对某些金属离子有亲和力的配体[89,90]。同一作者还使用ESI-MS来研究不同金属复合物在连续流配体交换反应器中的复合物形成情况[91]。在另一项研究中,用质谱监测酶产物形成的在线生化分析被用来筛选乙酰胆碱酯酶(AChE)抑制剂[92]。这种BCA用质谱监测乙酰胆碱的水解情况。从HPLC中洗脱出来的抑制剂会暂时抑制AChE,从而改变乙酰胆碱的水解,由质谱监测。MS同时测量洗脱的抑制剂的质谱。鉴定蛋白质的分析方法通常包括一个消化步骤,通常使用胰蛋白酶作为蛋白质分解酶。在大多数情况下,在LC-MS分析系统之前实施离线和在线消化方法,以获得高序列覆盖率,进行明确的蛋白质鉴定。对于氨基酸序列高度重叠的蛋白质,例如,治疗性蛋白质及其代谢物,在消化和随后的电喷雾质谱分析标记肽之前,必须分离蛋白质。Bruyneel等[93]展示了一种在线柱后溶液相消化方法,该方法是基于向纳米C_{18}反相柱下游连续注入蛋白酶的方式。根据蛋白质的保留时间,结合柱后消化物中形成的特定标记肽的检测,来鉴定蛋白质。

如果没有基于荧光的生化分析法,基于质谱检测的FA-BCD-HPLC检测是酶法生化分析系统的一个可能选择。然而,这些生化分析法由于最佳生化分析条件通常采用与质谱不兼容的缓冲液和辅助因子条件而受到严重阻碍。在这些生化分析中,酶促产物的形成被跟踪,必须选择与质谱兼容的生化分析条件。这很可能导致生化分析条件远离最佳状态或生化分析条件使生化分析不适合在线生化分析法。当研究基于质谱的FA-BCD-HPLC检测时,在监测结合相互作用时,需要高亲和力的相互作用以获得良好的信噪比。因此,这些生化分析法可能不能用于大多数受体系统,与基于抗体和阿维菌素的系统相比,这些系统的亲和力范围通常较低。另外,当与基于抗体和阿维菌素的系统相比,使用较低的受体浓度范围时,通常会导致信噪比较低。此外,质谱兼容的生化分析条件在大多数情况下不适合于受体生化分析,而膜结合的受体系统很可能会带来质谱相关的问题。因此,基于质谱的FA-BCD-HPLC格式的明显选择是非膜结合的酶系统,它能迅速将底物转化为其产物(通过质谱测量),并且没有荧光酶生化分析可用。

5.4 结论

流动注射分析或液相色谱与生化分析和质谱检测的结合已被证明是提高蛋白质-蛋白质和蛋白质-配体相互作用的亲和力分析过程中的选择性和敏感性以及样品产量的绝

佳方法。使用生化检测的传统荧光或辐射检测模式有一个缺点，即必须合成一个对受体或酶有显著亲和力的荧光探针或底物。另一个限制是，标记反应可能是费力和耗时的。另一方面，质谱检测只需要使用报告配体——一种与受体蛋白结合的化合物，并且可以很容易地被质谱仪检测到。

为了将流动系统与生化分析相结合，可以应用三种方法：在分析分离之前使用生化相互作用、在分析分离期间使用生化相互作用或在分析分离之后使用生化相互作用。所有三种程序在选择性（例如干扰基质成分的影响）和灵敏度（例如分馏过程中的潜在样品损失）方面各有优缺点。通过使用在线系统，可以获得高通量和可重复性。此外，这种类型的程序允许同时测定化学（例如定量和定性信息）和生物学（例如 IC_{50}）信息。

参考文献

［1］Hage, D. S. (1998) Survey of recent advances in analytical applications of immunoaffinity chromatography. *Journal of Chromatography. B, Biomedical Sciences and Applications*, 715, 3–28.

［2］Hennion, M. C. and Pichon, V. (1000) Immuno-based sample preparation for trace analysis. *Journal of Chromatography. A*, 2003, 29–52.

［3］Hage, D. S. (2002) High-performance affinity chromatography: a powerful tool for studying serum protein binding. *Journal of Chromatography. B, Analytical Technologies in the Biomedical and Life Sciences*, 768, 3–30.

［4］Phillipson, D. W., Milgram, K. E., Yanovsky, A. I., Rusnak, L. S., Haggerty, D. A., Farrell, W. P., Greig, M. J., Xiong, X. and Proefke, M. L. (2002) High-throughput bioassay-guided fractionation: a technique for rapidly assigning observed activity to individual components of combinatorial libraries, screened in HTS bioassays. *Journal of Combinatorial Chemistry*, 4, 591–599.

［5］Hewitt, L. M. and Marvin, C. H. (2005) Analytical methods in environmental effects-directed investigations of effluents. *Mutation Research*, 589, 208–232.

［6］Trubetskoy, O. V., Gibson, J. R. and Marks, B. D. (2005) Highly miniaturized formats for *in vitro* drug metabolism assays using vivid fluorescent substrates and recombinant human cytochrome P450 enzymes. *Journal of Biomolecular Screening*, 10, 56–66.

［7］Reineke, N., Bester, K., Huhnerffiss, H., Jastorff, B. and Weigel, S. (2002) Bioassay-directed chemical analysis of River Elbe surface water including large volume extractions and high performance fractionation. *Chemosphere*, 47, 717–723.

［8］Katta, V. and Chait, B. T. (1991) Observation of the heme-globin complex in native myoglobin by electrospray-ionization mass spectrometry. *Journal of the American Chemical Society*, 113, 8534.

［9］Ganem, B., Li, Y. and Henion, J. D. (1991) Detection of noncovalent receptor-ligand complexes by mass spectrometry. *Journal of the American Chemical Society*, 113, 6294–6296.

［10］Chowdhury, S. K., Katta, V. and Chait, B. T. (1990) Probing conformational changes in proteins by mass spectrometry. *Journal of the American Chemical Society*, 112, 9012–9013.

［11］Craig, T. A., Veenstra, T. D., Naylor, S., Tomlinson, A. J., Johnson, K. L. M. S., Juranic, N. and Kumar, R. (1997) Zinc binding properties of the DNA binding domain of the 1, 25-dihydroxyvitamin

D3 receptor. *Biochemistry*, 36, 10482-10491.

[12] Kheterpal, I., Cook, K. D. and Wetzel, R. (2006) Hydrogen/deuterium exchange mass spectrometry analysis of protein aggregates. *Methods in Enzymology*, 413, 140-166.

[13] Whitelegge, J., Halgand, F., Souda, P. and Zabrouskov, V. (2006) Top-down mass spectrometry of integral membrane proteins. *Expert Review of Proteomics*, 3, 585-596.

[14] Evers, T. H., van Dongen, J. L., Meijer, E. W. and Merkx, M. (2007) Ligand-induced monomerization of Allochromatium vinosum cytochrome c' studied using native mass spectrometry and fluorescence resonance energy transfer. *Journal of Biological Inorganic Chemistry*.

[15] Mazon, H., Gabor, K., Leys, D., Heck, A. J., van der Oost, J. and van den Heuvel, R. H. (2007) Transcriptional activation by CprKl is regulated by protein str uctural changes induced by effector binding and redox state. *Journal of Biological Chemistiy*, 282, 11281-11290.

[16] van Duijn, E., Bakkes, P. J., Heeren, R. M., van den Heuvel, R. H., van Heerikliuizen, H., van der Vies, S. M. and Heck, A. J. (2005) Monitoring macromolecular complexes involved in the chaperonin-assisted protein folding cycle by mass spectrometry. *Nature Methods*, 2, 371-376.

[17] Bovet, C., Wortmann, A., Eiler, S., Granger, F., Ruff, M., Gerrits, B., Moras, D. and Zenobi, R. (2007) Estrogen receptor-ligand complexes measured by chip-based nanoelectrospray mass spectrometry: an approach for the screening of endocrine disruptors. *Prolein Science*, 16, 938-946.

[18] van den Heuvel, R. H. and Heck, A. J. (2004) Native protein mass spectrometry from intact oligomers to functional machineries. *Current Opinion in Chemical Biology*, 8, 519-526.

[19] van Elswijk, D. A., Diefenbach, O., van der Berg, S., Irth, H., Tjaden, U. R. and van der Greef, J. (1020) Rapid detection and identification of angiotensin-converting enzyme inhibitors by on-line liquid chromatography-biochemical detection, coupled to electrospray mass spectrometry. *Journal of Chromatography A*, 2003, 45-58.

[20] Schenk, T., Appels, N. M., van Elswijk, D. A., Irth, H., Tjaden, U. R. and van der Greef, J. (2003) A generic assay for phosphate-consuming or-releasing enzymes coupled on-line to liquid chromatography for lead finding in natural products. *Analytical Biochemistry*, 316, 118-126.

[21] Schenk, T., Breel, G. J., Koevoets, P., van den Berg, S., Hogenboom, A. C., Irth, H., Tjaden, U. R. and van der Greef, J. (2003) Screening of natural products extracts for the presence of phosphodiesterase inhibitors using liquid chromatography coupled online to parallel biochemical detection and chemical characterization. *Journal of Biomolecular Screening*, 8, 421-429.

[22] van Elswijk, D. A., Schobel, U. P., Lansky, E. P., Irth, H. and van der Greef, J. (2004) Rapid dereplication of estrogenic compounds in pomegranate (Punica granatum) using on-line biochemical detection coupled to mass spectrometry. *Phytochemistry*, 65, 233-241.

[23] Schobel, U., Frenay, M., van Elswijk, D. A., McAndrews, J. M., Long, K. R., Olson, L. M., Bobzin, S. C. and Irth, H. (2001) High resolution screening of plant natural product extracts for estrogen receptor alpha and beta binding activity using an online HPLC-MS biochemical detection system. *Journal of Biomolecular Screening*, 6, 291-303.

[24] Hogenboom, A. C., de Boer, A. R., Derks, R. J. and Irth, H. (2001) Continuous-flow, on-line monitoring of biospecific interactions using electrospray mass spectrometry. *Analytical Chemistiy*, 73, 3816-3823.

[25] de Boer, A. R., Letzel, T., van Elswijk, D. A., Lingeman, H., Niessen, W. M. and Irth, H. (2004) On-line coupling of high-performance liquid chromatography to a continuous-flow enzyme assay based

on electrospray ionization mass spectrometry. *Analytical Chemistry*, 76, 3155-3161.

[26] Johnson, B. M., Nikolic, D. and van Breemen, R. B. (2002) Applications of pulsed ultrafiltration-mass spectrometry. *Mass Spectrometry Reviews*, 21, 76-86.

[27] van Breemen, R. B., Huang, C. R., Nikolic, D., Woodbury, C. P., Zhao, Y. Z. and Venton, D. L. (1997) Pulsed ultrafiltration mass spectrometry: a new method for screening combinatorial libraries. *Analytical Chemistry*, 69, 2159-2164.

[28] Geun Shin, Y., Bolton, J. L. and van Breemen, R. B. (2002) Screening drugs for metabolic stability using pulsed ultrafiltration mass spectrometry. *Combinatorial Chemistry & High Throughput Screening*, 5, 59-64.

[29] van Elswijk, D. A., Tjaden, U. R., van der Greef, J. and Irth, H. (2001) Mass spectrometry-based bioassay for the screening of soluble orphan receptors. *International Journal of Mass Spectrometry*, 210, 625-636.

[30] Annis, D. A., Nazef, N., Chuang, C. C., Scott, M. P. and Nash, H. M. (2004) A general technique to rank protein - ligand binding affinities and determine allosteric versus direct binding site competition in compound mixtures. *Journal of the American Chemical Society*, 126, 15495-15503.

[31] Annis, D. A., Athanasopoulosa, J., Currana, P. J., Felscha, J. S., Kalghatgia, K., Leea, W. H., Nasha, H. M., Orminatia, J. P. A., Rosnera, K. E., Shipps, G. W., Jr., Thaddupathyb, G. R. A., Tylera, A. N., Vilenchek, L., Wagner, C. R. and Wintner, E. (2004) An affinity selection-mass spectr-ometry method for the identification of small molecule ligands from selfencoded combinatorial libraries. Discovery of a novel antagonist of *E. coli* dihydrofolate reductase.

[32] Haake, H. M., de Best, L., Irth, H., Abuknesha, R. and Brecht, A. (2000) Label - free biochemical detection coupled on-line to liquid chromatography. *Analytical Chemistry*, 72, 3635-3641.

[33] Hsieh, Y. F., Gordon, N., Regnier, F., Afeyan, N., Martin, S. A. and Vella, G. J. (1997) Multidimensional chromatography coupled with mass spectrometry for targetbased screening. *Molecular Diversity*, 2, 189-196.

[34] Kaur, S., McGuire, L., Tang, D., Dollinger, G. and Huebner, V. (1997) Affinity selection and mass spectrometry-based strategies to identify lead compounds in combinatorial libraries. *Journal of Protein Chemistry*, 16, 505-511.

[35] Lenz, E., Taylor, S., Collins, C., Wilson, I. D., Louden, D. and Handley, A. (2002) Flow injection analysis with multiple online spectroscopic analysis (LIV, IR, 1H-NMR and MS). *Journal of Pharmaceutical and Biomedical Analysis*, 27, 191-200.

[36] Rich, R. L., Hoth, L. R., Geoghegan, K. F., Brown, T. A., LeMotte, P. K., Simons, S. P., Hensley, P. and Myszka, D. G. (2002) Kinetic analysis of estrogen receptor/ ligand interactions. *Proceedings of the National Academy of Sciences of the United States of America*, 99, 8562-8567.

[37] Schuffenhauer, A., Popov, M., Schopfer, U., Acklin, P., Stanek, J. and Jacoby, E. (2004) Molecular diversity management strategies for building and enhancement of diverse and focused lead discovery compound screening collections. *Combinatorial Chemistry & High Throughput Screening*, 7, 771-781.

[38] Liu, B., Li, S. and Hu, J. (2004) Technological advances in high-throughput screening. *American Journal of PhannacoGenomics*, 4, 263-276.

[39] Fura, A., Shu, Y. Z., Zhu, M., Hanson, R. L., Roongta, V. and Humphreys, W. G. (2004) Discovering drugs through biological transformation: role of pharmacologically active metabolites in drug discovery. *Journal of Medicinal Chemistry*, 47, 4339-4351.

[40] van Rhee, A. M., Stocker, J., Printzenhoff, D., Creech, C., Wagoner, P. K. and Spear, K. L. (2001) Retrospective analysis of an experimental high-throughput screening data set by recursive partitioning. *Journal of Combinatorial Chemistry*, 3, 267-277.

[41] Fura, A. (2006) Role of pharmacologically active metabolites in drug discovery and development. *Drug Discovery Today*, 11, 133-142.

[42] Ohno, K., Fukushima, T., Santa, T, Waizumi, N., Tokuyama, H., Maeda, M. and Imai, K. (2002) Estrogen receptor binding assay method for endocrine disruptors using fluorescence polarization. *Analytical Chemistry*, 74, 4391-4396.

[43] de Boer, T., Otjens, D., Muntendam, A., Meulman, E., van Oostijen, M. and Ensing, K. (2004) Development and validation of fluorescent receptor assays based on the human recombinant estrogen receptor subtypes alpha and beta, *Journal of Pharmaceutical and Biomedical Analysis*, 34, 671-679.

[44] Lim, H. K., Stellingweif, S., Sisenwine, S. and Chan, K. W. (1999) Rapid drug metabolite profiling using fast liquid chromatography, automated multiplestage mass spectrometry and receptorbinding. *Journal of Chromatography A*, 831, 227-241.

[45] Irth, H., Oosterkamp, A. J., van der Welle, W, Tjaden, U. R. and van der Greef, J. (1993) On-line immunochemical detection in liquid chromatography using fluorescein-labelled antibodies, *Journal of Chromatography*, 633, 65-72.

[46] Oosterkamp, A. J., Irth, H., Beth, M., Unger, K. K., Tjaden, U. R. and van de Greef, J. (1994) Bioanalysis of digoxin and its metabolites using direct serum injection combined with liquid chromatography and on-line immunochemical detection. *Journal of Chromatography. B, Biomedical Applications*, 653, 55-61.

[47] Schenk, T, Molendijk, A., Irth, H., Tjaden, U. R. and van der Greef, J. (2003) Liquid chromatography coupled on-line to flow cytometry for postcolumn homogeneous biochemical detection. *Analytical Chemistry*, 75, 4272-4278.

[48] Oosterkamp, A. J., Villaverde Herraiz, M. T., Irth, H., Tjaden, U. R. and van der Greef, J. (1996) Reversed-phase liquid chromatography coupled on-line to receptor affinity detection based on the human estrogen receptor. *Analytical Chemistry*, 68, 1201-1206.

[49] Rhee, I. K., Appels, N., Luijendijk, T., Irth, H. and Verpoorte, R. (2003) Determining acetylcholinesterase inhibitory activity in plant extracts using a fluorimetric flow assay. *Phytochemical Analysis*, 14, 145-149.

[50] Przyjazny, A., Kjellstrom, T. L. and Bachas, L. G. (1990) High-performance liquid chromatographic postcolumn reaction detection based on a competitive binding system. *Analytical Chemistry*, 62, 2536-2540.

[51] Przyjazny, A., Hentz, N. G. and Bachas, L. G. (1993) Sensitive and selective liquid chromatographic postcolumn reaction detection system for biotin and biocytin using a homogeneous fluorophore-linked assay. *Journal of Chromatography A*, 654, 79-86.

[52] Nuengchamnong, N., de Jong, C. F., Bruyneel, B., Niessen, W. M., Irth, H. and Ingkaninan, K. (2005) HPLC coupled on-line to ESI-MS and a DPPH-based assay for the rapid identification of antioxidants in Butea superba. *Phytochemical Analysis*, 16, 422-428.

[53] Oosterkamp, A. J., Irth, H., Villaverde Herraiz, M. T., Tjaden, U. R. and van der Greef, J. (1997) Theoretical concepts of online liquid chromatographic-biochemical detection systems. I. Detection systems based on labelled ligands. *Journal of Chromatography A*, 787, 27-35.

[54] Oostercamp, A. J., Irth, H., Tjaden, U. R. and van der Greef, J. (1997) Theoretical concepts of on-line liquid chromatographic-biochemical detection systems. II. Detection systems based on labelled affinity proteins. *Journal of Chromatography. A*, 787, 37-46.

[55] Miller, K. J. and Herman, A. C. (1996) Affinity chromatography with immunochemical detection applied to the analysis of human methionyl granulocyte colony stimulating factor in serum. *Analytical Chemistry*, 68, 3077-3082.

[56] Schenk, T., Irth, H., Marko-Varga, G., Edholm, L. E., Tjaden, U. R. and van der Greef, J. (2001) Potential of on-line micro-LC immunochemical detection in the bioanalysis of cytokines. *Journal of Pharmaceutical and Biomedical Analysis*, 26, 975-985.

[57] Oosterkamp, A. J., Irth, H., Heintz, L., Marko-Varga, G., Tjaden, U. R. and van der Greef, J. (1996) Simultaneous determination of cross-reactive leukotrienes in biological matrices using on-line liquid chromatography immunochemical detection. *Analytical Chemistry*, 68, 4101-4106.

[58] van Bommel, M. R., de Jong, A. P., Tjaden, U. R., Irth, H. and van der Greef, J. (1999) Enzyme amplification as detection tool in continuous-flow systems. I. Development of an enzyme-amplified biochemical detection system coupled on-line to flow injection analysis. *Journal of Chromatography A*, 855, 383-396.

[59] van Bommel, M. R., de Jong, A. P., Tjaden, U. R., Irth, H. and van der Greef, J. (1999) Enzyme amplification as detection tool in continuous-flow systems. II. On-line coupling of liquid chromatography to enzyme-amplified biochemical detection after pre-column derivatization with biotin. *Journal of Chromatography A*, 855, 397-409.

[60] van Bommel, M. R., de Jong, A. P., Tjaden, U. R., Irth, H. and van der Greef, J. (2000) High-performance liquid chromatography coupled to enzyme-amplified biochemical detection for the analysis of hemoglobin after pre-column biotinylation. *Journal of Chromatography A*, 886, 19-29.

[61] Lutz, E. S., Irth, H., Tjaden, U. R. and van der Greef, J. (1996) Applying hollow fibres for separating free and bound label in continuous-flow immunochemical detection. *Journal of Chromatography A*, 755, 179-187.

[62] Mazereeuw, M., de Best, C. M., Tjaden, U. R., Irth, H. and van der Greef, J. (2000) Free flow electrophoresis device for continuous on-line separation in analytical systems. An application in biochemical detection. *Analytical Chemistry*, 72, 3881-3886.

[63] Kool, J., Ramautar, R., van Liempd, S. M., Beckman, J., de Kanter, F. J., Meerman, J. H., Schenk, T., Irth, H., Commandeur, J. N. and Vermeulen, N. P. (2006) Rapid online profiling of estrogen receptor binding metabolites of tamoxifen. *Journal of Medicinal Chemistry*, 49, 3287-3292.

[64] van Liempd, S. M., Kool, J., Niessen, W. M., van Elswijk, D. E., Irth, H. and Vermeulen, N. P. (2006) On-line formation, separation, and estrogen receptor affinity screening of cytochrome P450-derived metabolites of selective estrogen receptor modvlators. *Drug Metabolism and Disposition*, 34, 1640-1649.

[65] Cooper, M. A. (2004) Advances in membrane receptor screening and analysis. *Journal of Molecular Recognition*: JMR, 17, 286-315.

[66] Ragno, P. (2006) The urokinase receptor: a ligand or a receptor? Story of a sociable molecule. *Cellular and Molecular Life Sciences*: CMLS.

[67] Oosterkamp, A. J., van der Hoeven, R., Glassgen, W., Konig, B., Tjaden, U. R., van der Greef, J. and Irth, H. (1998) Gradient reversed-phase liquid chromatography coupled on-line to receptor-

affinity detection based on the urokinase receptor. *Journal of Chromatography B, Biomedical Sciences and Applications*, 715, 331–338.

[68] Ingkaninan, K., de Best, C. M., van der Heijden, R., Hofte, A. J., Karabatak, B., Irth, H., Tjaden, U. R., van der Greef, J. and Verpoorte, R. (2000) High-performance liquid chromatography with on-line coupled UV, mass spectrometric and biochemical detection for identification of acetylcholinesterase inhibitors from natural products. *Journal of Chromatography. A*, 872, 61–73.

[69] Ingkaninan, K., Hazekamp, A., de Best, C. M., Irth, H., Tjaden, U. R., van der Heijden, R., van der Greef, J. and Verpoorte, R. (2000) The application of HPLC with on-line coupled UV/MS-biochemical detection for isolation of an acetylcholinesterase inhibitor from narcissus 'Sir Winston Churchill'. *Journal of Natural Products*, 63, 803–806.

[70] Hirata, J., de Jong, C. F., van Dongen, M. M., Buijs, J., Ariese, F., Irth, H. and Gooijer, C. (2004) A flow injection kinase assay system based on time-resolved fluorescence resonance energy-transfer detection in the millisecond range. *Analytical Chemistry*, 76, 4292–4298.

[71] Hirata, J., Chung, L. P., Ariese, F., Irth, H. and Gooijer, C. (1081) Coupling of sizeexclusion chromatography to a continuous assay for subtilisin using a fluorescence resonance energy transfer peptide substrate: testing of two standard inhibitors. *Journal of Chromatography A*, 2005, 140–144.

[72] de Boer, A. R., Bruyneel, B., Krabbe, J. G., Lingeman, H., Niessen, W. M. and Irth, H. (2005) A microfluidic-based enzymatic assay for bioactivity screening combined with capillary liquid chromatography and mass spectrometry. *Lab-on-a-Chip*, 5, 1286–1292.

[73] Guengerich, F. P. (2001) Common and uncommon cytochrome P450 reactions related to metabolism and chemical toxicity. *Chemical Research in Toxicology*, 14, 611–650.

[74] Nebert, D. W. and Gonzalez, F. J. (1987) P450 genes: structure, evolution, and regulation. *Annual Review of Biochemistry*, 56, 945–993.

[75] Vermeulen, N. P. (2003) Prediction of drug metabolism: the case of cytochrome P450 2D6. *Current Topics in Medicinal Chemistry*, 3, 1227–1239.

[76] Strege, M. A. (1999) High-performance liquid chromatographic-electrospray ionization mass spectrometric analyses for the integration of natural products with modern high-throughput screening. *Journal of Chromatography B, Biomedical Sciences and Applications*, 725, 67–78.

[77] Kool, J., van Liempd, S. M., Ramautar, R., Schenk, T., Meerman, J. H., Irth, H., Commandeur, J. N. and Vermeulen, N. P. (2005) Development of a novel cytochrome p450 bioaffinity detection system coupled online to gradient reversed-phase high-performance liquid chromatography. *Journal of Biomolecular Screening*, 10, 427–436.

[78] Kool, J., van Liempd, S. M., van Rossum, H., van Elswijk, D. A., Irth, H., Commandeur, J. N. and Vermeulen, N. P. (2007) Development of three parallel cytochrome P450 enzyme affinity detection systems coupled on-line to gradient high-performance liquid chromatography. *Drug Metabolism and Disposition*, 35, 640–648.

[79] Kool, J., Eggink, M., van Rossum, H., van Liempd, S. M., van Elswijk, D. A., Irth, H., Commandeur, J. N., Meerman, J. H. and Vermeulen, N. P. (2007) Online biochemical detection of glutathione-S-transferase P1-specific inhibitors in complex mixtures. *Journal of Biomolecular Screening*, 12, 396–405.

[80] de Boer, A. R., Alcaide-Hidalgo, J. M., Krabbe, J. G., Kolkman, J., van Emde Boas, C. N., Niessen, W. M., Lingeman, H. and Irth, H. (2005) High-temperature liquid chromatography coupled on-

line to a continuous-flow biochemical screening assay with electrospray ionization mass spectrometric detection. *Analytical Chemistry*, 77, 7894-7900.

[81] Sugita, O., Ishizawa, N., Matsuto, T., Okada, M. and Kayahara, N. (2004) A new method of measuring the antioxidant activity of polyphenols using cumene hydroperoxide. *Annals of Clinical Biochemistry*, 41, 72-77.

[82] Manzocco, L., Calligaris, S. and Nicoli, M. C. (2002) Assessment of pro-oxidant activity of foods by kinetic analysis of crocin bleaching. *Journal of Agricultural and Food Chemistry*, 50, 2767-2771.

[83] Cardenosa, R., Mohamed, R., Pineda, M. and Aguilar, M. (2002) On-line HPLC detection of tocopherols and other antioxidants through the formation of a phosphomolybdenum complex. *Journal of Agricultural and Food Chemistry*, 50, 3390-3395.

[84] Cano, A., Alcaraz, O., Acosta, M. and Arnao, M. B. (2002) On-line antioxidant activity determination: comparison of hydrophilic and lipophilic antioxidant activity using the ABTS* + assay. *Redox Report*, 7, 103-109.

[85] Kool, J., Van Liempd, S. M., Harmsen, S., Schenk, T., Irth, H., Commandeur, J. N. and Vermeulen, N. P. (2007) An on-line post-column detection system for the detection of reactive-oxygen-species-producing compounds and antioxidants in mixtures. *Analytical and Bioanalytical Chemistry*, 388, 871-879.

[86] Fabel, S., Niessner, R. and Weller, M. G. (1099) Effect-directed analysis by high-performance liquid chromatography with gas-segmented enzyme inhibition. *Journal of Chromatography A*, 2005, 103-110.

[87] Derks, R. J., Hogenboom, A. C., van der Zwan, G. and Irth, H. (2003) On-line continuous-flow, multi-protein biochemical assays for the characterization of bioaffinity compounds using electrospray quadrupole time-of-flight mass spectrometty. *Analytical Chemistry*, 75, 3376-3384.

[88] Krabbe, J. G., Lingeman, H., Niessen, W. M. and Irth, H. (2003) Ligand-exchange detection of phosphorylated peptides using liquid chromatography electrospray mass spectrometry. *Analytical Chemistry*, 75, 6853-6860.

[89] Krabbe, J. G., Lingeman, H., Niessen, W. M. and Irth, H. (1093) Screening for metal ligands by liquid chromatography-ligand-exchange-electrospray mass spectrometry. *Journal of Chromatography A*, 2005, 36-46.

[90] Krabbe, J. G., Gao, F., Li, J., Ahlskog, J. E., Lingeman, H., Niessen, W. M. and Irth, H. (2006) Selective detection and identification of phosphorylated proteins by simultaneous ligand-exchange fluorescence detection and mass spectrometry. *Journal of Chromatography A*, 1130, 287-295.

[91] Krabbe, J. G., de Boer, A. R., van der Zwan, G., Lingeman, H., Niessen, W. M. and Irth, H. (2007) Metal-complex formation in continuous-flow ligandexchange reactors studied by electrospray mass spectrometry. *Journal of the American Society for Mass Spectrometry*, 18, 707-713.

[92] de Jong, C. F., Derks, R. J., Bruyneel, B., Niessen, W. and Irth, H. (2006) High-performance liquid chromatography-mass spectrometry-based acetylcholinesterase assay for the screening of inhibitors in natural extracts. *Journal of Chromatography A*, 1112, 303-310.

[93] Bruyneel, B., Hoos, J. S., Smoluch, M. T., Lingeman, H., Niessen, W. M. and Irth, H. (2007) Trace analysis of proteins using postseparation solution-phase digestion and electrospray mass spectrometric detection of marker peptides. *Analytical Chemistry*, 79, 1591-1598.

6 流动分析中的原子光谱

José L. Burguera 和 Marcela Burguera

6.1 引言

任何分析方法中最关键和最耗时的步骤是样品预处理。分析人员必须经常处理以所有三种状态（固体、液体和气体）出现的各种生物、环境和地质样品，这需要应用最合适的制备技术，例如稀释、过滤、溶解、消化、提取等。大多数样品处理程序仍然是手动执行的。它们的在线实施包括从样品制备到数据管理的所有程序，并在分析物转移到检测器的微妙阶段发挥着重要作用。为此，包括流动进样（FI）在内的连续流动系统已经很容易与原子光谱（AS）技术相连接，例如：原子吸收光谱（AAS）或火焰原子吸收光谱（FAAS）或电热（ET AAS）雾化，电感耦合等离子体发射光谱法（ICP-OES）或ICP-质谱法（ICP-MS）和原子荧光光谱法（AFS）。

文献检索显示，许多书籍[1-4]和综述[5-13]中有整章内容专门介绍与原子光谱相结合的在线样品制备技术，它们批判性地介绍了到2001年为止所发表的最重要的进展。定期发表在《分析原子光谱学杂志》上的年度评论（参考文献部分给出了一个例子）在很大程度上涵盖了这个问题[14]。本章回顾了过去六年中文献中描述的不同和最相关的在线原子光谱样品制备技术，强调了在线分离/预浓缩程序，如化学气相色谱（CVG）、柱子或结式反应器（KR）上的吸附，以及一些用于标本分析的连带技术，如高效液相色谱（HPLC）、毛细管电泳（CE）或微透析。与ICP-OES和ICP-MS有关的内容在本书中另有一章介绍。这里引用的论文的选择是基于以下标准：①涉及在线分离/预浓缩系统；②在2001—2007年期间发表在高影响因子的分析杂志上。我们也避免引用在分析过程中没有什么创新的论文。有趣的是，在线微波辅助样品消化已经不再是20世纪90年代的研究课题[6-8, 11, 12]，可能是因为它已经成为一种常规技术，可以被广泛使用；或者是因为它对于复杂的固体基质难以处理。

6.2 火焰原子吸收光谱法

火焰原子吸收光谱（FAAS）加上流动进样系统仍然是常规测定各种样品中许多无机元素的最流行方法之一。这可能是由于其成本低，而且容易将任何流动进样系统与气动雾化器的吸气效应产生的直接吸气相连接，以便将样品引入火焰中。尽管火焰原子吸收光谱具有单一元素的特点，但它是一种稳健的技术，在大多数临床和分析实验室都可以进行mg/L水平的元素分析。因此，对于许多样品，火焰原子吸收光谱在灵敏度方面不能满足特定的要求，各种在线分离及（或）预浓缩程序常常被用来提高性能。其中，在特定的柱子上分离后的化学气相色谱是最近最广泛使用的，这样分析物离子在柱子上在线浓缩，而潜在的干扰物被丢弃。由于用于测定成氢元素的仪器也足以用于冷蒸气生成，我们在本节中包括了与汞测定有关的部分论文，尽管火焰不用于雾化。

6.2.1 使用火焰原子吸收光谱检测的预浓缩/分离系统

6.2.1.1 化学蒸气的生成

化学气相色谱（CVG）是一种多功能技术，通常用于从复杂的基质中分离某些元素，也可用于其预浓缩，以及在受控条件下，用于标本分析的目的。冷蒸气或氢化物生成（HG）管线与原子光谱检测器在线连接，然后在装有特定吸附剂材料的迷你柱上对被分析物进行在线分离/预浓缩，这已成为测定汞[15,16]、砷[17,18]或铋[19]等元素的一种高选择性和灵敏度的方法，原则上，迄今为止开发的程序都非常相似。相应的无机离子被在线预浓缩及（或）分离，并用酸性溶液（主要是 HCl 和/或 HNO_3）洗脱，与还原剂（$SnCl_2$ 或 $NaBH_4$）混合，产生特定的挥发性物质，通过氮气或氩气流剥离到 QC 雾化器中。电化学（EC）产生的氢化物也可以直接在加热的 QC[20]或迷你柱[21]上进行预浓缩，这类系统是专门为提高灵敏度和去除其余基质而设计的，它们的一些特点见表 6.1[15-21]。在这些情况下，样品类型的选择受制于样品消耗量，可能是 25mL 或者更多，这显然取决于系统中使用的样品流速和要实现的预浓缩。报告的增强因子（EF）也显示出很大的变化（从 10~100 倍），并且取决于预浓缩时间，而预浓缩时间的变化是以样品产量为代价的，预浓缩时间仅仅达到 25/h 或者更少。所开发的程序之间的主要区别在于包装材料的类型和所测定的真实样品的性质，尽管大多数应用涉及天然水域。在所有情况下，检测限（DL）都达到了 ng/L 水平。

表 6.1　在小型色谱柱上进行预浓缩后的在线化学气相法测定的部分应用

确定的元素	样品类型	吸附材料	保留的元素种类	洗脱液	检测限/（ng/L）	参考文献
汞总量	自来水和头发	活性炭	Hg^{2+}	HNO_3	10	[15]
汞（Ⅱ）	湖水和深井水	Cyanex 923	Hg^{2+}	HNO_3	0.2	[16]
砷（Ⅲ）	自然水体	聚四氟乙烯车床切屑	As-PDC	HCl	20	[17]
砷总量	矿物水、饮用水、河水和泉水	活性氧化铝	As^{3+}	HCl	150	[18]
铋	尿液	Amberlite	$BiCl_4^-$	HCl	225	[19]
锑总量（EC-HG）	饮用水、土壤/沉积物	Heated QC	SbH_3	加热+H_2	53	[20]
锑总量（EC-HG）	堆肥、海洋沉积物	Chelex-100	Fe（Ⅱ）	HCl	—	[21]

还有一些其他 CVG 开发虽然不使用任何吸附剂材料进行分析物分离/预浓缩，但由于其新颖的程序而值得一提。例如，位于气液分离器（GLS）和 QC 之间的 7cm 长硅胶管用于捕集，从而预浓缩传统 HG-FI 系统中在 500℃下产生的氢化铅[22]。当捕集器被进一步加热到 750℃时，分析物被释放出来。氢气需要重新挥发，氧气也必须存在。在 60s 的捕集时间内达到的检测限是 19ng/L，但总体效率只有 49%。这一原理最近被用于将铋直接捕集到一个改良的 QC 雾化器中，该雾化器被外部加热到 900℃[23]。在这种情况下，收集/挥发效率为 100%±2.5%，预浓缩比率为 530，收集时间为 300s，对应的样品量为 20mL。在这些条件下，检测限为 3.9ng/L。作者声明，同样的方法在分析上也适用于锑化氢，但不适用于砷化氢。

除了少数例外，上述系统没有能力区分分析物的不同存在形态，也从来没有对各自的有机化合物进行标定。

然而，其他实验安排在 CVG 之前结合了分离技术，如 HPLC。HPLC 与传统的在线 HG-AAS 系统的联用已证明适用于此类形态研究。这些耦合方法大多数已用于 As 的存在形态研究[24-27]，尽管包含各自有机和无机化合物的样品中的 Se[27] 也被分离。在将含砷的样品在微波炉中进行碱性过二硫酸盐氧化后，进行了与物质存在形态无关的量化[24]。此类应用针对环境和生物样品中具有重要毒理学意义的物质。

6.2.1.2 微量元素固相萃取预浓缩

鉴于 FAAS 的低灵敏度，许多在线吸附剂萃取/预浓缩系统已被优化，以测定各种性质的样品中存在的 μg/L 水平的微量元素。为此，研究者最近测试了天然物质（高岭土、沸石、氧化铝、稻壳）和合成材料（琥珀石、硅胶、聚氨酯泡沫）在被化学固定化或被不同化合物功能化后作为吸附材料的性能，用于将金属离子从基体中分离出来并预浓缩，以便随后用 FAAS 测定。然后用稀释的酸或有机溶剂洗脱吸附的复合物，并将其带到检测器中。表 6.2[28-72] 所示为与此相关的一些出版物的分析数据、一些具体特征和当前的应用。在这种情况下，这些应用是针对更复杂的基质，如食品、生物和环境样品。对于 60~90s 的预浓缩时间，采样频率很少超过 40/h。

在柱子上吸附的同时，在结式反应器上保留络合剂也可以选择性地保留一种分析物。这种方法包括在结式反应器上预涂螯合试剂以吸附分析物，然后引入固定的小体积洗脱液，对吸附的螯合物进行定量洗脱。Cerutti 等[73] 回顾了采用结式反应器作为有机金属络合物吸附介质的原子光谱技术的 FI 在线分离和预浓缩系统的进展，其中讨论了利用不同的有机和无机试剂测定不同样品中的不同元素。

表 6.2 用于测定痕量元素的在线 FAAS 柱或 KR 吸附/预浓缩系统

元素	样品类型	吸附材料	络合剂	洗脱液	富集因子	检测限/($\mu g/L$)	参考文献
Cd、Pb	CRM：猪肾、榉树叶	改性后的稻壳	NaOH	HNO_3	72.4 46	1.14 14.1	[28]
Co	水、维生素 B_{12}、复合维生素 B 注射液	表面活性剂涂层的氧化铝	2-亚硝基-1-萘酚	乙醇	125	0.02	[29]
Ag	自来水、井、雨、海水、放射学胶片	表面活性剂涂层的氧化铝	Ag-DDTC, pH3~4	乙醇	125	1.7	[30]
Cu、Pb	水、维生素片、合金	表面活性剂涂层的氧化铝	Salen I, pH9	HNO_3	100、75	0.3、2.6	[31]
Cr^{6+} Cr^{3+}	污水	活性氧化铝	—	NH_4OH HNO_3	—	42 81	[32]
Mn	复合材料	高岭土	5-Br-PADAP pH8.5~10.0	H_2SO_4	—	4.3	[33]

续表

元素	样品类型	吸附材料	络合剂	洗脱液	富集因子	检测限/(μg/L)	参考文献
Cu、Zn	水 CRM	Natrolite 沸石	5-Br-PADAP pH 7.5~9.5	HNO_3	—	0.03、0.006	[34]
Pb	虾、牡蛎、蟹、鱼、贻贝	聚氨酯泡沫	BTAC pH	HCl	26	1.0	[35]
Pb	酒	聚氨酯泡沫	BTAC,pH7	HCl	26	1.0	[36]
Cu、Pb、Cr^{6+}	天然水域生物材料	聚氨酯泡沫	Me-PDC	甲基异丁酮	170 131 28	0.2 1.8 2.0	[37]
Ga	铝合金、天然水、尿液	聚氨酯泡沫	氯化镓络合物	甲基异丁酮	40	6	[38]
Zn	自然水	聚氨酯泡沫	Me-BTABr,pH6.5~9.2	HCl	23	0.37	[39]
Zn	饮用水,来自炼油厂的盐水废物	聚氨酯泡沫	聚芳基酸酯 pH8.3	HCl	91.23	0.28	[40]
Pb	饮用水,来自炼油厂的盐水废物	聚氨酯泡沫	聚芳基酸酯 pH8.3	HCl	51	0.4	[41]
Cd	饮用水	聚氨酯泡沫	聚芳基酸酯 pH8.2	HCl	158	0.02	[42]
Cr^{6+}	自然水、沉积物	聚四氟乙烯车床切屑	Cr-PDC, pH0.8~1.4	甲基异丁酮	80	0.8	[43]
Pb	水、沉积物、鱼类组织	聚四氟乙烯车床切屑	Pb-PDC, pH1.4~3.2	甲基异丁酮	330	0.8	[44]
Cu、Cd	CRM 和水	改性硅胶	氧化铌(V)	HNO_3	34.2 33	0.4 0.1	[45]
Zn	尿液、血浆、红细胞	改性硅胶	氧化铌(V)	HCl	77	0.77	[46]
Cr^{3+} Cr^{6+}	水	改性硅胶	磷化锆(IV) 氧化锆(IV)	HNO_3 THAM	20.8 24.9	1.9 2.3	[47]
Cu	水溶液	硅胶	3-(1-咪唑基)丙基	HNO_3	19.5~25.8	0.4	[48]
Cu、Pb	水	聚氯三氟乙烯粒子	Me-DDPA	甲基异丁酮	250	0.07、2.7	[49]
Cr^{6+} Pb	水	聚氯三氟乙烯粒子	Me-PDC, pH1~1.6 Cr^{6+}, pH1.5~3.2, Pb	甲基异丁酮	94 220	0.4 1.2	[50]
Pb	地表水、地下水、污水、土壤	梅里菲尔德氯甲基化树脂粒子	Me-DDTC pH8.0~9.0	酸化甲醇	48	1.3	[51]

续表

元素	样品类型	吸附材料	络合剂	洗脱液	富集因子	检测限/($\mu g/L$)	参考文献
Pd	合成地质、颗粒型废弃的汽车催化剂	聚胺 Metalfix-Chelamine 树脂	Pd^{2+}	硫脲盐酸盐	20	9	[52]
Mn	水、CRM	Amberlite IRA-904	5,10,15,20-四（4-羧基苯）卟啉	HNO_3	30	12	[53]
Cd、Mn、Ni、Zn、Fe	水溶液	装有焦儿茶酚紫的 Amberlite XAD-4	苄基二甲基十四烷基氯化铵 pH7.4~8.2	HNO_3	85	—	[54]
Pb	生物	改性 Amberlite XAD-2	BTAC, pH6.5~8.5	HCl	27	3.7	[55]
Ni	米粉、菠菜、果园和桃叶	负载 Amberlite XAD-2	BTAC pH	HCl	30	1.1	[56]
Cu	米粉、菠菜和苹果叶	负载 Amberlite XAD-2	2-（2-噻唑基偶氮）-5-二甲氨基苯酚	HCl	62	0.23	[57]
Cu	米粉和淀粉	功能化 AmberliteXAD-2	3,4-二羟基苯甲酸	HCl	33	0.27	[58]
Cd、Co、Cu、Ni	米粉、菠菜、红茶	用邻苯二酚锚定的 Amberlite XAD-2	—	HNO_3	36~69	0.31~1.64	[59]
Cr	注射液	Amberlite-XAD-16	TAR	乙醇	50	0.02	[60]
Cd	生物样品	Amberlite XAD-4	Me-DDTP	HCl	20	5~1	[61]
Cd	海水	Chelite P	氨基甲基膦基	HCl	1556	2.7	[62]
Cr^{6+}、Co、Ni、Cu	水溶液	KR	Me-PDC	乙醇	44、78、65、75	0.40、0.33、0.31、0.26	[63]
Cd	水溶液	KR	Me-PAN	HCl	18	0.1	[64]
Pb	水溶液	KR	Me-PAN	HCl	26.5	0.43	[65]
Cu	尿液、血清	螯合胺	Cu^{2+}	HNO_3	—	35、0.67	[66]
Cu	水溶液	苯乙烯-二乙烯基苯树脂	(S)-2-[羟基-双-(4-乙烯基-苯基)-甲基]-吡咯烷-1-羧酸乙酯 pH9	HCl	43	1.1	[67]
Co、Cu、Ni	食物（红茶、米粉）	AmberliteX AD-2	Me-BTAC	HCl	19、12、12	—	[68]
Cd、Cu	饮用水、湖水和海水	壳聚糖微球	磺酸 8-羟基喹啉 pH7 (Cd), pH10 (Cu)	HNO_3	19.1、13.9	0.2、0.3	[69,70]

续表

元素	样品类型	吸附材料	络合剂	洗脱液	富集因子	检测限/(μg/L)	参考文献
Pd	街道/风扇叶片灰尘和岩石	膨胀石墨	Pd-DADC	甲醇	—	1.0	[71]
Pb	水性标准品	多壁碳纳米管	用 HNO_3 氧化的 Pb(Ⅱ) 在 pH 为 4.7 时被吸收	HNO_3	44.2	2.6	[72]

注：THAM—三（羟甲基）胺；Salen I—NX-双（水杨基）乙二胺；5-Br-PADAP—2-（5-Br-2-吡啶偶氮）-5-二乙氨基苯酚；BTAC—2-（2-苯并噻唑基偶氮）-对甲酚；TAR—4-（2-thiazolylazo）-间苯二酚；PAR—4-（2-吡啶偶氮）-间苯二酚；DADC—二乙基二硫代氨基甲酸铵。

6.2.1.3 用于火焰原子吸收光谱检测的其他预浓缩系统

其他预浓缩方法也被用于在用 FAAS 测定前对不同真实样品中的金属离子进行预浓缩。其中，值得一提的是沉淀/溶解[74]和溶剂萃取[75,76]。例如，作为四碘酸盐和奎宁的离子对，通过沉淀将 Cd 预浓缩 32 倍，然后用乙醇溶解沉淀物[74]。该方法被用于测定贻贝样品中含量为 μg/g 水平的 Cd。另外，通过顺序注射（SI）-FAAS，以 18/h 的采样率，顺序测定佐餐葡萄酒中的 Fe^{3+} 和总 Fe 含量[75]。用甲基异丁酮提取 Fe^{3+} 和硫氰酸盐之间形成的复合物，而总的 Fe 含量是在没有溶剂提取的情况下测定的。新设计的重力相分离器用于在线提取 Cd-PDC 络合物到甲基异丁酮中，采样频率为 33/h，富集因子为 155，检测限为 20ng/L[76]，在线连续超声辅助提取被用于火焰原子吸收光谱测定海产品样品中的 Cu[77]和 Mn[78]，检测限分别为 0.06μg/g 和 0.4μg/g。

6.2.1.4 间接测定

根据某些阴离子物种如氰化物、碘化物、亚硝酸盐等的特定化学特性，也可以用火焰原子吸收光谱间接测定它们。

一个简单的 FI-FAAS 系统被用于间接测定工业废水中的氰化物，方法是将样品溶液通过一个装有悬浮在硅胶珠上的碳酸镉的柱子[79]。作为洗脱剂的氢氧化钠溶液（pH 为 10）将分析物以氰化镉络合物的形式带到火焰中。吸光度与样品中的氰化物浓度成正比。该系统提供的检测限为 0.2mg/L，相对标准偏差为 1.22%，采样率为 72/L。食品和废水中的亚硝酸盐在填充有二氧化铅的微柱中被在线氧化后，也可以用 FAAS 间接测定[80]。样品的流动将 PbO_2 固相试剂还原成 Pb^{2+}，用火焰原子吸收光谱测量，注入 0.4mL 样品的检测限为 0.11mg/L，采样率约为 80/h。

除了使用固相萃取外，上述所有案例中的检测限都在 mg/L 范围内，这样的范围太大，无法应用于环境或生物样品，报告的采样率是 >70/h。

6.3 样品稀释

到目前为止，与流动进样分流板结合的在线稀释的原子光谱技术是那些连续检测系统，如 FAAS[1,2,81,82]和 ICP-MS[83]以及 ET AAS[84]。一般来说，稀释的目的是提高样品产量和精度，扩大 FI-FAAS 系统的分析动态范围，使分析物浓度在仪器工作范围内，并减少样品预处理步骤，从而将污染的风险降到最低。

在这些系统中，在线稀释是在基于不同原理的歧管中实现的，如微体积的区域渗透和分散、级联方法的实施或使用不同的设备，如稀释室、再循环回路等。精度是稀释系统的一个重要标准，它在很大程度上取决于推进装置的性能和传输管道的几何形状。当使用活塞泵而不是蠕动泵时（蠕动泵会出现损坏的脉动），当导管打结或盘绕而不是直的时候，可以获得更好的精度，从而有利于样品区与载体的径向混合。通过选择适当的泵和采样时间，可以获得宽泛的稀释度（2～130000）。根据所构建的歧管的目的，采样频率也显示出宽泛的变化（60～200/h），在优化的条件下，重复性保持<3%，一些系统与常规应用高度兼容。

6.4 电热原子吸收光谱法

ET AAS 是一种强大的分析技术，在大多数实验室中都可以使用，并且具有足够的灵敏度，可以常规地应用于测定 $\mu g/L$ 或 $\mu g/g$ 水平的大多数微量元素。尽管 ET AAS 仪器取得了进展，但由于严重的基质干扰，复杂的样品不能直接处理，尽管开发了高效的背景校正装置，但仍未将基质干扰降到最低。

与 ET AAS 耦合的在线样品处理系统的实施受到了限制，因为像 ET AAS 这样不连续的技术很难耦合到任何流动进样系统。复杂接口的发展使在线 ET AAS 成为一种快速发展的技术，这值得出版一些综述，提供各种在线 ET AAS 系统的主要进展和最新趋势[5, 7, 9-11, 13, 85]。迄今为止开发的样品处理系统有：沉淀和共沉淀、柱上吸附、溶剂萃取、微透析、电化学反应、化学气相色谱、乳化和稀释。

6.4.1 用于 ET AAS 检测的分析物预浓缩

6.4.1.1 沉淀或共沉淀/溶解反应

沉淀或共沉淀过程的目的是预浓缩分析物或将其从烦琐的基质中分离出来。下面的溶解过程用于溶解含有分析物或其他基质成分的沉淀物，以便将分析物引入雾化器或将选定的成分转移到废物中。使用在线沉淀和共沉淀反应的优势是 2001 年之前发表的一系列文章的主题，Vereda[10]、Burguera 和 Burguera 等[11]的综述中涉及了这些文章。从那时起，这类系统的发展就减少了，可能是因为它们不像基于固相萃取的系统那样具有多功能性。

Nakajima 等[86]描述了海水中 Pb 的测定，首先通过与氢氧化铁在线共沉淀，然后将沉淀物溶解在 HNO_3 中，并通过一种铅选择性树脂（Pb-Spec）固相萃取将分析物与 Fe 分离，吸附的 Pb 用乙二胺四乙酸溶液洗脱。将对应于最高分析物浓度区的 $30\mu L$ 洗脱液部分注入石墨炉。

Burguera 等[87]描述了一种沉淀/溶解系统，用于在完整的在线流动进样微波辅助矿化后测定血清和全血中的钼。在样品在线暴露于微波辐射后，钼用亚铁氰化钾沉淀。将亚铁氰化钼沉淀物用稀释的氢氧化钠溶解，然后在取样臂组件的毛细管中收集子样品。最后，将该溶液的等分试样通过正空气置换引入雾化器。该方法适用于范围在 0.2~20.0μg/L 的钼，检测限为 0.1μg/L。该系统的优点是完全由软件控制，无需修改仪器软件。

化学改性的 C_{18} 粒子用于在阀上实验室（LOV）系统中收集氢氧化镉沉淀物，并结合 ETAAS[88]。在用 HNO_3 洗脱沉淀物后，得到的富集因子为 28、检测限为 1.7ng/L，采样频率为 13/h。通过将样品量从 600μL 增加到 1200μL，富集因子可以进一步提高到 44。

上述论文最突出的优点是对共存的干扰有很高的容忍度，这允许在相当复杂的基质中测定分析物，如海水[86]、血清和全血[87]、河流沉积物、莴苣和冷冻牛血[88]。然而，在流动进样系统中对相对大量的沉淀物进行在线操作，给生产稳健的系统带来一些困难。这一点可以通过使用阀上实验室系统来规避[88]，该系统在定义明确的管道中提供可控的收集能力，并提供更多的流程。

6.4.1.2 柱吸附

最常用的 FI-ET AAS 系统涉及在流动进样歧管中加入填料微柱，这些微柱要么取代了 ET AAS 自动进样器臂的尖端，尽管就获得的分析性能而言，两种设计都没有明显的优劣之分。然后，通过空气置换或使用选择阀将所需体积引入雾化器。

各种各样的吸附剂材料都被用来填充用于预浓缩目的的微型色谱柱，如疏水性的聚四氟乙烯车床切屑[89]、活性炭[90,91]、硅胶[92-97]、新开发的螯合树脂[98,99]，尽管最广泛使用的是 Amberlite[93] Dowex[100,101] 和 Muromac[102] 系列的市售树脂。在结式反应器上耦合流动进样吸附和 ET AAS 的结果表明，柱中柱系统可以替代不同的分析目的。结式反应器的体积比柱子大，这给雾化器的有限容量带来了严重的问题。解决这个问题的方法是利用装满聚四氟乙烯粒子的柱状反应器，或者利用空气分割的离散区洗脱。通过这种方式，被引入雾化器的洗脱体积被限制在气泡之间，其分散性被最小化，以获得更好的富集因子并提高结果的可重复性。表 6.3[89-106] 中总结了最近涉及这些方法的论文。

表 6.3　带有 ET AAS 检测器的填充柱中的在线吸附预浓缩

分析物	样品	吸附材料	洗脱液	富集因子	检测限/(ng/L)	参考文献
Co	水体、贻贝	聚四氟乙烯车屑	甲基异丁酮	87 倍	4	[89]
Co	饮用水	活性炭（pH 9.5）	HNO_3	190 倍	5	[90]
Cr^{3+}、Gr^{6+}	饮用水	活性炭（pH 5）	HNO_3	35	3	[91]
Cr^{3+}、Gr^{6+}	天然水域	Amberlite 或功能化硅胶	HNO_3	7.4、5.6	140、80	[92]
Cd	海水	功能化硅胶	HNO_3	2	60	[93]
Sb^{3+}	环境介质	固定化硅胶	HNO_3	—	300	[94]
Rh	催化剂	硅胶	HNO_3	—	300	[95]

续表

分析物	样品	吸附材料	洗脱液	富集因子	检测限/(ng/L)	参考文献
Pt	催化剂	硅胶	HNO_3	40~42	800	[96,97]
Pt	气溶胶	螯合剂	乙醇	42	100	[98]
Pb	海水	螯合树脂	乙醇	20.5	140	[99]
Se^{4+}、Se^{6+}	饮用水	Dowex	HCl	82	10	[100]
Rh	环境介质	负载 PSTH 的 Dowex	HNO_3	—	300	[101]
Bi、Cd、Pb	尿液	Muromak	HNO_3		13、2、4.5	[102]
Hg^{2+} 有机汞	鱼类组织	香烟滤嘴	乙醇	75	6.8 3.4ng/g	[103]
Pd	血液、灰尘	KR [ion pair: K^+-18-crown-6 and Pd$(SCN)_4^{2-}$]	甲醇	29	16	[104]
As^{3+}	海水	KR (As^{3+}-PDC)	乙醇	44	8	[105]
Co	标准化的海水	KR (氢醌类)	—	10~34	—	[106]

注：PSTH—1,5-双（2-吡啶基）-3-磺基苯基亚甲基硫代碳酰肼。

尽管用上述预浓缩程序评估的检测限很低（表 6.3），但当样品具有复杂的基质时，它们仍然是不够的。为此，Hansen 小组[107-114]使用了粒子填充的微柱和具有可再生反相表面的 SI-LOV 方案，这增强了 SI-ET AAS 分析更复杂样品的潜力。这些作者声称，在吸附步骤中，顺序注射的流动方向逆转增加了分析物-吸附材料的相互作用，提高了富集因子，因为样品塞子一个接一个地堆叠起来，以便进行物理操作。

可再生反相表面涉及使用：①含有垂悬的十八烷基分子的聚（苯乙烯-二乙烯基苯）粒子（Ci8-PS/DVB）[109,113]，浸渍有选择性的有机金属螯合剂；②含有二乙基二硫代磷酸铵（ADDP）的聚（四氟乙烯）（PTFE）-颗粒状 Algoflon 粒子[110]；③亲水螯合的 Sepharose[111,112]，在阀上实验室单元的微孔导管中操作这些粒子。粒子的尺寸均匀性和完美的球性提高了重现性、保留效率、增强因子和线性动态范围[109]。这些 SI-LOV 方案的潜力通过测定环境（自然和/或海水[109,112]和河流沉积物[113]）中的 Cd[108,110,111]、Cr^{3+}[112]、Cr^{6+}[112,113]、Pb^{2+}[111]、Ni^{2+}[111]和生物样品（尿液）[111]而得到证明，无需任何稀释步骤。一般来说，更宽的动态范围、高富集因子（20 以上）和大于等于 10 的样品通量都很容易获得。

6.4.1.3 溶剂萃取

在某些情况下，人工进行的液-液萃取过程意味着要进行大量连续的萃取。另一种方法是在两个液相之间进行在线传质，在使用适当的装置时，可以用最少的样品量和提取溶剂进行有效的提取。

钢中的砷是根据 As^{3+} 与碘化物离子在浓盐酸介质中反应生成三碘化砷，使用重力相分离器将其提取到苯中，然后反萃取到水中[115]。检测限为 $0.2\mu g$。同一研究小组[116]使用该萃取系统来测定钢中的砷或锡，但使用硫酸溶液进行反萃取。在这种情况下，砷和锡的检测限分别为 $0.2\mu g$ 和 $0.1\mu g$。

水溶液[117]、天然水和尿液[118]中的镉首先分别与吡咯烷二硫代氨基甲酸铵和偶氮二甲酰二哌啶络合，并将相应的螯合物提取到甲基异丁酮中，通过重力相分离器与水相分离。二烷基二硫代磷酸镉络合物被回抽到含有 Hg^{2+} 的稀硝酸中作为剥离剂，而二烷基二硫代磷酸镉络合物的一个等分点被直接注入石墨炉中。得到的富集因子为 21.4 和 24.6，检测限为 2.7ng/L 和 2.8ng/L，同时采样频率分别为 13/h 和 30/h。

Nan 和 Yan[119]开发了一种动态的二维纳米管（由十六烷基三甲基溴化铵作为阳离子表面活性剂）溶剂萃取结合 ET AAS 测定饮用水中的 Cr^{6+}，分析物与 APDC 在线络合，并被溶解在硅胶填充的微柱纳米管中。然后用乙腈洗脱柱子，用 ET AAS 测定。胶束的形成显示了可逆性，允许进行 22 次重复测定。富集因子为 32，采样量为 31/h，检测限为 3.0ng/L。

正确地评估上述液-液 FI-ET AAS 系统的分析潜力，相对于其他预浓缩方法，增强因子是相当适中的。操作者要有良好的经验来确保正确的操作和可靠的分析结果需要，特别是在传质步骤（可能影响灵敏度）和使用的相分离器种类（影响重现性）方面[115-118]。在常规分析中使用重力相分离器进行平稳操作是非常困难的[115-118]，因为其坚固性极低。为了尽量减少这些缺点，可以采用以下替代方案：①回抽到水溶液中[115-117]；②将有机复合物溶解在用硅胶包装的微柱中[119]。

6.4.2 ET AAS 检测前的分析物分离

在某些情况下，在通过 ET AAS 进行测定之前，分析物只需要从基质中分离（不一定是预浓缩的）。有几种新颖的方法，例如，使用微透析膜、电化学过程或化学蒸汽发生来评估分离情况。

基于流动透析的装置的突出特点是通过膜将分析物与基质成分进行高分辨率的分离。在流动进样中，为了降低试剂消耗和提高产量，需要小型化的系统。Tseng 等[120]开发了一种在线微透析采样和 FI-ET AAS，用于活体大鼠大脑中细胞外扩散性锰的体内监测。微透析样品通过植入的探针进行灌注，用在线注射阀的样品环收集，并由流动进样系统直接引入雾化器中。这个系统的检测限为 0.43ng Mn/L，能够在给麻醉大鼠注射氯化锰后监测锰。25min 的时间分辨率是可以实现的。同样的方法被应用于活体兔子在服用硫酸镁后血液中可扩散的 Mg 的体内监测，时间分辨率为 1.5min[121]，检测限为 0.53mg/L。

Promchan 和 Shiowatana[122]描述了一种通过模拟胃肠道消化体外测定铁生物利用度的类似方法，用于估计矿物质生物利用度的方法是基于批次的胃消化系统和动态连续流肠道消化，这是在透析袋中进行的，透析袋放置在含有流动的透析液流的腔室内。肠道消化阶段的矿物质浓度和透析矿物质被用来评估不同种类牛乳中铁的透析能力。最近，Tseng 等[123]使用在线微透析结合 SIMAAS 对盐水和细胞悬浮液中的多元素（Cu、

Mn、Ni 和 Se）进行了连续监测。由于使用超纯 NaCl 溶液作为灌注液，因此样品的基质中含有高浓度的盐。使用 Pd-Mg（NO_3）$_3$ 作为基体改性剂提高了热解温度，防止目标元素在 NaCl 之前蒸发。Miro 等[124]提出了一种新的概念，可以在对采样点干扰最小的情况下对土壤中的金属离子进行在线微量采样和连续监测。它涉及一个中空纤维微透析器，该微透析器作为微型传感装置被植入土壤中。本应用中微透析背后的想法是模拟被动采样器的功能，以预测痕量金属的实际流动性和生物利用度，而不是潜在的流动性和生物利用度。水性模型溶液具有较长的毛细管，在 2.0μL/min 的流动速率下，足以获得较高的铅透析回收率。在研究中发现，惰性土壤基质将会使金属的吸收量减少 30%。对微透析分析仪在土壤污染风险评估和金属分配研究方面的潜力进行了初步研究，干净的微量透析液被在线注入雾化器。植入式微透析装置的一个显著特点是，能够在模拟的自然场景或人为行为下跟踪金属释放的动力学。

在线电解技术可能是由于在线电解过程与 ET AAS 同步的复杂性，因而几乎没有被使用。据我们所知，在过去五年中只有两篇论文描述了这种方法。Knápek 等[125]通过 ET AAS 测定海水中的镉，使用微电池中的电荷分离。该方法包括四个步骤：①在微电池中进行电解；②在流动模式下，在控制电流的条件下将分析物电沉积在石墨电极上；③将分析物溶解到洗脱液（稀释的 HNO_3 溶液）中；④将整个体积（40L）直接注入雾化器。检测限为 25ng/L。用 ET AAS 作为检测器，采用在线双向电堆积的方法对天然水体中的砷进行标本分析和预浓缩[126]，当施加 750V 的电压并持续 20min 时，检测限为 0.35μg/L，预浓缩系数为 4.8。用 $KMnO_4$ 将所有价态的砷氧化成 As^{5+}，以测定砷总量。

与 ET AAS 连接的在线 CVG 方法可以提供低至 ng/L 水平的检测限、基质改性、灵敏度提高、选择性测定和干扰物的抑制[13]。CVG 装置和雾化器之间的传输线应尽可能短，以尽量减少传输过程中分析物的损失。

Ma 等[127]将矿石样品消化液中产生的铜挥发物捕获在石墨炉的内壁上。在捕集时间为 40s，采样频率为 50/h 的情况下，得到了 0.8ng/mL 的检测限。

Chuachuad 和 Tyson[128]首次使用四氢硼酸盐形式的阴离子交换器对天然水、葡萄酒、人类唾液、橙汁和人类头发中的镉进行了 FI-CVG 测定。所有样品的检测限都在 40~50ng/L 的范围内。这个系统后来使用雾化器内捕集技术进行了改进[129]。在这种情况下，AmberliteIRA-400 柱首先装入样品的硼氢化物溶液，然后将稀释的 HNO_3 流经该柱。洗脱后，分析物被截留在涂有锆-铱的石墨管内。在 L-半胱氨酸、硫脲和钴的作用下，信号被增强。天然水域的检测限被极大地改善，达到 3~15ng/L，其他样品的检测限改善到 90~180ng/L。

最近，Petrov 等[130]描述了一个通过 FI-HG 测定尿液中有毒的砷种类的程序，在一个横向加热的石墨雾化器的综合平台上收集生成的无机和甲基氢化物。在这种情况下，使用了一个大的（20cm^3）定制版 GLS，它对水浸和气溶胶的形成表现出良好的耐受性。石英吸头被调整为在用锆和铱处理的平台上方约 1.3mm 处提供氢化物，这既是有效的氢化物封存介质，也是永久的化学改性剂，允许多达 700~800 次的燃烧。在 10 倍和 25 倍的稀释下，特征质量、特征浓度和检测限分别为 39pg、0.078μg/L 和 0.038μg/L，样品通过率为 25/h。

从上面可以得出结论，原位捕集方法适用于所有情况，这可能是因为在这种方式下，雾化与雾化器表面的微小变化和用于传输挥发性物质的气态介质的组成无关。虽然 L-半胱氨酸被用来提高灵敏度[128-130]，但阴离子交换器允许在相当复杂的基质中测定分析物[129]。

6.4.3 其他

乳液是两种或更多液体的混合物，其中一种以微观或超微观大小的液滴存在[131, 132]。乳液的使用让各种在线 FI-ET AAS 可以自动测定高黏性样品中的金属[132]。

Burguera 等[131]将顺序注射技术用于在线制备"水包油"微乳液，以测定新旧润滑油中的铝含量。乙氧基化的壬基酚表面活性剂、样品和辅助表面活性剂（仲丁醇）溶液的混合物被依次吸入一个贮存盘管中。超声处理和重复改变流动方向提高了乳剂的稳定性。校准图从 7.7μg/L 到 120μg/L 的铝呈线性，检测限为 2.3μg/L。表面活性剂/油/水方法学的最佳相行为配方大大简化了这些高黏度样品的制备时间。此外，由于微乳剂降低了黏度对测定精度的影响，因此，以一种非常好的方式，解决了那些与标准品黏度差异很大的样品溶液的分析问题。

最近有人描述了一种 SI-ET AAS 方法，通过 ET AAS 对样品进行在线稀释后，在实验过程中不改变样品传输管道，综合测定人体唾液中的锌[84]。稀释程序和将溶液注入雾化器是由计算机控制的，并与温度程序的运行同步。通过对真实唾液样品的重复分析测试，其精度优于 3%，检测限为 0.35μg/L。这种方法的潜力刚刚出现，因为用于校准目的的储备溶液的受控稀释，可以导致标准添加方法和其他标准化程序的实施，从而显著减少燃烧时间和燃烧次数。

6.5 原子荧光光谱法

原子荧光光谱法是一种常规的检测系统，用于环境和毒理学重要样品中氢化物生成和循环伏安法的测定。然而，与 ET AAS 和 ICP-OES 相比，原子荧光光谱法具有较低的运行成本和类似的检测限。它在临床实践中的巨大应用潜力，需要在医生和分析化学家的共同努力下实现。

6.5.1 用于原子荧光光谱法检测的分析物预浓缩

原子荧光光谱法（AFS）已被广泛地应用于从基质中分离出氢化物形成的元素，并通过在填料柱或结式反应器上的保留、共沉淀、选择性萃取或络合物的形成进行预浓缩。这些系统中的大多数都允许进行形态研究和满足较低的检测限，尽管除了少数例外，其仅限应用于环境样品。表 6.4 所示为一些选定系统的分析特征[133-139]。

可以使用连续萃取结合 FI-HG-AFS[140]确定土壤样品中的无机砷，萃取剂的顺序是：水、磷酸一钾、HCl 和 KOH 以及在柠檬酸介质中分析的 As^{3+}，同时使用 L-半胱氨酸在线还原 As^{5+} 以确定砷总量。As^{3+} 和 As^{5+} 的检测限分别为 0.11μg/L 和 0.07μg/L，10μg/L

水平的 RSD 为 1.43%（$n=11$）。遵循相同的想法，Dong 和 Yan[141] 开发了一种与 HG-AFS 耦合的在线顺序提取方法，用于快速和自动分离土壤中的砷。填充到微柱中的土壤样品通过连续泵送单独的萃取剂（水、KOH 和 HCl 溶液），通过柱子进行动态萃取。提取的砷与过硫酸钾溶液合并，用于将所有价态的砷在线氧化为 As^{5+}。使用在线标准添加校准策略对提取的砷进行定量。这个方法已成功应用于认证土壤标准物质中砷的分馏。

HG-AFS 技术也已扩展到镉[139] 和金[142] 等元素，这些元素不能形成氢化物。使用可编程间歇流动反应器，在 HCl 介质中用硼氢化钾还原，在室温下产生金的挥发性物质，然后通过地质认证参考材料中的 AFS 测定[142]，微量二乙基二硫代氨基甲酸钠的存在大大增强挥发性物质的产生效率，而载气的流速对 Au 的信号强度有显著影响。此外，高表面积/体积比和快速 GLS 对于挥发性金的产生至关重要。在优化条件下，可以达到 0.23μg/L 的检测限，20μg/L 的金的测量精度为 1.7%。在 Co^{2+} 的存在下，镉挥发性物质的生成效率可以大大提高。

表 6.4　　　　　　　　　　　用于 AFS 检测的在线预浓缩系统

元素	样品	柱吸附	洗脱液	检测限/(ng/L)	富集因子	取样/h	相对标准偏差/%	参考文献
Se^{4+}	河水、湖水、海水	聚四氟乙烯纤维上的 Se^{4+}-PDC	$NaBH_4$	4	67%	26	1.5	[133]
	生活饮用水	Se^{4+} 与 La(OH)$_2$ 共沉淀在聚四氟乙烯粒上	HCl	5	11%	38	1.2	[134]
	生活饮用水	Se^{4+} 与 La(OH)$_2$ 在 KR 上共沉淀	HCl	14	18	24	2.5	[135]
Ge	水溶液	Ge^{4+} 与 Ni(OH)$_2$ 在 KR 上共沉淀	磷酸	110	11	24	5.6	[136]
Hg^{2+}、MeHg	生活饮用水	样品在紫外辐射下分解后在硅胶上的吸附-2-巯基苯并咪唑	KCN、HCl	0.07、0.05	13、24	—	8.8、10	[137]
	生活饮用水	KR 上的 Hg^{2+}-DDP 和甲基汞-双硫腙	HCl	3.6、2.0	13、24	30、20	2.2、2.8	[138]
Cd	茶	在 Cyanex 923 上的吸附	H_2SO_4	10.8	—	40	0.97	[139]

与 FAAS 的情况一样，上述系统最多只能区分同一元素的两种价态 [如 As^{3+} 和 As^{5+}]。对于含有成氢元素的有机化合物的标本，在 AFS 检测之前，强大的分离技术被连接到 HG 歧管上。

6.5.2　用 AFS 检测进行形态研究的连带技术

有人将任何 HG-AFS 系统与高效液相色谱或反相液相色谱（RPLC）等分离技术结

合起来，将标本研究扩展到复杂的含有机物分子，如硒胺酸、单甲基胂酸（MMAA）、二甲基胂酸（DMAA）、亚砷酸（AsB）、砷胆碱（AsC）、甲基汞（MeHg）、乙基汞（EthHg）或苯基汞（PhHg）化合物。

使用含有甲醇和双十烷基二甲基溴化铵的醋酸铵缓冲液（pH 为 4），通过在线偶联 HPLC 和 HG-AFS 分离选定的硒氨基酸，如硒代胱氨酸（SeCys）、硒代甲硫氨酸（SeMet）、硒代乙硫氨酸（SeEt）以及 Se（IV）[143]。SeCys、SeMet、SeEt 和 Se^{4+} 的检测限，分别为 18pg/L、70pg/L、96pg/L 和 16pg/L。还在基于糖肽替考拉宁的手性固定相（ChirobioticT）上使用 HPLC 与 AFS 和 ICP-MS 联用，在手性分离后测定母乳和配方乳中的 SeMet[144]。这样的仪器组合需要在线的柱后分析处理，如微波辅助消化、通过离心和超滤消除脂肪和蛋白质，来进行严格地样品清理。未衍生化的 DL-SeMet 对映异构体，在无缓冲水流动相下 10min 内可以完全分离。有良好的选择性和灵敏度 L-SeMet 和 D-SeMet 硒的检测限分别 3.1ng/mL 和 3.5ng/mL，这种方法具有稳健性和简单性，并且 AFS 检测器成本很低，因此，适用于婴儿牛乳的常规分析。

有人开发了一种 HPLC-HG-AFS 装置，用于对奶牛饲草补充不同种类的硒（有机硒为硒化酵母，无机硒为亚硒酸钠）后，测定牛乳中的硒形态[145]。在流动相中使用四乙基氯化铵（作为带正电荷的离子对试剂）在 μBondapack C_{18} 柱中进行分离。洗脱级分中不同硒化合物的预还原，通过紫外线辐射和加热块进行改善。补充有机饲料后获得的牛乳样品中含有 SeCys、SeMet 和 Se^{4+}，而补充无机饲料后获得的牛乳样品中仅存在 SeCys 和 Se^{4+}。

离子交换液相色谱通过氢化物生成与原子荧光光谱法在线耦合，用于测定土壤、沉积物和污水污泥的正磷酸提取物中的无机 [As^{3+} 和 As^{5+}] 和有机（MMAA 和 DMAA）砷化合物[146]。该方法显示了在环境固体中进行常规砷标本分析的良好潜力，所有基质中所有物种的检测限都非常低，从 0.02mg/kg 到 0.04mg/kg。对萃取程序的效率和萃取物的稳定性进行了详细的研究。同样，12 种无机和有机砷化合物在 Dionex AS7 柱中被分离，使用醋酸缓冲液和硝酸作为流动相[147]。分离后，这些化合物通过紫外线照射被在线氧化，通过氢化物生成挥发，并由氩气流带到检测器中。检测限在 4~22pg 范围内变化，该方法被应用于海洋样品。在所有样品中都检测到 AsB 是主要的物种。在另一期刊上发表了一种在 20min 内分离 AsB、AsC、三甲基氧化砷和四甲基砷离子的色谱条件[148]。

四种含汞物质（Hg^{2+}、MeHg、EthHg 和 PhHg 氯化物）在 RP C_{18} 色谱柱中使用含有四丁基溴化铵和氯化钠的甲醇流动相在 13min 内进行基线分离[149]。为了提高有机汞化合物向无机汞的转化效率，在酸化过硫酸钾的存在下应用了柱后微波消解。为避免水蒸气和甲醇进入原子荧光光谱法检测器，使用冰水混合浴来冷却消解物。上述方法准确地应用于海鲜样品中的汞形态分析。Hg^{2+} 和 MeHg 的检测限得到改善，在氢化物生成或乙基化低温聚焦气相色谱分离后由原子荧光光谱法同时测定[150]。同样，基于与 CV-AF 在线耦合的 RPC，可以分析某些金属硫蛋白[151]、植物螯合素（PC）[152] 和挥发性硫醇，包括硫化氢[153]。首先，硫醇蛋白在含有尿素和对羟基苯甲酸酯（p-HHgB）的磷酸盐缓冲液中进行柱前变性和衍生化，然后在 C_4 Vydac 反相柱上分离相应的复合物。在 HCl 培养基中由 KBr/KBrO$_3$ 原位生成的衍生化的变性蛋白质与溴的柱后在线反应，允许游离和蛋白质结合的 p-HHgB 快速转化为无机 Hg，然后由 AFS 选择性检测降低到 Hg^0。使用这种方法，可以确定能够与 p-HHgB 复合的兔肝脏中含有金属硫蛋白的—SH 基团的数量。在 phitochelatins 的情况下，合成溶液（apo-PCs 和 Cd^{2+}-complexed PCs）被分析：①通过尺寸排阻色谱（SEC）；②通过用 p-衍生化 SEC 部分中的 PC-SH 基团 HHgB，通过 RPC 分离复合物，然后通过 CV-AFS 间接测定。从生长在含镉营养液中的三角褐指藻的细胞培养物提取物中测定 PC。来自受污染空气的挥发性硫化合物（H$_2$S、甲硫醇、乙硫醇和丙硫醇）被捕获并预先浓缩在含有 p-HHgB 的碱性溶液中，形成强的 Hg-SH 共价键，在室温下至少稳定 12h，如果在 -20℃ 冷冻保存，则稳定 3 个月。因此，通过 AFS 间接测定了沼气和原油分馏厂空气中的不同含硫化合物。用 CV-AFS 检测得到的 CH$_3$SH、C$_2$H$_5$SH 和 C$_3$H$_7$SH 的检测限分别为 9.7μg/L、13.7μg/L 和 17.7μg/L。完全相同的柱前和柱后衍生程序被用于 CV-AFS 对半胱氨酸、青霉胺和谷胱甘肽复合物形式的无机汞和有机汞（MeHg、EthHg 和 PhHg）进行标本化和直接测定[154]，分离的汞-硫醇复合物也在室温下，在 30cm 的针织线圈中，在不到 2.5s 内被溴氧化为无机汞。在优化的络合和洗脱条件下，Hg^{2+}、MeHg、EthHg 和 PhHg 的检测限分别为 16pg、18pg、18pg 和 20pg（以汞的形式）。

在用 AFS 测定当地收集的水样中的 As^{3+} 和 As^{5+} 之前，一种新的方法是将基于微流控芯片的 CE 与挥发性形态生成系统连接起来[155]，使用 H$_3$BO$_3$ 和 CTAB（pH8.9）的混合物作为缓冲剂，在 54s 内实现了不同价态砷的分离。在 3.0mg/L 的水平上，As^{3+} 和 As^{5+} 的检测限分别为 76μg/L 和 112μg/L，精度（RSD，$n=5$）变化范围为 1.4~207%（As 总量）。同样的系统被用来测定无机汞（Hg^{2+}）和有机汞（MeHg$^+$）[156]，这些汞在 64s 内被分离为它们的 L-半胱氨酸络合物，对于 2mg/L 的 Hg^{2+} 和 4mg/L 的 MeHg，精确度（RSD，$n=5$）分别为 0.7% 和 0.9%，Hg^{2+} 和 MeHg 的检测限分别为 53μg/L 和 161μg/L。相同的作者最近在不同的杂志社发表了同样的工作，在系统设计上有微小的变化[157]。奇怪的是，除了检测限严重降低外，其余特征数是相同的。

6.6 结论和展望

涉及样品预处理系统与原子光谱技术在线耦合的全自动歧管的进一步开发应更好地发挥其样品消耗少以及单位时间内样品分析频率较高的优势。微量样品引入系统、预浓缩和化学蒸气发生程序将用于提高对高溶解固体、有机溶剂和高黏性样品的耐受性。在这种情况下，未来的研究应该涉及使用超声波和微波辐射来去除一些基质成分，而用于提高灵敏度的可用预浓缩歧管需要进一步研究。

应进一步提高用于消除基质成分和预浓缩分析物的吸附柱的效率，也许最好是将所有关于"自制"微柱的现有信息进行汇编，并利用现有的经认证的参考材料，尝试在重现性和长期稳定性方面对特定用途的方法进行标准化。顺序注射和阀上实验室装置将允许化学反应和/或物理操作在不断小型化的装置中进行，这些装置最终可用于现场研究。

如果有足够的物质存在形态方面的信息，可以更好地理解与环境和生物系统相关的许多相关问题。鉴于生物系统的复杂性，与现有的环境系统，尤其是天然水体相比，物种形成的数据仍然很少。物种形成结果的验证需要有足够的标准参考材料。例如，与原子光谱技术结合的透析歧管允许对可扩散物质进行原位和体外监测，以便更多地了解它们在生物系统中的生物利用度。从神经化学应用和药代动力学研究改编的刺激反应方案可以扩展到其他高度特异性的分离技术，如电化学过程或毛细管电泳，它们很容易在线连接，前提是通过耦合能够获得足够灵敏的技术来规避灵敏度限制，如FT AAS和ICP-MS。这将对未来痕量水平未知元素物种的临床和环境监测计划产生重大影响。

最后，希望在下一阶段，在线原子光谱系统在临床和环境实验室中的使用频率将会增加，特别是在常规分析中。

参考文献

[1] Burguera, J. L. (ed.) (1989) *Flow Injection Atomic Spectroscopy*, Marcel Dekker, New York.

[2] Fang, Z. (1995) *Flow Injection Atomic Spectrometry*, John Wiley, New York.

[3] Sanz-Medel, A. (ed.) (1999) *Flow Analysis with Atomic Spectrometric Detectors*, Elsevier, Amsterdam.

[4] Ruzucka, J. and Hansen, E. H. (1988) *Flow Injection Analysis*, 2nd edn, Wiley, New York.

[5] Fang, Z. and Tao, G. (1996) New developments in flow injection separation and preconcentration techniques for electrothermal atomic absorption spectrometry. *Fresenius' Journal of Analytical Chemistry*, 355, 576-580.

[6] Burguera, J. L. and Burguera, M. (1997) Flow injection for automation in atomic spectrometry. *Journal of Analytical Atomic Spectrometry*, 12, 643-651.

[7] Burguera, J. L. and Burguera, M. (1998) On-line sample pre-treatment systems interfaced to electrothermal atomic absorption spectrometry. *Analyst*, 123, 561-569.

[8] Burguera, M. and Burguera, J. L. (1998) Microwave-assisted sample decomposition in flow analysis. *Analytica Chimica Acta*, 366, 63-80.

[9] Fang, Z. L. (1998) Trends and potentials in flow injection on-line separation and preconcentration techniques for electrothermal atomic absorption spectrometry. *Spectrochimica Acta Part B-Atomic Spectroscopy*, 53, 1371-1379.

[10] Vereda, A., García de Torres, A. and Cano Pavón, J. M. (2001) Flow injection on-line electrothermal atomic absorption spectrometry. *Talanta*, 55, 219-232.

[11] Burguera, J. L. and Burguera, M. (2001) Flow injection-electrothermal atomic absorption spectrometry configurations: recent developments and trends. *Spectrochimica Acta Part B-Atomic Spectroscopy*, 56, 1801-1829.

[12] Burguera, J. L. and Burguera, M. (2001) On-line flow injection-atomic spectroscopic configurations: road to practical environmental analysis. *Quím Analítica*, 20, 255-273.

[13] Burguera, J. L. and Burguera, M. (2001) Volatile species generation in flow injection for the on-line determination of species in environmental samples by ET AAS. *Journal of Flow Injection Analysis*, 18,

5-12.

[14] Evans, E. H., Day, J. A., Palmer, C., Price, W. J., Smith, C. M. M. and Tyson, J. F. (2006) Analytical spectrometry update. Advances in atomic emission, absorption and fluorescence spectrometry, and related techniques. *Journal of Analytical Atomic Spectrometry*, 21, 592-625.

[15] Ferrúa, N., Cerutti, S., Salonia, J. A., Olsina, R. A. and Martinez, L. D. (2007) On-line preconcentration and determination of mercury in biological and environmental samples by cold vapor-atomic absorption spectrometry. *Journal of Hazardous Materials*, 141, 693-699.

[16] Duan, T., Song, X., Xu, J., Guo, P., Chen, H. and Li, H. (2006) Determination of Hg (II) in waters by on-line preconcentration using Cyanex 923 as a sorbent-cold vapor atomic absorption spectrometry. *Spectrochimica Acta Part B-Atomic Spectroscopy*, 61, 1069-1073.

[17] Anthemidis, A. N. and Martavaltzoglou, E. K. (2006) Determination of arsenic (III) by flow injection solid-phase extraction coupled with on-line hydride generation atomic absorption spectrometry using PTFE turnings-packed micro column. *Analytica Chimica Acta*, 57, 413-418.

[18] Bartoleto, G. G. and Cadore, S. (2005) Determination of total inorganic arsenic in water using on-line preconcentration and hydride generation atomic absorption spectrometry. *Talanta*, 67, 169-174.

[19] Carrero, P., Gutierrez, L., Rondón, C., Burguera, J. L., Burguera, M. and Petit de Peña, Y. (2004) Flow injection determination of bismuth in urine by successive retention of Bi (III) and tetrahydroborate (III) on an anion-exchange resin and hydride generation atomic absorption spectrometry. *Talanta*, 64, 1309-1316.

[20] Menemenlioglu, I., Korkmaz, D. and Yavuz Ataman, O. (2007) Determination of antimony by using a quartz atom trap and electrochemical hydride generation atomic absorption spectrometry. *Spectrochimica Acta Part B-Atomic Spectroscopy*, 62, 40-47.

[21] Bolea, E., Arroyo, D., Laborda, F. and Castillo, J. R. (2006) Determination of antimony by electrochemical hydride generation atomic absorption spectrometry in samples with high iron content using chelating resins as on-line removal system. *Analytica Chimica Acta*, 569, 227-233.

[22] Korkmaz, D. K., Ertas, N. and Yavuz Ataman, O. (2002) A novel silica trap for lead determination by hydride generation atomic absorption spectrometry. *Spectrochimica Acta Part B - Atomic Spectroscopy*, 57, 571-580.

[23] Kratzer, J. and Dedina, J. (2006) In situ trapping of bismuthine in externally heated quartz tube atomizers for atomic absorption spectrometry. *Journal of Analytical Atomic Spectrometry*, 21, 208-210.

[24] Villa-Lojo, M. C., Alonso-Rodriguez, E., López-Mahía, P., Muniategui-Lorenzo, S. and Prada-Rodriguez, D. (2002) Coupled high performance liquid chromatography-microwave digestion hydride generation-atomic absorption spectrometry for inorganic and organic arsenic speciation in fish tissue. *Talanta*, 57, 741-750.

[25] Sur, R. and Dunemann, L. (2004) Method for the determination of five toxicologically relevant arsenic species in human urine by liquid chromatography-hydride generation atomic absorption spectrometry. *Journal of Chromatography*, 807, 169-176.

[26] Tseng, W. C., Cheng, G. W., Lee, C. F., Wu, H. L. and Huang, Y. L. (2005) On-line coupling of microdialysis sampling with high performance liquid chromatography and hydride generation atomic absorption spectrometry for continuous *in vivo* monitoring of arsenic species in the blood of living rabbits. *Analytica Chimica Acta*, 543, 38-45.

[27] Niedzielski, P. (2005) The new concept of hyphenated analytical systems: simultaneous

determination of inorganic arsenic (III), arsenic (V), selenium (IV) and selenium (VI) by high performance liquid chromatography – hydride generation – (*fast sequential*) atomic absorption spectrometry during single analysis. *Analytica Chimica Acta*, 551, 199-206.

[28] Tarley, C. R. T., Ferreira, S. L. C. and Arruda, M. A. Z. (2004) Use of modified rise husks as a natural solid adsorbent of trace metals: characterization and development of an on-line preconcentration system for cadmium and lead determination by FAAS. *Microchemical Journal*, 77, 163-175.

[29] Aji Shabani, A. M., Dadfarnia, S. and Dehghan, K. (2003) On-line preconcentration and determination of cobalt by chelating microcolumns and flow injection atomic spectrometry. *Talanta*, 59, 719-725.

[30] Dadfarnia, S., Haji Shabani, A. M. and Gohari, M. (2004) Trace enrichment and determination of silver by immobilized DDTC microcolumn and flow injection flame atomic absorption spectrometry. *Talanta*, 64, 682-687.

[31] Dadfarnia, S., Haji Shabani, A. M., Tamaddon, F. and Rezaei, M. (2005) Immobilized salen (*N*, *N'* – bis (salicylidene) ethylenediamine) as a complexing agent for on-line, sorbent extraction/preconcentration and flow injection – flame atomic absorption spectrometry. *Analytica Chimica Acta*, 539, 69-75.

[32] Marquéz, M. J., Morales-Rubio, A., Salvador, A. and de la Guardia, M. (2001) Chromium speciation using activated alumina microcolumns and, sequential injection analysis-flame atomic absorption spectrometry. *Talanta*, 53, 1229-1239.

[33] Afzali, D., Taher, M. A., Mostafavi, A. and Mobarakeh, S. Z. M. (2005) Thermal modified Kaolinite as useful material for separation and preconcentration of trace amounts of manganese ions. *Talanta*, 65, 476-480.

[34] Mostafavi, A., Afzali, D. and Taher, M. A. (2006) Atomic absorption spectrometric determination of traces amounts of copper and zinc after simultaneous solidphase extraction and preconcentration onto modified natrolite zeolite. *Analytical Sciences*, 22, 849-853.

[35] Lemos, V. A. and Ferreira, S. L. C. (2001) On-line preconcentration system for lead determination in seafood simples by flame atomic absorption spectrometry using polyurethane foam loaded with 2- (2-benzothiazolylazo) -2-p-cresol. *Analytica Chimica Acta*, 441, 281-289.

[36] Lemos, V. A., de la Guardia, M. and Ferreira, S. L. C. (2002) An on-line system for preconcentration and, determination of lead in wine samples by FAAS. *Talanta*, 58, 475-480.

[37] Anthemidis, A. N., Zachariadis, G. A. and Stratis, J. A. (2002) On-line preconcentration and determination of copper, lead and chromium (VI) using unloaded polyurethane foam packed column by flame atomic absorption spectrometry in natural waters and biological samples. *Talanta*, 58, 831-840.

[38] Anthemidis, A. N., Zachariadis, G. A. and Stratis, J. A. (2003) Gallium trace on-line preconcentration/separation and determination using polyurethane foam minicolumn and flame atomic absorption spectrometry. *Talanta*, 60, 929-936.

[39] Lemos, V. A., dos Santos, W. N. L., Santos, J. S. and de Carvalho, M. B. (2003) On-line preconcentration system using a minicolumn, of polyurethane foam loaded with Me – BTABr for zinc determination by flame atomic absorption spectrometry. *Analytica Chimica Acta*, 481, 283-290.

[40] dos Santos, W. N. L., Santos, C. M. C. and Ferreira, S. L. C. (2003) Application of three-variables Doehlert matrix for optimization o fan on-line preconcentration system for zinc determination in natural water samples by flame atomic absorption spectrometry. *Microchemical Journal*, 75, 211-221.

[41] dos Santos, W. N. L., dos Santos, C. M. M., Costa, J. L. O., Andrade, H. M. C. and Ferreira, S. L. C. (2004) Multivariate optimization and validation studies in online preconcentration system for lead determination in drinking water and saline waste from oil refinery. *Microchemical Journal*, 77, 123-129.

[42] dos Santos, W. N. L., Costa, J. L. O., Araujo, R. G. O., de Jesus, D. S. and Costa, A. C. S. (2006) An on-line preconcentration system for determination of cadmium in drinking water using FAAS. *Journal of Hazardous Materials*, 137, 1357-1361.

[43] Anthemidis, A. N., Zachariadis, G. A., Kougoulis, J. S. and Stratis, J. A. (2002) Flame atomic absorption spectrometric determination of chromium (VI) by on-line preconcentration system using a PTFE packed column. *Talanta*, 57, 15-22.

[44] Zachariadis, G. A., Anthemidis, A. N., Bettas, P. G. and Stratis, J. A. (2002) Determination of lead by on-line solidphase extraction using a PTFE microcolumn and flame atomic absorption spectrometry. *Talanta*, 57, 919-927.

[45] da Silva, E. L., Ganzarolli, E. M. and Carasek, E. (2004) Use of $Nb_2O_5-SiO_2$ in an automated on-line preconcentration system for determination of copper and cadmium by FAAS. *Talanta*, 62, 727-733.

[46] Dutra, R. L., Maltez, H. F. and Carasek, E. (2006) Development of an on-line preconcentration system for zinc determination in biological samples. *Talanta*, 69, 488-493.

[47] Maltez, H. F. and Carasek, E. (2005) Chromium speciation and preconcentration using Zr (IV) oxide and Zr (IV) phosphate chemically immobilized onto silica gel surface using a flow system and FAAS. *Talanta*, 65, 537-542.

[48] da Silva, E. L., Martins, A. O., Valentini, A., de Favere, V. T. and Carasek, E. (2004) Application of silica gel organofunctionalized with 3 (1-imidazolyl) propyl in an on-line preconcentration system for the determination of copper by FAAS. *Talanta*, 64, 181-189.

[49] Anthemidis, A. N. and Ioannou, K. I. G. (2006) Evaluation of polychloro trifluoroethylene as sorbent material for on-line solid-phase extraction systems: determination of copper and lead by flame atomic absorption spectrometry in water samples. *Analytica Chimica Acta*, 575, 126-132.

[50] Anthemidis, A. N. and Koussoroplis, S. J. V. (2007) Determination of chromium (VI) and lead in water samples by on-line sorption preconcentration coupled with flame atomic absorption spectrometry using a PCTFE-beads packed column. *Talanta*, 71, 1728-1733.

[51] Praveen, R. S., Naidu, G. R. K. and Prasada Rao, T. (2007) Dithiocarbamate functionalized or surface sorbed Merrifield resin beads as column materials for on-line flow injection-flame atomic absorption spectrometry determination of lead. *Analytica Chimica Acta*, 600, 205-213.

[52] Iglesias, M., Antico, E. and Salvado, V. (2003) On-line determination of trace level of palladium by flame atomic absorption spectrometry. *Talanta*, 59, 651-657.

[53] Knap, M., Kilian, K. and Pyrzynska, K. (2007) On-line enrichment system for manganese determination in water samples using FAAS. *Talanta*, 71, 406-410.

[54] Mostafavi, A., Afzali, D. and Taher, M. A. (2006) Simultaneous separation and preconcentration of traces amounts of Cd^{+2}, Mn^{+2}, $Ni^{+2} \cdot Zn^{+2}$ and Fe^{+2} onto modified amberlite XAD-4 resin loaded with, pyrocatechol violet. *Asian Journal of Chemistry*, 18, 2303-2309.

[55] Ferreira, S. L. C., Lemos, V. A., Santelli, R. E., Ganzarolli, E. and Curtius, A. J. (2001) An automated on-line flow system for the preconcentration and determination of lead by flame atomic absorption spectrometry. *Microchemical Journal*, 68, 41-46.

[56] Ferreira, S. L. C., dos Santos, W. N. L. and Lemos, V. A. (2001) On-line preconcentration

system for nickel determination, in food simples by flame atomic absorption spectrometry. *Analytica Chimica Acta*, 445, 145-151.

[57] Ferreira, S. L. C., Bezerra, M. A., dos Santos, W. N. L. and Neto, B. B. (2003) Application of Doehlert designs for optimization o fan on–line preconcentration system for copper determination by flame atomic absorption spectrometry. *Talanta*, 61, 295-303.

[58] Lemos, V. A., Baliza, P. X., Yamaki, R. T., Rocha, M. E. and Alves, A. P. O. (2003) Synthesis and application of a functionalized resin in on–line system for copper preconcentration and determination in food by flame atomic absorption spectrometry. *Talanta*, 61, 675-682.

[59] Lemos, V. A., da Silva, D. G., de Carvalho, A. l., de Andrade Santana, D., dos Santos Novaes, G. and dos Passos, A. S. (2006) Synthesis of amberlite XAD-2-PC resin for, preconcentration and determination of trace elements in food simples by flame atomic absorption spectrometry. *Microchemical Journal*, 84, 14-21.

[60] Wuilloud, G. M., Wuilloud, R. G., de Wuilloud, J. C. A., Olsina, R. A. and Martinez, L. D. (2003) On-line preconcentration and determination of chromium, in parenteral solutions by flow injection-flame atomic absorption spectrometry. *Journal of Pharmaceutical and Biomedical Analysis*, 31, 117-124.

[61] dos Santos, E. J., Herrmann, A. B., Ribeiro, A. S. and Curtius, A. J. (2005) Determination of Cd in biological simples by flame AAS following on–line preconcentration by complexation with O, O–diethyldithiophosphate and solidphase extraction with Amberlite, XAD-4. *Talanta*, 65, 593-597.

[62] Yebra-Biurrun, M. C., Moreno-Cid, A. and Puig, L. (2004) Minicolumn field preconcentration and flow injection flame atomic absorption spectrometric determination of cadmium in seawater. *Analytica Chimica Acta*, 524, 73-77.

[63] Li, Y, Jiang, Y. and Yan, X. P. (2004) Further study on a flow injection on–line multiplex sorption preconcentration coupled with flame atomic absorption spectrometry for trace element determination. *Talanta*, 64, 758-765.

[64] Souza, A. S. dos Santos, W. N. L. and Ferreira, S. L. C., (2005) Application of Box-Behnken design in the optimization, of an on–line preconcentration system using knotted reactor for cadmium determination by flame atomic absorption spectrometry. *Spectrochimica Acta Part B-Atomic Spectroscopy*, 60, 737-742.

[65] Souza, A. S., Brandao, G. C., dos Santos, W. N. L., Lemos, V. A., Ganzarolli, E. M., Bruns, R. E. and Ferreira, S. L. C. (2007) Automatic on-line preconcentration system using, knotted reactor for the FAAS determination of lead in drinking water. *Journal of Hazardous Materials*, 141, 540-545.

[66] Lopes, C. M. P. V., Almeida, A. A., Santos, J. L. M. and Lima, J. L. F. C. (2006) Automatic flow system for the sequential determination of copper in serum and urine by flame atomic absorption spectrometry. *Analytica Chimica Acta*, 555, 370-376.

[67] Cassella, R. J., Magalhaes, O. I. B., Couto, M. T., Lima, E. L. S., Neves, M. A. F. S. and Cautinho, F. M. B. (2005) Synthesis and application of a functionalized resin for flow injection/FAAS copper determination in waters. *Talanta*, 67, 121-128.

[68] Lemos, V. A., David, G. T. and Santos, L. N. (2006) Synthesis and application of XAD-2/Me-BTAP resin for n-line solid-phase extraction and determination of trace metals in biological simples by FAAS. *Journal of Brazilian Chemical Society*, 17, 697-704.

[69] Martins, A. O., da Silva, E. L., Carasek, E., Laranjeira, M. C. M. and de Favere, V. T. (2004) Sulphoxine immobilized onto chitosan microspheres by, spray drying: application for metal ions

preconcentration by flow injection analysis. *Talanta*, 63, 397-403.

[70] Martins, A. O., da Silva, E. L., Carasek, E., Goncalves, N. S., Laranjeira, M. C. M. and de Favere, V. T. (2004) Chelating resin from functionalization of chitosan with, complexing agent 8 - hydroxyquinoline: application for metal ions on-line preconcentration system. *Analytica Chimica Acta*, 521, 157-162.

[71] Praveen, R. S., Daniel, S., Prasada Rao, T., Sampath, S. and Sreenivasa Rao, K. (2006) Flor injection on-line solid phase extractive preconcentration of palladium (II) in dust and rock samples using exfoliated graphite packed microcolumns and determination by flame atomic absorption spectrometry. *Talanta*, 70, 437-443.

[72] Barbosa, A. F., Segatelli, M. G., Pereira, A. C., de Santana Santos, A., Kubota, L. T., Luccas, P. O. and Tarley, C. R. T. (2007) Solid-phase extraction system for Pb (II) ions enrichment, based on multiwall carbon nanotubes coupled on-line to flame atomic absorption spectrometry. *Talanta*, 71, 1512-1519.

[73] Cerutti, S., Martínez, L. D. and Wuilloud, R. G. (2005) Knotted reactors and their role in flow injection on-line preconcentration systems coupled to atomic spectrometry-based detector. *Applied Spectroscopy Reviews*, 40, 71-101.

[74] Yebra, M. C., Enriquez, M. F. and Cespon, R. M. (2000) Preconcentration and flame atomic absorption spectrometry determination of cadmium in mussels by an on-line continuous precipitationdissolution flow system. *Talanta*, 52, 631-636.

[75] de Campos Costa, R. C. and Araujo, A. N. (2001) Determination of Fe (III) and total Fe in wines by sequential injection analysis and flame atomic absorption spectrometry. *Analytica Chimica Acta*, 438, 227-233.

[76] Anthemidis, A. N., Zachariadis, G. A., Farastelis, C. G. and Stratis, J. A. (2004) On-line liquid-liquid extraction system using a new phase separator for flame atomic absorption spectrometric determination of ultra-trace cadmium in natural waters. *Talanta*, 62, 437-443.

[77] Moreno-Cid, A. and Yebra, M. C. (2002) Flow injection determination of copper in mussels by flame atomic absorption spectrometry after on-line continuous ultrasound-assisted extraction. *Spectrochimica Acta Part B-Atomic Spectroscopy*, 57, 967-974.

[78] Yebra, M. C. and Moreno-Cid, A. (2003) On-line determination of manganese in solid seafood samples by flame atomic absorption spectrometry. *Analytica Chimica Acta*, 477, 149-155.

[79] Naroozifar, M., Khorasani-Motlagh, M. and Hosseini, S. N. (2005) Flow injection analysis-flame atomic absorption spectrometry system for indirect determination of cyanide using cadmium carbonate as a new solid-phase reactor. *Analytica Chimica Acta*, 528, 269-273.

[80] Naroozifar, M., Khorasani-Motlagh, M., Taheri, A. and Homayoonfard, M. (2006) Indirect determination of nitrite by flame atomic absorption spectrometry using lead (IV) dioxide oxidant microcolumn. *Bulletin of the Korean Chemical Society*, 27, 875-880.

[81] Tao, G., Fang, Z.-L., Baasner, J. and Welz, B. (2003) Flow injection on-line dilution for flame atomic absorption spectrometry by micro-sample introduction and dispersion using syringe pumps. *Analytica Chimica Acta*, 481, 273-281.

[82] López García, I., Kozak, J. and Hernández Córdoba, M. (2007) Use of membrane micropumps for introducing the simple solution in flame atomic absorption spectrometry. *Talanta*, 71, 1369-1374.

[83] Wang, J., Hansen, E. H. and Gammelgaard, B. (2001) Flow injection on-line dilution for

multielement determination in human urine with detection by ICP-MS. *Talanta*, 55, 117-126.

[84] Burguera-Pascu, M., Rodriguez-Archilla, A., Burguera, J. L., Burguera, M., Rondón, C. and Carrero, P. (2007) Flow injection on-line dilution for zinc determination in human saliva with ETAAS detection. *Analytica Chimica Acta*, 600, 214-220.

[85] Wang, J. and Hansen, H. E. (2005) Trends and perspectives of flow injection/ sequential injection on-line sample-pretreatment schemes coupled to ETAAS. *Trends in Analytical Chemistry*, 24, 1-8.

[86] Nakajima, J., Hirano, Y. and Oguma, K. (2003) Determination of lead in seawater by flow injection on-line preconcentration-electrothermal atomic absorption spectrometry after coprecipitation with iron (Ⅲ) hydroxide. *Analytical Science*, 19, 585-588.

[87] Burguera, J. L., Burguera, M. and Rondón, C. (2002) An on-line flow injection microwave-assisted mineralization and a preconcentration/dissolution system for the determination of molybdenum in blood serum and whole blood by electrothermal atomic absorption spectrometry. *Talanta*, 58, 1167-1175.

[88] Wang, Y., Wang, J.-H. and Fang, Z.-L. (2005) Octadecyl immobilized surface for precipitate collection with a renewable microcolumn in a lab-on-valve coupled to an electrothermal atomic absorption spectrometer for ultratrace cadmium determination. *Analytical Chemistry*, 77, 5396-5401.

[89] Anthemidis, A. N., Zachariadis, G. A. and Stratis, J. A. (2002) Cobalt ultra-trace online preconcentration and determination using a PTFE turnings packed column and electrothermal atomic absorption spectrometry. Applications in natural waters and biological samples. *Journal of Analytical Atomic Spectrometry*, 17, 1330-1334.

[90] Cerutti, S., Moyano, S., Gásquez, J. A., Stripeikis, J., Olsina, R. A. and Martínez, L. D. (2003) On-line preconcentration of cobalt in drinking water using a minicolumn packed with activated carbon coupled to electrothermal atomic absorption spectrometric determination. *Spectrochimica Acta Part B-Atomic Spectroscopy*, 58, 2015-2021.

[91] Gil, R. A., Cerutti, S., Gásquez, J. A., Olsina, R. A. and Martinez, L. D. (2006) Preconcentration and speciation of chromium in drinking water samples by coupling of on-line sorption on activated carbon to ET AAS determination. *Talanta*, 68, 1065-1070.

[92] Cordero, M. T. S., Alonso, E. I. V., García de Torres, A. and Pavón, J. M. C. (2004) Development of a new system for the speciation of chromium in natural waters and human urine samples by combining ion exchange and ETA-AAS. *Journal of Analytical Atomic Spectrometry*, 19, 398-403.

[93] Vereda Alonso, E. I., Gil, L. P., Siles Cordero, M. T., García de Torres, A. and Cano Pavón, J. M. (2001) Automatic online column preconcentration system for determination of cadmium by electrothermal atomic absorption spectrometry. *Journal of Analytical Atomic Spectrometry*, 16, 293-295.

[94] Ojeda, C. B., Rojas, F. S., Pavón, J. M. C. and Martín, L. T. (2005) Use of 1, 5-bis (di-2-pyridyl) methylene thiocarbohydrazide immobilized on silica gel for automated preconcentration and selective determination of antimony (Ⅲ) by flow injection electrothermal atomic absorption spectrometry. *Analytical and Bioanalytical Chemistry*, 382, 513-518.

[95] Sánchez Rojas, F., Bosch Ojeda, C. and Cano Pavón, J. M. (2005) Application of flow injection on-line electrothermal atomic absorption spectrometry to the determination of rhodium. *Annali di Chimica*, 95, 437-445.

[96] García, M. M., Rojas, F. S., Ojeda, C. B., García de Torres, A. and Pavón, J. M. C. (2003) On-line ion-exchange preconcentration and determination of traces of platinum by electrothermal atomic absorption spectrometry. *Analytical and Bioanalytical Chemistry*, 375, 1229-1233.

[97] Bosch Ojeda, C., Sánchez Rojas, F., Cano Pavón, J. M. and García de Torres, A. (2003) Automated on-line separation preconcentration system for platinum determination by electrothermal atomic absorption spectrometry. *Analytica Chimica Acta*, 494, 97–103.

[98] Limbeck, A., Rudolph, E., Hann, S., Koellensperger, G., Stingeder, G. and Rendl, J. (2004) Flow injection on-line pre-concentration of platinum coupled with electrothermal atomic absorption spectrometry. *Journal of Analytical Atomic Spectrometry*, 19, 1474–1478.

[99] Vereda, A. E., Siles Cordero, M. T., Garcia de Torres, A. and Cano Pavon, J. M. (2006) Lead ultra-trace on-line preconcentration and determination using selective solid-phase extraction and electrothermal atomic absorption spectrometry: Applications in seawaters and biological samples. *Analytical and Bioanalytical Chemistry*, 385, 1178–1185.

[100] Stripeikis, J., Pedro, J., Bonivardi, A. and Tudino, M. (2004) System coupled to electrothermal atomic spectrometry with permanent chemical modifiers. *Analytica Chimica Acta*, 502, 99–105.

[101] Sánchez Rojas, F., Bosch Ojeda, C. and Cano Pavón, J. M. (2004) On-line preconcentration of rhodium on an anion-exchange resin loaded with 1,5-bis(2-pyridyl)-3-sulphophenyl methylene thiocarbonohydrazide and its determination in environmental samples. *Talanta*, 64, 230–236.

[102] Sung, Y.-H. and Huang, S.-D. (2003) Online preconcentration system coupled to electrothermal atomic absorption spectrometry for the simultaneous determination of bismuth, cadmium, and lead in urine. *Analytica Chimica Acta*, 495, 165–176.

[103] Yan, X.-P., Li, Y. and Jiang, Y. (2003) Selective measurement of ultratrace methylmercury in fish by flow injection on-line microcolumn displacement sorption preconcentration and separation coupled with electrothermal atomic absorption spectrometry. *Analytical Chemistry*, 75, 2251–2255.

[104] Dimitrova, B., Benkhedda, K., Ivanova, E. and Adams, F. (2004) Flow injection online preconcentration of palladium by ionpair adsorption in a knotted reactor coupled with electrothermal atomic absorption spectrometry. *Journal of Analytical Atomic Spectrometry*, 19, 1394–1396.

[105] Herbello-Hermelo, P., Barciela-Alonso, M. C., Bermejo-Barrera, A. and Bermejo-Barrera, P. (2005) Flow on-line sorption preconcentration in a knotted reactor coupled with electrothermal atomic absorption spectrometry for selective As(III) determination in sea-water samples. *Journal of Analytical Atomic Spectrometry*, 20, 662–664.

[106] Tsakovski, S., Benkhesdda, K., Ivanova, E. and Adams, F. C. (2002) Comparative study of 8-hydroxyquinoline derivatives as chelating reagents for flow injection preconcentration of cobalt in a knotted reactor. *Analytica Chimica Acta*, 453, 143–154.

[107] Wang, J., Hansen, E. H. and Miró, M. (2003) Sequential injection-bead injection-lab-on-valve schemes for on-line solid-phase extraction and preconcentration of ultra-trace levels of heavy metals with determination by electrothermal atomic absorption spectrometry and inductively coupled plasma mass spectrometry. *Analytica Chimica Acta*, 499, 139–147.

[108] Wang, J. and Hansen, E. H. (2002) Sequential injection on-line matrix removal and trace metal preconcentration using a PTFE beads packed column as demonstrated for the determination of cadmium with detection by electrothermal atomic absorption spectrometry. *Journal of Analytical Atomic Spectrometry*, 17, 248–252.

[109] Miró, M., Jóncyk, S., Wang, J. and Hansen, E. H. (2003) Exploiting the beadinjection approach in the integrated sequential injection lab-on-valve format using hydrophobic packing materials for on-line matrix removal and preconcentration of trace levels of cadmium in environmental and biological samples

via formation of non-charged chelates prior to EETAAS detection. *Journal of Analytical Atomic Spectrometry*, 18, 89-98.

[110] Long, X., Chomchoei, R., Gala, P. and Hansen, E. H. (2004) Evaluation of a novel PTFE material for use as a means for separation and preconcentration of trace levels of metal ions in sequential injection (SI) and sequential injection lab-on-valve (SI-LOV) systems: Determination of cadmium (II) with detection by electrothermal atomic absorption spectrometry (ETAAS). *Analytica Chimica Acta*, 523, 279-286.

[111] Long, X., Hansen, E. H. and Miró, M. (2005) Determination of trace metal ions via on-line separation and preconcentration by means of chelating Sepharose beads in a sequential injection lab-on-valve (SI-LOV) system coupled to electrothermal atomic absorption spectrometric detection. *Talanta*, 66, 1326-1332.

[112] Long, X., Miró, M. and Hansen, E. H. (2005) An automatic micro-sequential injection bead injection Lab-on-Valve, (μSI-BI-LOV) assembly for speciation analysis of ultra trace levels of Cr (III) and Cr (VI) incorporating on-line chemical reduction and employing detection by electrothermal atomic absorption spectrometry (ETAAS). *Journal of Analytical Atomic Spectrometry*, 20, 1203-1211.

[113] Long, X., Miró, M. and Hansen, E. H. (2005) Universal approach for selective trace metal determinations via sequential injection-bead injection-lab-on-valve using renewable hydrophobic bead surfaces as reagent carriers. *Analytical Chemistry*, 77, 6032-6040.

[114] Wang, J. and Hansen, E. H. (2001) On-line ion exchange preconcentration in a sequential injection lab-on-valve microsystem incorporating a renewable column with ETAAS for trace level determination of bismuth in urine and river sediments. *Atomic Spectroscopy*, 22, 312-318.

[115] Sakuragawa, A., Taniai, T. and Uzawa, A. (2003) Determination of arsenic in steel by metal furnace-AAS and flow injection analysis based on method with on-line iodide trace extraction system. *Tetsu-To-Hagane/Journal of the Iron and Steel Institute of Japan*, 89, 927-934.

[116] Taniai, T., Sakuragawa, A. and Uzawa, A. (2004) Determination of arsenic or tin in steels by the automated extraction system with a recycled solvent and an improved gravity phase separation column. *ISIJ Interntional*, 44, 1852-1858.

[117] Wang, J. and Hansen, E. H. (2002) Development of an automated sequential injection on-line solvent extraction-back extraction procedure as demonstrated for the determination of cadmium with detection by electrothermal atomic absorption spectrometry. *Analytica Chimica Acta*, 456, 283-292.

[118] Anthemidis, A. N., Zachariadis, G. A. and Stratis, J. A. (2003) Development of online solvent extraction system for electrothermal atomic absorption spectrometry utilizing a new gravitational phase separator. Determination of cadmium in natural waters and urine samples. *Journal of Analytical Atomic Spectrometry*, 18, 1400-1403.

[119] Nan, J. and Yan, X.-P. (2005) On-line dynamic two-dimensional admicelles solvent extraction coupled to electrothermal atomic absorption spectrometry for determination of chromium (VI) in drinking water. *Analytica Chimica Acta*, 536, 207-212.

[120] Tseng, W.-C., Sun, Y.C., Yang, M.-H., Chen, T.-P., Lin, T.-H. and Huang, Y.-L. (2003) On-line microdialysis sampling coupled with flow injection electrothermal atomic absorption spectrometry for *in vivo* monitoring extracellular manganese in brains of living rats. *Journal of Analytical and Atomic Spectrometry*, 18, 38-43.

[121] Tseng, W.-C., Sun, Y.C., Lee, C.-F., Chen, B.-H., Yang, M.-H., Chen, T.-P. and

Huang, Y. -L. (2005) On-line microdialysis sampling coupled with flame atomic absorption spectrometry for continuous *in vivo* monitoring of diffusible magnesium in the blood of living animals. *Talanta*, 66, 740-745.

[122] Promchan, J. and Shiowatana, J. (2005) A dynamic continuous-flow dialysis system with on-line electrothermal atomic-absorption spectrometric and pH measurements for-in-vitro determination of iron bioavailability by simulated gastrointestinal digestion. *Analytical and Bioanalytical Chemistry*, 382, 1360-1367.

[123] Tseng, W.-C., Chen, P.-H., Tsay, T.-S., -H-Chen, B. and Huang, Y.-L. (2006) Continuous multi-element (Cu, Mn, Ni, se) monitoring in saline and cell suspension using on-line microdialysis coupled with simultaneous electrothermal atomic absorption spectrometry. *Analytica Chimica Acta*. 576, 2-8.

[124] Miró, M., Jimoh, M. and Frenzel, W. (2005) A novel dynamic approach for automatic microsampling and continuous monitoring of metal ion release from soils exploiting a dedicated flow-through microdialyser. *Analytical and Bioanalytical Chemistry*, 382, 396-404.

[125] Knàpek, J., Komàrek, J. and Kràsensky, P. (2005) Determination of cadmium by electrothermal atomic absorption spectrometry using electrochemical separation in a microcell. *Spectrochimica Acta Part B-Atomic Spectroscopy*, 60, 393-398.

[126] Coelho, L. M., Coelho, N. M. M., Arruda, M. A. Z. and de la Guardia, M. (2007) Online bi-directional electrostacking for As speciation/preconcentration, using ET AAS. *Talanta*, 71, 353-358.

[127] Ma, H., Fan, X., Zhou, H. and Xu, S. (2003) Preliminary studies on flow injection *in situ* trapping of volatile species of gold in graphite furnace and atomic absorption spectrometric determination. *Spectrochimica Acta Part B-Atomic Spectroscopy*, 58, 33-41.

[128] Chuachuad, W. and Tyson, J. F. (2004) Determination of cadmium by electrothermal atomic absorption spectrometry with flow injection chemical vapor generation from a tetrahydroborate-form anion-exchanger and in-atomizer trapping. *Canadian Journal of Analytical Science and Spectroscopy*, 49, 362-373.

[129] Chuachuad, W. and Tyson, J. F. (2005) Determination of cadmium by flow injection atomic absorption spectrometry with cold vapor generation by a tetrahydroborate-form-anion-exchanger. *Journal of Analytical Atomic Spectrometry*, 20, 273-281.

[130] Petrov, P. K., Serafimovski, I., Dtafilov, T. and Tsalev, D. L. (2006) Flow injection hydride generation electrothermal atomic absorption spectrometric determination of toxicologically relevant arsenic in urine. *Talanta*, 69, 1112-1117.

[131] Burguera, J. L., Burguera, M., Anton, R.-E., Salager, J. L., Arandia, M. A., Rondón, C., Carrero, P., Petit de Peña, Y., Brunetto, R. and Gallignani, M. (2005) Determination of aluminum by electrothermal atomic absorption spectroscopy in lubricating oils emulsified in a sequential injection analysis system. *Talanta*, 68, 179-186.

[132] Burguera, J. L. and Burguera, M. (2004) Analytical applications of organized assemblies for on-line spectrometric determinations: present and future. *Talanta*, 64, 1099-1108.

[133] Lu, C.-Y., Yan, X.-P., Zhang, Z.-P., Wang, Z.-P. and Liu, L.-W. (2004) Flow injection on-line sorption preconcentration coupled with hydride generation atomic fluorescence spectrometry using a polytetrafluoroethylene fiber packed microcolumn for determination of Se (IV) in natural waters. *Journal of Analytical Atomic Spectrometry*, 19, 277-281.

[134] Tang, X., Xu, Z. and Wang, J. (2005) A hydride generation fluorescence spectrometric procedure for selenium determination after flow injection on-line coprecipitation preconcentration.

Spectrochimica Acta Part B-Atomic Spectroscopy, 601, 580-1585.

［135］Wu, H., Jin, Y., Shi, Y. and Bi, S. (2007) Online organoselenium interference removal for inorganic selenium species by flow injection coprecipitation preconcentration coupled with hydride generation atomic fluorescence spectrometry. *Talanta*, 71, 1762-1768.

［136］Jianbo, S., Zhiyong, T., Chunhua, T., Quan, C. and Zexiang, J. (2002) Determination of trace amounts of germanium by flow injection hydride generation atomic fluorescence spectrometry. *Talanta*, 56, 711-716.

［137］Bagheri, H. and Gholami, A. (2001) Determination of very low levels of dissolved mercury (II) and methylmercury in river waters by continuous flow with on-line UV decomposition and cold vapor atomic fluorescence spectrometry after preconcentration on a silica gel-2-mercaptobenzimidazol sorbent. *Talanta*, 55, 1141-1150.

［138］Wu, H., Jin, Y., Han, W., Miao, Q. and Bi, S. (2006) Non-chromatographic speciation analysis of mercury by flow injection on-line preconcentration in combination with chemical vapor generation atomic fluorescence spectrometry. *Spectrochimica Acta Part B-Atomic Spectroscopy*, 61, 831-840.

［139］Duan, T., Song, X., Jin, D., Li, H., Xu, J. and Chen, H. (2005) Preliminary results on the determination of ultratrace amounts of cadmium in tea samples using a flow injection on-line solid-phase extraction separation and preconcentration technique to couple with a sequential injection hydride generation atomic fluorescence spectrometry. *Talanta*, 67, 968-974.

［140］Bo Shi, J., Tang, Z.-Y., Jin, Z.-X., Chi, Q., He, B. and Jiang, G.-B. (2003) Determination of As (III) and As (V) in soils using sequential extraction combined with flow injection hydride generation atomic fluorescence spectrometry. *Analytica Chimica Acta*, 477, 139-147.

［141］Dong, J.-M. and Yan, X.-P. (2005) On-line coupling of flow injection sequential extraction to hydride generation atomic fluorescence spectrometry for fractionation of arsenic in soils. *Talanta*, 65, 627-631.

［142］Li, Z. (2006) Studies on the determination of trace amounts of gold by chemical vapor generation non-dispersive atomic fluorescence spectrometry. *Journal of Analytical Atomic Spectrometry*, 21, 435-438.

［143］Ipolyi, I., Stefanka, Z. and Fodor, P. (2001) Speciation of Se (IV) in the selenoaminoacids by high performance liquid chromatography-direct hydride generation atomic fluorescence spectrometry. *Analytica Chimica Acta*, 435, 367-375.

［144］Gómez-Ariza, J. L., Bernal-Daza, V. and Villegas-Portero, M. J. (2004) Comparative study of the instrumental couplings of high performance liquid chromatography with microwave-assisted digestion hydride generation atomic fluorescence spectrometry and inductively coupled plasma mass spectrometry for chiral speciation of selenomethionine in breast milk. *Analytica Chimica Acta*, 520, 229-235.

［145］Muñiz-Naveiro, O., Dominguez-Gonzalez, R., Bermejo-Barrera, A., Bermejo Barrera, P., Cocho, J. A. and Fraga, J. M. (2007) Selenium speciation in cow milk obtained after supplementation with different selenium forms to the cow feed using liquid chromatography coupled with hydride generation-atomic fluorescence. *Talanta*, 71, 1587-1593.

［146］Gallardo, M. V., Bohari, Y., Astruc, A., Potin-Goutier, M. and Astruc, M. (2001) Speciation analysis of arsenic environmental solids reference materials by high-performance liquid chromatography-hydride generation-atomic fluorescence spectrometry following orthophosphoric acid extraction. *Analytica Chimica Acta*, 441, 257-268.

［147］Simón, S., Tran, H., Pannier, F. and Potin-Gautier, M. (2004) Simultaneous determination of twelve inorganic and organic arsenic compounds by liquid chromatography-ultraviolet irradiation-hydride generation atomic fluorescence spectrometry. *Journal of Chromatography*, 1024, 105-113.

[148] Simón, S., Lobos, G., Pannier, F., de Gregori, I., Pinochet, H. and Potin-Gautier, M. (2004) Speciation analysis of organoarsenical compounds in biological matrices by coupling ion chromatography to atomic fluorescence spectrometry photooxidation and hydride generation. *Analytica Chimica Acta*, 521, 99-108.

[149] Liang, L. N., Jiang, G.-B., Liu, J.-F. and Hu, J.-T. (2003) Speciation analysis of mercury in seafood by using high-performance liquid chromatography on-line coupled with cold vapor atomic fluorescence spectrometry via a post column microwave digestion. *Analytica Chimica Acta*, 477, 131-137.

[150] Stoichev, T., Rodriguez, R. C., Tessier, E., Amouroux, D. and Donard, O. F. X. (2004) Improvement of analytical performances for mercury speciation byon-line derivatization, cryophocusing and atomic fluorescence spectrometry. *Talanta*, 62, 433-438.

[151] Bramanti, E., Lomonte, C., Galli, A., Onor, M., Zamboni, R., Raspi, G. and D'Ulivo, A. (2004) Characterization of denatured metallothioneins by reverse phase coupled with on-line chemical vapor generation and atomic fluorescence spectrometric detection. *Journal of Chromatography*, 1054, 285-291.

[152] Bramanti, E., Toncelli, D., Morelli, E., Lampugnani, L., Zamboni, R., Miller, K. E., Zemewtra, J. and D'Ulivo, A. (2006) Determination and characterization of phytochelatins by liquid chromatography coupled with on-line chemical vapor generation and atomic fluorescence spectrometric detection. *Journal of Chromatography*, 1133, 195-203.

[153] Bramanti, E., D'Ulivo, L., Lomonte, C., Galli, A., Onor, M., Zamboni, R., Raspi, G. and D'Ulivo, A. (2006) Determination of hydrogen sulfide and volatile thiols in air samples by mercury probe derivatization coupled with liquid chromatography-atomic fluorescence spectrometry. *Analytica Chimica Acta*, 579, 38-46.

[154] Bramanti, E., Lomonte, C., Galli, A., Onor, M., Zamboni, R., D'Ulivo, A. and Raspi, G. (2005) Mercury speciation by liquid chromatography coupled with on-line chemical vapor generation and atomic fluorescence spectrometric detection. *Talanta*, 66, 762-768.

[155] Li, F., Wang, D.-D., Yan, X.-P., Su, R.-G. and Lin, J.-M. (2005) Speciation analysis of inorganic arsenic by microchip capillary electrophoresis coupled with hydride generation atomic fluorescence spectrometry. *Journal of Chromatography*, 1081, 232-237.

[156] Li, F., Wang, D.-D., Yan, X.-P., Lin, J.-M. and Su, R.-G. (2005) Development of a new hybrid technique for rapid speciation analysis by directly interfacing capillary electrophoresis system to atomic fluorescence spectrometry. *Electrophoresis*, 26, 2261-2268.

[157] Wang, D.-D., Li, F. and Yan, X.-P. (2006) On-line hyphenation of flow injection miniaturized capillary electrophoresis and atomic fluorescence spectrometry for high throughput speciation analysis. *Journal of Chromatography*, 1117, 246-249.

7 流动进样质谱法

Maria Fernanda Ciné

7.1 引言

7.1.1 流动注射分析质谱的作用和重要性

流动注射分析为在线溶液管理提供了多功能性，其可以在计算机控制的条件下执行不同的单元操作和过程[1]，当与具有高检测能力的分析技术结合时，例如质谱（MS）[2]，这些特性具有重要价值。

在载体溶液中注入离散的样品会产生通过 FIA-MS 接口的瞬时通道，这使得基线可以有条不紊地恢复，并避免注入溶液之间的交叉污染。流动注射分析与质谱耦合的效率可以使得样品和标样的进样或同位素溶液的加标按顺序进行，从而简化定量过程并最大限度地减少系统误差和污染[3,4]。流动注射分析以高度重现的方式运行小体积（例如，<100μL）的样品、标样和试剂在质谱中得到了大量应用[5]。流动进样系统已用于通过质谱检测进行分子和元素分析。这些应用提高了将样品引入电离源的效率，从而克服了基质效应，并为执行在线校准提供了便利。

7.1.2 质谱

质谱是一种具有强大离子分离、鉴定和定量能力的仪器技术。质谱系统包括一个样品引入装置、一个电离源和一个隔室，用于将离子传输到质量分析器，在分析器中它们按质荷比（m/z）进行分离并发送到检测器系统。广义质谱仪的隔室呈现在图 7.1 的方案中。样品引入可以将流动注射分析与膜探针结合起来，这些探针通过渗透蒸发被有机物质渗透。通过气相色谱（GC）[6]和液相色谱（LC）[7]、毛细管电泳（CE）[8]、场流分级（FFF）[9]进行的分析物质瞬态分离已与 MS 结合使用。在线蒸气生成，或者更具体地说，氢化物生成（HG）以及不同类型的雾化器已被用于样品引入[10]。

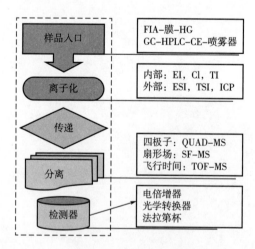

图 7.1 质谱分析步骤
样品入口和外部电离源均处于大气压下，其他组件在真空下运行。

通过质谱分析气相中有机或无机样品的分子或元素离子。离子通过不同的机制产生，包括能量转移以形成气相分子或元素离子，具体取决于分析目的。电离源可以位于质谱内部、真空下或真空外部，并通过接口连接。对于大多数分子离子，电离过程涉及真空下内腔中的电子碰撞（EI）、化学电离（CI）或热电离（TI）[11]，对于分子分析，样品雾化和电离组合可以在通过电喷雾（ESI）[12]和热喷雾电离（TSI）[13]测定大气压。露天直接电离技术，如解吸电喷雾电离（DESI）[14]和实时直接分析（DART）[15]，已成为有机质谱中发展最快的领域之一。电感耦合等离子体是用于多元素无机分析的大气压电离源[16]。对于ICP-MS，样品通常以溶液的方式通过雾化器引入。

来自离子源的代表性原子和分子离子被引导通过质量分析器中真空下的电场和磁场，并用检测器系统收集。通常用于检测来自瞬态流动注射分析离子的质量分析器的一些特性将在后续部分中描述。

7.1.2.1 四极杆质谱仪（QMS）

四极杆质谱仪（QMS）是由四个棒状电极组成的阵列，向其施加交替（r_f）和直流电压的组合，如图7.2[16-19]所示。电极棒定义了一个内部双曲线带电空间，在离子源之后由适当的离子光学器件处理后，离子被引入带电场空间的中心。满足可以稳定通过这一标准的离子遵循明确的路径通过四极杆到达检测器。其他离子不稳定并且丢失。定义的m/z离子的稳定轨迹取决于带电场参数，这些参数确定了用于传输的质荷离子的电气组合。通过改变施加的四极杆电压参数，可以以快速序列在质量范围内扫描离子传输。四极杆的特点是分辨率低（例如，$R = m/\Delta m = 300$），它们的质量范围高达4000uma。

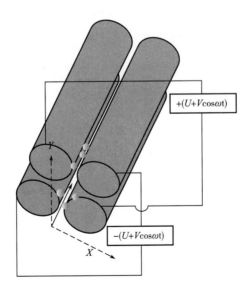

图7.2 四极杆质量分析器的示意图，显示施加到相对杆上的直流电和交流电 箭头表示离子进入方向。

7.1.2.2 扇形场质谱仪

扇形场质谱仪（SFMS）包括多种配置的磁性和/或静电分析仪[17-19]。

进入 SFMS 的离子由恒定电压 V 加速，该电压取决于它们的电荷 z 和电位，由下式给出：

$$E = zV$$

该公式还可以转化为动能方程：

$$zV = \frac{1}{2}mv^2 \tag{7.1}$$

以速度 v 进入扇形磁场 B 的离子获得垂直于两者的轨迹，这是由于洛伦兹力 $F = Bzv$ 和离心力导致的圆形轨迹：

$$F = \frac{mv^2}{R}$$

比较影响轨迹 R 的两种力：

$$Bzv = \frac{mv^2}{R} \text{ 和 } BR = \frac{mv}{z} \tag{7.2}$$

从关系式（7.1）和（7.2）推导出 m/z 对施加的加速电压 V 的依赖性，从而得出等式（7.3）：

$$\frac{m}{z} = \frac{B^2R^2}{2V} \tag{7.3}$$

位于焦平面中的一个或多个检测器可以收集穿过磁场和/或静电场的离子。

静电扇区由两个弯曲的板构成，它们的间距为 d，曲率比为 r。在对板施加电压差 ΔE（通常为 $0.5\sim 1.0$ kV）时，具有 m/z 和速度 v 的阴离子将聚焦在出射窗口，如果：

$$\frac{mv^2}{z} = \frac{Er}{d} \tag{7.4}$$

对于那些进入静电扇区的能量在 E_1 到 E_2 范围内且聚焦在出口狭缝处的离子，给出了稳定轨迹 $r=2v/E$。在质谱中使用了区分离子能量的能力并采用了双扇形排列以获得高分辨率（例如，$R=m/\Delta m=400\sim 10000$）。SF-MS 具有在宽范围（高达 15000uma）内分离质量的能力，并且可以在低、中和高分辨率（例如，R = 400、3000、10000）下进行质量分离。

7.1.2.3 多接收器质谱仪

多接收器质谱仪（MCMS）是一种双聚焦质量分析器的组合，将基于极点的质量分离与扇形磁场质谱相结合，该质谱仪与 ICP 电离源相连接，可以同时进行多组分检测并提高检测限[19,21]。与 ICP 电离源耦合的安装座旨在获得多种元素的精确同位素比率，这足以确定放射成因或稳定同位素的自然变化。该质谱仪主要应用于地球科学和地质科学。一旦瞬态信号中的同位素比率受到漂移和测量偏差的影响，通常可以从连续样品引入中获得良好的性能[20]。

7.1.2.4 飞行时间质谱仪（TOFMS）

飞行时间（TOF）法分离是通过脉冲静电场实现的，该静电场通过评估的无场线型管道加速离子包。飞行时间取决于离子的 m/z，这是通过快速电子检测系统记录的[19,22]。

进入飞行时间光谱仪的离子被脉冲电压加速,脉冲电压通常为2kV,持续约5ns,并被输送到圆柱形管内的无场真空中。加速电位脉冲转化为动能[方程(7.1)]。离子包在真空中非常快速地飞行,其中具有不同质荷比的离子通过它们的速度来区分。从方程(7.1):

$$zV = \frac{1}{2}mv^2$$

在定义的时间t内穿过管的线性距离d的速度可以通过以下替换方程式表示:

$$\frac{m}{z} = \frac{2V}{d^2}t^2 \tag{7.5}$$

由同一脉冲加速的不同m/z检测离子的时间间隔计算如下:

$$t_1 - t_2 = \sqrt{\frac{d^2}{2zV}}(\sqrt{m_2} - \sqrt{m_1}) \tag{7.6}$$

相关文献已经描述了具有不同ICP界面对齐的两种ICP-TOFMS配置[23],正交和同轴ICP-TOFMS中的离子轨迹如图7.3所示。采集区(排斥电极)前面的电极板通过正交排列促进离子的采集。偏转轨迹是分开的,因此与同轴方案(采集电极与飞行管同轴)相比,其分辨率更高。此外,现代飞行时间质谱仪在线型漂移管的末端包含一个反射离子透镜(反射器)。离子被电场反射,具有相同m/z的离子云被压缩,从而提高分辨率。同轴TOFMS中铅同位素的分辨率为1400~1500,正交设计的分辨率可达2300。

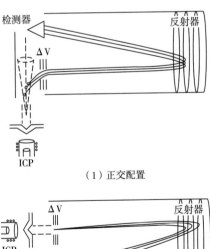

(1)正交配置

(2)轴向配置

图7.3 ICP-TOF质谱仪的方案

在快速序列中检测不同m/z的离子。例如,考虑到1m的飞行管和2kV的脉冲,检测H^+的时间为1.7μs,而对于m/z为2500的分子、离子检测时间为50μs。由脉冲电压加速的对封装中携带离子的快速采集（每秒约 20000~30000 个质谱）可以沿瞬态信号检测多个点，从而最大限度地减少光谱偏斜误差。当通过扫描可变时间-浓度峰值进行检测时，误差来源于失真。图 7.4 有助于理解频谱偏斜误差。TOFMS 在瞬态采样技术（例如，激光烧蚀、电热气化）和流动注射分析[10,23-25]中展现出了出色的性能。在某些情况下，时间分辨率消除了质谱的重叠[24]。

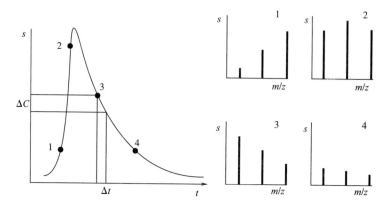

图 7.4　偏斜效应误差示意图

信号(5)的瞬态轮廓对应于浓度的时间变化。点 1~4 表示 3 种丰度之间的不同可能性。

7.2　FIA-MS 样品引入装置

当流动注射分析应用于元素和有机物的质谱分析时，需要将样品引入装置与电离源和质谱仪连接起来。图 7.5 总结了已经描述的用于耦合 FIA-MS 进行元素和分子分析的仪器设备。

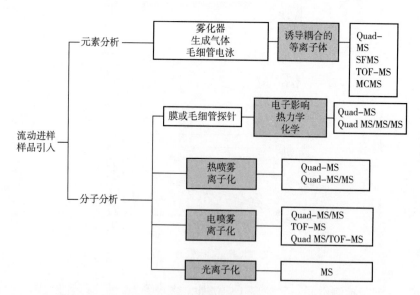

图 7.5　用于耦合 FIA-MS 进行元素和分子分析的仪器流程图

灰色阴影区域表示电离技术。

监测来自 FIA-MS 的瞬态信号的能力取决于液体分析物向电离源转移的动力学和效率以及转化为气态离子的速率、效率和离子采样的动力学和传输到光谱仪的速率、质量分离和 m/z 检测的速度。

7.2.1 将样品瞬时引入质谱电离室

具有内部电离源（即 EI、TI 和 CI）的经典质谱仪的特点是仅引入和分析气态样品。气态化合物通过不同的程序作为稳态流引入，例如，差异扩散、通过探针直接插入、加热毛细管和隔膜。1957 年文献[6]中描述的第一台气相色谱质谱仪开始将瞬态气体物质引入电离室。在高压下从色谱柱中排出的部分气体通过毛细管传送到处于轻度真空状态的质谱电离室。频谱由示波器监测。此后，设计了几种 GC-MS 接口设备，例如，膜探针、加热毛细管、毛细管探针和扩散射流，以将气态和液态样品引入质谱电离源。为了提高质谱检测灵敏度和选择性，人们提出了不同的方法将液相色谱与质谱耦合[11]。将离散样品溶液转移到探针尖端，在那里将真空条件下闪蒸产生的气态化合物引入电离室。在 LC-MS 中，柱后溶液首先进行闪蒸，然后在三个同心硅膜管装置中进行气相分离。分子分析物在增加的真空条件下通过膜进行传输[7]。

在 GC-MS 和 LC-MS 接口之后，带有膜渗透的 FIA-MS 被用于将挥发性有机化合物引入质谱电离室。膜引入质谱仪（MIMS）被修改为直接膜探针（DIMP）并在真空下被引入质谱电离室[26]。渗透性挥发性化合物的分子离子通过电子碰撞、热电离和/或化学电离产生。通常使用的质量分析器是 QMS 和串联质谱，其分两个或三个阶段，以获取分段信息。

选择性高并且带有特定膜的流动注射分析的优点是可以通过水进行在线分割，进而清洁膜和促进基线的恢复[27]。通过图 7.6 中的歧管可以依次注入样品、水、标样从而进行在线定量。

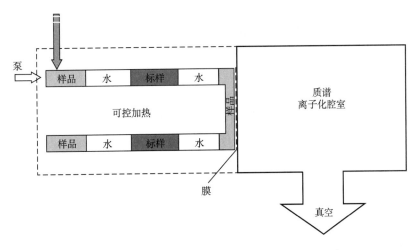

图 7.6 通过膜渗透 FIA-MS 引入样品和标样的方案

（经许可来自参考文献 [27]。）

水分割通过清洁膜表面避免了样品以及标样的残留。将200μL由水分割的有机样品泵入直接插入的膜探针中，可以仅将挥发性物质注入质谱电离室。渗透蒸发分离需要从加热（70~95℃）膜的高压侧选择性吸附分析物，在真空下渗透和解吸成蒸气状态。有机化合物从流动注射分析分段流系统到电离室的可重复渗透已经被证明。在使用膜引入的FIA-MS中分析物传输的动力学已被建模，以研究流动参数对跨膜传质曲线的影响[28]。

膜和毛细管探针可用于实施FIA-MS，以在线监测发酵和生物反应器中的挥发性产物。

当需要高通量时，在流动条件下的在线监测过程中对薄膜进行了测试[29]。其他FIA-MS应用与石化衍生物的在线监测相关，具有处理实时化学反应（如光解）以及提供自动反馈控制[30]的优势。

FIA-MS采用了不同的质谱仪组合，例如单四极杆QMS和串联四极杆（QMS/MS/MS）。流动注射分析众所周知的液-液分散分布特性在通过气体环境的路径中被放大，同时需要有效的联用装置来避免污染。达到这一条件是为了在线监测乙醇制造工厂的生物反应，从而控制发酵化合物的主要产物和挥发性代谢物的选择性[31]。然而，对其他类型样品的分析应用受到膜的特异性的限制。

7.2.2 将流动进样样品引入外部电离源-质谱

FIA-MS的耦合接口是为了在质谱检测之前加强对样品溶液的管理。传统的质谱样品制备过程受产生挥发性或气态化合物的要求的限制，这限制了几种材料的质谱分析（主要是元素分析）。然而，人们已经开发出了不同的外部电离源以促进从液体和固体样品中产生代表性的分子和原子离子。

大多数FIA-MS应用是通过将在线液体管理程序的多功能性与引入的高效的外部质谱电离源相结合而开发的。液体溶液直接进入外部电离源，如等离子体[32]、热喷雾[33]和电喷雾[12,34]。FIA-ICP-MS在元素分析中的适用性通过在线稀释、校准、添加、分析物分离、预浓缩和其他化学程序得到了验证[2]。

相关文献报道了流动注射分析与样品引入/电离源的耦合，例如电喷雾电离[34]、热喷雾电离[35]、大气压化学电离（APCI）[36]、大气压光离子化（APPI）[37]和激光诱导多光子或用于分子分析的脉冲样品引入接口（PSI）[38]。使用流动注射分析进样的原因是可以获得无污染的高样品通量，进行在线稀释和进行标准化处理。带有热喷雾电离和大气压化学电离的流动注射分析校准具有两个数量级的宽线性范围[39]，使用FIA-MS通过标样添加法对污水处理厂流出物中不可生物降解的污染物进行定量比LC-MS快25~30倍[33]。

7.3 与外部电离源耦合的流动系统-质谱

7.3.1 FIA-ICP-MS

在ICP-MS元素分析中广泛使用了FIA-MS。通过将流动注射分析与多同位素能力、高检测能力、快速采集速度和 ICP-MS 样品引入的兼容性相结合，实现了高精度的痕量元素测定。FIA-ICP-MS 与在线程序耦合的便利性已经发表了大约 300 篇论文。大多数将流动注射分析与 ICP-MS 结合的研究旨在通过在线基质分离、分析物预浓缩或蒸气生成来提高痕量元素的检测限。通过精确的分析物在线添加或同位素加标，改善了标样添加和同位素稀释的定量过程。在接下来的章节中将会描述用于产生离子的等离子体源配置以及用于采集离子并将它们引导至光谱仪的接口。

7.3.1.1 ICP 来源

ICP 是一种在受振荡磁场影响的氩气氛围中产生的常压电离源，该振荡磁场由流过感应线圈的射频（RF）电源维持。射频在 27~40MHz 下以 1.0~1.5kW 的功率运行。在大气压下，等离子体在同心石英管（即炬管）组件的末端形成，释放热量、辐射和离子。炬管包括三个同心圆柱管，其中氩气以 12~15L/min 的速度流过外部部分以维持等离子体。ICP 处于非热力学平衡状态，其特征是焦耳效应产生的热量呈环形分布，在感应线圈区域的等离子体核心处温度达到最高（10000K）。由大约 2% 的电离氩提供的等离子体的电子密度为 10^{13}~10^{15} 个电子/cm^3[18,19]。液体样品通常以气溶胶（由雾化器产生）的形式引入。气溶胶的均匀性取决于几个参数，例如液滴尺寸和分布以及雾化室对液滴尺寸的区分。样品传输的效率取决于云均匀性，该均匀性受液体表面张力、溶液密度、雾化器气体流速和雾化室几何因素的影响。在流动注射分析中由液体载体传输的注入样品具有层流分散的特征，当通过流通式检测器检测时其呈现出典型的峰轮廓。该峰被雾化器的湍流气溶胶传输完全改变了，这主要是由于雾化室内大部分分析物的损失。这会使得瞬时的峰展宽接近于高斯分布。

等离子体能量转移到样品上并导致样品组分的干燥、蒸发、解离、原子化和电离。产生离子的效率取决于样品基质、提供的射频功率、样品气溶胶的产生、传输流速和等离子气体成分。氩等离子体中的可用能量为 15.7eV，该能量足以电离元素周期表中的大部分元素。

用于引入小体积样品的直接注射雾化器系统（DIN）与流动注射分析系统相类似[40]。歧管由一个三通道电磁阀组成，该阀带有可更换的回路以注入不同体积的样品。注入的样品由液压气体置换泵（GDP）通过炬管中心管末端附近的毛细管泵送。注入的样品形成气溶胶，该气溶胶直接注入等离子体中并进行气化、雾化和电离。产生的瞬态峰呈现出平滑的峰值和面积，具体取决于注入量和载体溶液的泵送压力。具有大平台期的峰形显示出较小的分散效应，主要是在注入样品塞和高效引入等离子体的极端处，最大限度地减少死体积。

流动注射分析液体样品可以通过蒸气发生器[41]、热喷雾（TS）[42] 和氢化物发生器（HG）[43] 引入 ICP。从液态到气态的转化提高了分析物传输至等离子体的效率并且使基质分离更加便利，从而降低了检测限。

7.3.1.2 用于减少基质效应的 FIA-ICP-MS

基质效应是与复杂样品相关的非质谱干扰，呈现超出仪器工作条件范围的特征。用于 ICP 的进样雾化器通常受基质效应（例如，溶剂黏度引起的物理基质效应）的影响，并且将雾化器与去溶剂化装置耦合的设备在减少等离子体放电的溶剂负载方面非常有效。

在 ICP-MS 中，除了雾化器以外，离子采样和传输到质谱仪的接口也受样品基质的影响。离子采样接口由两个金属锥体（即采样锥体和截取锥）组成，依次放置有直径分别为 1mm 和 0.7mm 的小中心孔。采样锥孔会被盐沉积物堵塞。因此，样品固含量限制在 2% 以下。

在通过采样锥之后，大部分中性组分通过锥之间的真空条件（100Pa）去除，而离子被分离锥上的低电位所吸引。穿过截取锥的离子由真空条件（10Pa）下的电场和磁场引导通过圆柱形透镜到达质谱仪隔室。

一些仪器设备和隔室的尺寸不适合管理复杂的溶液，离子可能会受到质量差异的影响。仪器质量的偏差可以通过分析由认证同位素组成的参比溶液进行校正，例如，由美国国家标准研究院 NIST-951 制备的硼酸[19]。然而，由于在离子通过透镜区域的过程中产生的样品基质效应，光同位素的传导会受到质量偏差的影响。重离子优先沿中心轴传输，轻离子会因库仑排斥而丢失。这种效应取决于重离子和轻离子的相对浓度及其电离电位，其也被认为是 ICP-MS 中基质效应的主要原因。轻同位素透射中的同位素质量差异会影响含有重元素的样品基质中 B 和 Li 的同位素比。可以通过常见的分析物内标算法来校正由于样品基质引起的质量偏差效应[44]。

当引入微升体积的溶液时，首次结合 FIA-ICP-MS 的实验为痕量金属的多元素测定提供了很有前景的同位素数据[45]，这些结果预示了 FIA-ICP-MS 在减少基质效应方面的优势。在载体溶液之间注入定义的样品体积，以可重现的方式在线稀释，可以使基线快速恢复并降低锥体上盐沉积的风险。

除了流动注射分析在样品引入方面的优势外，其他在质谱分析之前管理溶液的设施也得到了改进。耦合 ICP-MS 的流动系统与不同类型的雾化器一起使用，例如，用于引入纳升体积的微流雾化器[46]。流动注射分析与高效雾化器相结合用于注入含有低浓度分析物的小体积样品[47]。使用具有在线基质去除[48-50]、溶剂萃取/反萃取预浓缩或液固相萃取[51]以及添加内标[52]的流动注射分析系统克服了基质效应。基质干扰的诊断可以通过 FIA-ICP-MS 中的梯度稀释进行，其类似于梯度校准方法[53]。在 FIA-ICP-TOFMS 中进行梯度洗脱的排列以表征基质干扰[54]。相关研究描述了通过在线添加标样进行定量的流动注射分析设计，该设计是为了减少基质效应。应用标样添加法（SAM）需要在将已知量的标样添加到样品之前和之后测量样品中的分析物，以获得没有基质效应的校准曲线。相关文献已经描述了几种在与原子光谱仪耦合的流动系统中执行在线添加标样的方案[55]。

通过将标准溶液连续添加到分散的样品区，利用汇流流体的流动注射分析配置已被使用。此外，合并区域流动方案已与 ICP-MS 相结合，在该方案中分散的样品和标样区域在下游通过汇流合并。只注入一种标样（该标样的峰与样品峰重叠）的简单程序就足以应用 SAM。最近通过峰模拟程序对用于 FIA-ICP-MS 的浓度-时间曲线进行改善，进而计算在添加标样前后样品峰的面积[57]。

使用模块化流动注射分析歧管进行在线样品稀释和添加标样并与 ICP-MS 联用的校准技术被用于直接分析酒和尿液[58]。在准确性和灵活性方面对不同的铅定量方法（包括标样添加法、同位素稀释法和外部校准）作出了比较。将 Tl 在线添加作为内标的外部校准在大规模分析应用中具有简易性，但其准确度极度依赖于基质。同位素稀释和标样添加可提供准确的 Pb 结果，但灵活性较差。在 FIA-ICP-MS 中通过测定 Pt 的热喷雾接口对相同校准技术作出了比较[59]。通过使用气动雾化或热喷雾样品引入对流动注射分析瞬态峰进行比较呈现出有趣的特征，与使用雾化器引入相比，使用热喷雾引入具有更高的分析物传输效率，但精度较低。流动注射分析设计促进了样品的稀释，并且由于 Pt 的浓度低，最佳校准的选择不是决定性的。

7.3.1.3 执行同位素稀释（ID）的流动注射分析系统

同位素稀释（ID）是一种绝对定量方法，使用质量比测量作为主要计量程序[60]，它是一种准确的定量方法，适用于那些至少有两种同位素（即 A 和 B）的样品测量，并且这些同位素在样品中不会产生特殊的干扰。向体积明确的样品中（含有质量 M_s 未知的元素）加入质量 m_{sp} 已知的带有高丰度同位素 A 的加标样。样品（A_s 和 B_s）和加标样（A_{sp} 和 B_{sp}）的同位素组成为已知。加标了的样品充分混合以达到同位素平衡，然后测量同位素比 A/B。ICP-MS 测得的同位素比值是平衡后的结果，如下关系式所示：

$$R = \frac{M_s A_s + m_{sp} A_{sp}}{M_s B_s + m_{sp} B_{sp}}$$

流动进样程序的开发可以使加标过程更准确，降低了污染的风险并减少了样品和加标样的消耗。通过不同的流动注射分析程序以可重现的方式将含水加标样添加到样品溶液中，同时可以通过 ICP-MS 测量稀释过程涉及的两种同位素。FIA-ICP-MS 的在线同位素稀释于 1989 年首次报道[61]。其使用了基于时间的样品进样和合并区域流动配置中的加标。作者证明了通过 FIA-ID-ICP-MS 测定铅的方法可用于血液、灰尘、油漆、海洋和土壤沉积物的常规分析[61,62]。

用于执行同位素稀释的 r-流动注射分析或反向流动注射分析配置包括在恒定流量的加标溶液中注入样品。在连续流动的加标溶液中，采用了逆流进样和基于时间的可编程样品进样以获得最佳同位素比[63]。通过改变稀释率可以轻松地对不同的样品与加标样的比率进行编程。r-流动注射分析配置的主要缺陷来自加标溶液的连续泵送。当在加标溶液中注入明确体积的样品时，会在富集同位素加标的 m/z 中获得瞬态稀释峰，并根据样品中分析物的浓度增加其他同位素。该应用需要对样品进行初步分析，当分析物浓度在一个直线浓度范围内时是可行的[64]，分析物浓度的错误预测会导致稀释度很低。在这些情况下，仅测量一个同位素不能满足计数统计，同位素比受到精度差的影响。夹层配置考虑在两次样品注射之间注射加标样。在这种情况下，样品区的加标在到达 ICP-MS 之前被分散。表 7.1 总结了用于在线标样添加和同位素稀释的流动系统的配置及其应用。

表 7.1　　FIA-ICP-MS 在在线标样添加和同位素稀释中的应用

描述和分析目的	分析物	样品	参考文献
ICP-SFMS 中的 n-流动注射分析	U、Pu		[46]

续表

描述和分析目的	分析物	样品	参考文献
在线脱盐器最大限度地减少对 Cr 形态的干扰	Cr^{6+} 总 Cr	尿液、水	[50]
控制基质效应的流动策略：梯度稀释	Co、Ni、Cd、Pb	盐水溶液	[54]
用于测量 ICP-MS 中瞬态信号的流动注射分析标样添加校准	Mn、Co、Se、Mo、Cd、Pb、Sb、Tl	尿液	[57]
校准技术的比较：外部校准、标样添加、同位素稀释	Pb	葡萄酒、尿液	[58]
带有热喷雾接口的流动注射分析标样添加和同位素稀释	Pt	玉米叶、小鼠肝脏、$Ca_3(PO_4)$	[59]
使用混合的加标样和样品溶液执行同位素稀释	Pb	生物样品沉积物	[61, 62]
用于加标样品和标样的逆流进样，以最大限度地减少同位素残留	Ni、Mo	河水和咸水	[63]
加标样的夹层注射产生加标溶液的浓度梯度	Mo	盐水	[64]
可编程引入样品和加标溶液以获得精确的同位素比率	Cd	水溶液	[68]
在线透析，同位素稀释	Cu、Zn、Se、Pb	血清	[69]
纳米 HPLC-ID	S	肽	[72]
用于量化电解溶解质量的同位素稀释	Pb	铜	[78]
冷蒸气生成和同位素稀释	Hg	水	[79]
在阳离子交换色谱之前合并样品和加标样	Pd	尿液	[80]
^{26}Mg 的加标样应用伪同位素稀释测量 $^{26}Mg/^{23}Na$	Na	血清	[81]
使用超声波雾化器直接分析有机基质同位素稀释	Ag、Cd、Co、Cr、Fe、Mn、Ni、Pb	燃料乙醇	[82]
使用和不使用基质分离的同位素稀释比较	Cu、Cd、Pb	尿液	[83]
在线基质的电解分离	B	钢材	[84]
排除尺寸的色谱并结合有金属硫蛋白的金属同位素稀释	Cu、Cd、Zn	鲤鱼肾磷脂	[86]

通过同位素稀释获得的结果的准确性取决于测量同位素比的精度。使用 ICP-MS 进行同位素比测量的精度取决于质谱仪和来自样品引入装置、样品基质和数据采集参数等其他仪器因素[65]。

然而，测量同位素比率的精度的提高取决于同位素 σA 和 σB 测量的相对精度。精确的同位素比率测量反映了对样品中元素的准确测定。同位素比测量的最高精度取决于以下条件：①同位素 A 和 B 的测量满足以低标准偏差为特征的计数统计；②计数水平必须通过检测器死时间因子进行校正；③两种同位素测量精度的相似性[60]。

ICP-MS 中同位素比率测量的精度取决于样品引入、等离子体稳定性、离子采样、离子传输和质谱仪。样品引入系统的效率和基质效应会影响等离子体稳定性和代表性离子到质量分析器的传输。ICP-QMS 等序贯光谱仪中对两种同位素的测量均受等离子体波动的影响，同位素比率的精度以 RSD >0.5% 为特征。这种精度是通过以较短的停留时间快速测量两种同位素来实现的[57]。在瞬态分析中，以良好的精度测量同位素比（IR）的主要限制是与时间相关的曲线和有限的峰宽。通过使用 DIN-ICP-QMS 并进样 50μL 对 B 进行同位素比测量，测量的精度<1.4%。植物消解溶液中含有 300μg/L 的 B。图 7.7 中的 DIN 瞬态信号对应于注射的不同体积浓度为 100μg/L 的溶液，其峰的最大稳态值便于提高同位素比的测量[66]。

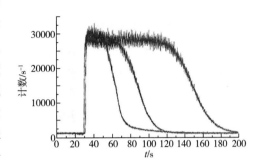

图 7.7 通过 DIN-ICP-MS 注入 20μL、50μL 和 100μL 的 100μg/L 溶液所获得的瞬态峰（m/z 10）

（资料来源：经许可，摘自参考文献 [66]。）

通过改进光谱仪的设计可在多接收器（MC）光谱仪[20]中以高分辨率同时检测少量同位素或通过 TOFMS[10]实现 ICP 离子的实时分析。通过 FIA-ICP-TOFMS 三次同时测定 22 种稀土元素的峰轮廓如图 7.8[67]所示。耦合 FIA-ICP-TOFMS 在通过同位素比检测持续 6s 的 FIA 信号时可以将精度控制在 0.5% 以内。通过将 CE 和 HPLC 与 ICP-TOFMS 耦合也实现了同位素比测量瞬态峰时的高精度，该方法用于对有机金属进行形态分析[24]。

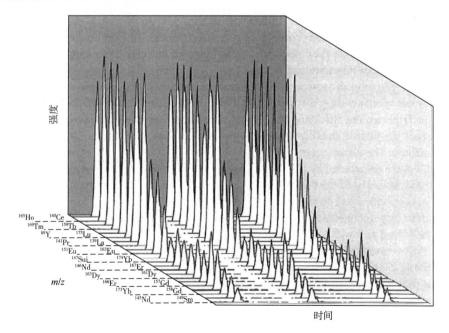

图 7.8 通过 FIA-TOFMS 获得的 22 种稀土元素的瞬态峰（一式三份）

（来自参考文献 [67]。）

计算机控制的流动注射分析歧管旨在提高同位素稀释的性能，可以为ICP-MS的同位素稀释应用执行可编程的在线加标。流动注射分析系统包括一组用于引入样品和加标样的电磁阀，以确保它们在到达雾化器之前在线混合。相关文献描述了通过不同分析物浓度的程序化在线加标来对精确的同位素比进行预测[68]。调整后的样品和每个样品的加标样体积在编程的和发现的同位素比之间呈现出良好的相关性。

7.3.1.4 ID-ICP-MS的流动注射分析应用

同位素稀释是一种定量程序，适用于部分回收分析物的过程，加标可被视为完美的内标。该应用程序仅限于没有大规模差异的过程。一些人工制品，如脱盐器和透析器，用于在线减少样品基质元素。这种策略有助于以超过90%的效率降低Na、K和样品中的Ca、Mg[69]。

FIA-ID对于小体积样品的多元素分析非常有用。同位素稀释的应用可以测定唾液中的多种元素[47]，以及咖啡叶[66]和体液[70]中的亚细胞组分。通过DIN-ICP-MS注入和引入样品。

对不同氧化态或与不同有机分子结合的金属同位素的稀释定量可以通过物质特异性或物质非特异性加标模式进行[71]。

物质特异性同位素稀释模式包括在分析程序开始之前用标记物质加标样品。通过不同的过程（例如透析、固相萃取和氢化物生成）在消解或样品基质分离之前加标样品可以对分析物进行准确定量，尽管它们的回收率可能不是定量的。该过程无需进行体积评估，从而避免了污染。该程序是在用加标金属孵育的血清样品中进行的，然后通过FIA-ICP-MS进行透析[69]。

物质非特异性在线同位素稀释模式已应用于在HPLC或毛细管电泳条件下对物质形成过程进行定量分析[85]。柱后分离的物质在流向ICP-MS时加标，同步的同位素比测量可以对元素进行定量。加标溶液被连续泵入等离子体，并获得分析物的反向瞬态峰。反向同位素稀释瞬态峰值受之前描述的相同效应所影响[62]。最近人们证明了柱前加标同位素稀释可用于通过S同位素稀释对半胱氨酸和甲硫氨酸肽进行定量。执行柱前同位素稀释的仪器系统包括分析物的在线预浓缩和仪器漂移的补偿[72]。

同位素稀释广泛适用于部分回收的在线过程，例如，使用不同的在线工件去除基质元素或保留/洗脱分析物。这些程序对于地质和环境样品通常是必要的，尤其是海水样品[73]。通过带有同位素稀释的FIA-ICP-MS可以测定小体积血清、脑和药物代谢物中的痕量元素。

7.3.1.5 流动注射分析与ICP-MS联用技术相结合

流动注射分析经常报道的应用之一包含用在线基质/分析物分离[47,48]和预浓缩[66]或在打结反应器[74]中的流通填充柱。用于基质分离、分析物预浓缩和元素形态分析的主要流通装置见表7.2。

几种校准策略利用流动注射分析设施来进行一些方法的自动化程序，例如内标法、标样添加和具有高重现性的同位素稀释法[54]。在流动系统中引入蒸气发生器设备提高了将样品引入质谱的效率[43]。用于管理小体积样品的流动注射分析设施在特定应用中被强调，例如识别与病毒感染相关的金属[75]，或生物样品的细胞部分中的B[66]。与雾化器和去溶剂化系统相关的纳米体积的流动进样有效地减少了氧化物的形成，其用于测定消化生物组织中的痕量镧系元素[76]。

表 7.2　FIA ICP-MS 在在线基质分离和/或分析物预浓缩中的应用

描述和分析目的	分析物	样品	参考文献
使用 8-HQS 色谱柱去除基质	Pb	海水	[73]
编结反应器-IDQ	Ag、Cd、Co、Ni、Pb、U	天然水域	[74]
在亚氨基乙酸酯树脂柱中去除 Ca、Na、Mg	Cu、Ni、Zn、Co、Pb、Cd、Fe	海水鱼虾	[48]
去除阴离子以避免基质干扰	B、Al、Cu、Mn、Fe、Cr、Cd、Pb、S	植物	[49]
溶剂/反萃取 DIN	Cu、Pb	尿液、水、沉淀物	[51]
使用 IGP-TOFMS 在编结反应器中进行在线预浓缩	REE	海水	[67]
样品加 EDTA 在 pH10.6 中在 Ln 树脂（HDEHP）中进行预浓缩	Ra-226	水	[98]
预浓缩/基质分离方法，使用 TRU 树脂	Am-241	尿液	[96]
阴离子交换树脂中的预浓缩和反应池中 CO 的使用	Pu	植被	[88]
萃取层析树脂	U、Th	血尿	[90]
体外微量透析-固相萃取	Cu、Zn、Mn	脑细胞外液	[91]
含有改性碳纳米纤维的色谱柱	REE	生物	[87]
将 8-氢醌固定化	V、Mn、Co、Ni、Cu、Zn、Mo、Cd、Pb、U	海水	[94]
亚氨基二乙酸酯树脂、Muromac A-1	REE	海水	[95]
Ag-PDC 的固相萃取	Pd	岩石	[89]
编结反应器二乙基二硫脲复合物	Pt、Rh、Pd	尿液、血清、道路灰尘	[92]
SIA-螯合树脂	Rh、Pt、Pd	尿液	[93]
FIA-生成氢化物避免基质干扰	As、Ge、Se	尿液	[43, 10]
在树脂中进行预浓缩，使用高灵敏度装置进行去溶剂雾化	Pu	尿液	[97]
利用 SF-ICP-MS ICP-TOF-MS 对生成的冷蒸气进行在线预还原	Hg	水	[79]

FIA-CE-ICP-MS 耦合系统被用于在线分析 Cr^{6+} 和 Cr^{3+} 的形成。在电解质溶液中使

用流动注射分析进行样品注射有助于将样品引入毛细管电泳（CE）中，并且避免了更换管路和溶液[77]。采用样品电溶法与 ICP-MS 在线测定电解铜中的 Pb 杂质。同位素稀释用于量化电溶解铅[78]。建议使用 FIA-HG-ICP-TOF-MS 组合来测定尿样中的 Ge、Sb、Sn、Hg 和 Bi[10]，通过分别扫描五次注入的 200ng/mL 溶液中的 Hg 同位素而得到的五个峰轮廓如图 7.9 所示。

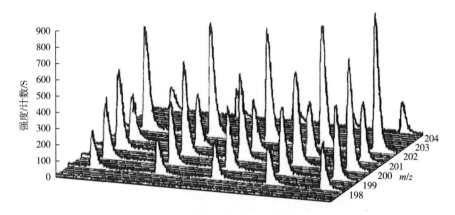

图 7.9 通过 FI-HG-ICP-TOF-MS 获得的六种汞同位素
（198、199、200、201、202 和 204）的流动进样信号的重复性
（来自参考文献 [10]。）

通过在流动条件下产生冷蒸气，使用流动注射分析系统测定水中超痕量水平的 Hg。产生的 Hg^0 在气液装置中分离，在液体捕集器中纯化，并在与 ICP-MS 相连的金箔中合并。加热金柱时产生的瞬态 Hg 信号通过同位素稀释进行量化。将 ICP-SF-MS 的性能与正交 ICP-TOF-MS 测定 Hg 的性能进行了比较。在相同条件下对产生的 Hg 冷蒸气进行直接分析，结果表明 ICP-SF-MS 的红外线测量精度很高。同时，当 Hg 通过在金中汞齐化进行预浓缩时，通过 ICP-TOF-MS 测量的瞬态信号中的同位素比拥有更高的精度，计算出的检测限为 0.9pg/g[79]。

7.3.2 流动注射分析-热喷雾质谱仪

热喷雾（TSI）电离是通过从直径小于 0.1mm 的毛细管尖端进入加热室的微滴产生超声气溶胶来实现的。含有溶剂化分析物分子的液滴在穿过加热室时开始失去溶剂，最后它们变成干燥的溶质颗粒。溶质粒子通过离子分子或电荷转移机制与离子碰撞而被电离，也可以使用替代电离装置，例如放电、电子轰击和裂变粒子轰击。另一种方法是向液体样品中添加电解质，例如 0.1mol/L 乙酸铵，以促进液相中的电荷转移。在这种情况下，离子在溶液中产生，并且在液滴干燥期间发生离子蒸发。通常会产生离子物质 MH^+，由热喷雾产生的离子通过离子路径被引导到光谱仪。

全自动 FIA-TSI-MS 用于测定和表征农业化学品。FIA-TSI 与串联 QMS/MS 耦合用于测定除草剂和杀虫剂[99]。FIA-TSI-MS/MS 与 LC-TSI-MS/MS 相比，其对测定肉类和血液中的磺胺显示出良好的灵敏度和可接受的检测限。流动注射分析能够每小时分析 50~70 次提取物而没有污染[100]。文献对比了 FIA-TSI-MS 和 LC-TSI-MS 两种方法测定来自生物废水处理厂的不可生物降解的极性化合物[101]。

7.3.3 电喷雾电离（ESI-MS）

7.3.3.1 电喷雾电离

电喷雾的电离过程将带电液滴的气溶胶（包含溶液中的离子）转化为气相中的离子。电喷雾装置由金属毛细管和连接到 2~3kV 电源的反电极组成。承受连续电流的液体介质中的电荷分离过程和电泳的机制相类似，并在毛细管尖端产生带电液滴。样品液滴穿过 0.1mm 内径的孔洞，从金属毛细管尖端出现并形成带电液滴的气溶胶，这些液滴在加热环境中逐渐蒸发溶剂。液滴变得更小且带电很高。当气体离子被释放时，增加的电荷排斥力最终克服了溶剂表面张力。毛细管尖端周围的电场具有很高的电场强度，高达 10^6V/m。反电极包含一个 0.3~0.5mm 的孔，用于将离子转移到低压区域。在不含孔的射流膨胀发生后，中性粒子被真空泵去除。截取锥在顶端有一个小孔（0.6~1.4mm），穿过孔的离子被磁场引导至质谱仪[12]。

7.3.3.2 FIA-ESI-MS

流动注射分析系统很容易连接到 ESI 的毛细管中，载流子流速低于 1mL/min。将样品注入极性溶剂混合物中，例如甲醇、乙醇水或丙酮、乙腈或二甲亚砜。流动注射分析通常用于在 FIA-ESI-MS 应用中引入小体积样品。

将同位素标记的内标噻菌灵[13]C 通过流动进样添加到杀真菌剂中，可以将其电离并通过 FIA-ESI-MS/MS 进行测定。该方法是液相色谱和气相色谱的替代方法，但由于不需要烦琐的样品净化，因此该方法易于执行[102]。相同的联用装置被用来量化抗肿瘤化合物中的 ^{10}B，^{10}B 应用于硼中子捕集疗法。只需注射 1μL，即可在尿液和血浆中定量测定含 ^{10}B 的药物化合物[103]。流动进样在 FIA-ESI-MS 的一些应用见表 7.3。

表 7.3　流动进样在电喷雾电离质谱（FIA-ESI-MS）中的应用

描述和分析目的	分析物	样品	参考文献
FIA-ESI-MS	杀真菌剂	柑橘	[102]
FIA-ESI-MS/MS 硼疗法	B-10 分子化合物	生物样本	[103]
FIA-CE-ESI-MS 毒理学	生物胺	葡萄酒	[104]
SIA-CE-ESI-MS 用于 CE 分离的样品预处理	几种胺	水	[105]
FIA-ESI-MS 生化标志物	胍基乙酸盐和肌酐	干血	[106]
FIA-ESI-MS/MS 医疗用途	琥珀酰丙酮	尿液、干血斑	[107]
FIA ESI-MS 提高样品通量	未衍生的氨基酸	酵母	[108]
FIA-HPLC ESI-TOF-MS 柱后标准化	碎片和杂质	毒品	[109]
FIA-ESI-MS 色谱分离的替代品	糖类	啤酒	[110]

续表

描述和分析目的	分析物	样品	参考文献
HPLC-FIA-ESI-MS 标准化	生物碱	天然保健品	[111]
直接引入	咖啡因肌酸	药物制剂	[112]
在线光衍生化 FIA-MS 药物	吲哚	磺胺类药物	[113]
FIA-ESI-MS 可重现且快速	细菌	粗细胞提取物	[114]
FIA-ESI-MS 流动性	铁载体	环境	[115]

联用的 FIA-CE-ESI-MS 用于测定葡萄酒中的生物胺。样品在注入毛细管电泳之前进行过滤以进行胺分离。耦合需要选择毛细管电泳电解质以确保与 ESI 的稳定电喷雾。在 CE-ESI-MS 接口中，加入鞘液以关闭毛细管电泳回路并将分离的分析物带到电喷雾中。与紫外检测相比，质谱的优势在于能够以高灵敏度识别不同的胺[104]。

自动顺序注射分析（SIA）与 CE-ESI-MS 联用，用于分析水样中的几种胺。顺序注射分析系统可以在毛细管电泳分离前添加试剂进行样品预处理和清理[105]。

在 FIA-ESI-MS/MS 中通过 ID 对干血斑提取物进行分析，以确定新生儿原发性肌酸紊乱的生化标志物（胍基乙酸盐和肌酸）。流动注射分析可以管理小体积样品，两种化合物均在 1min 内完成测定[106]。类似的，一种用于诊断肝肾疾病的代谢物琥珀酰丙酮（SA），被确定为 Girard T 衍生物，其通过 ESI 形成离子物质，通过 ^{13}C SA 和 FIA-ESI-MS-MS 进行同位素稀释被用于对血斑和尿样进行定量分析[107]。

相关文献提出了一种 FIA-ESI-MS 程序，该程序用于对通过液相色谱分离的化合物进行精确的质量测量[108]。引入柱后流动注射分析瞬态标准溶液的目的是叠加参比洗脱液峰并部分叠加洗脱的分析物峰。热喷雾电离独立发生，没有离子抑制效应，流动注射分析分析物可作为质量测量的参照物。

在 FIA-ESI-MS 系统中测定过氧化氢。流动注射分析系统通过促进过氧化氢与双核三氧化四氮-Fe（III）络合物的接触，形成过氧化物加成物。通过 ESI-MS 分析反应产物。在 m/z 251.5 和 240.5 处监测到两个加成物，过氧化氢 LOD 为 10^{-7} mol/L[109]。

7.4 结论

流动注射分析已与不同的质谱仪相结合用于有机和无机分析。流动系统在促进质谱检测之前的样品和标样管理方面做出了重大贡献。其主要应用集中在通过依次注入样品和标样来执行在线校准程序，以克服仪器漂移对定量过程的影响。通过在线标准化进行的校准程序，无论是通过执行标样添加还是通过使用不同的质谱仪器配置进行同位素稀释，都被证明是流动注射分析系统可以获得准确结果的原因。基质去除或分析物预浓缩的在线程序是 FIA-MS 应用的重要组成部分。固相树脂柱或汽化装置的插入是避免基质效应和获得较低检测限的决定性因素。使用与高效电离源相结合的小型样品引入装置的流动进样可以在减少干扰的情况下对小体积样品进行痕量分析。

参考文献

[1] Ruzicka, J. and Hansen, E. H. (1988) *Flow Injection analysis*, 2nd edn, Wiley, New York.

[2] Luque de Castro, M. D. and Tena, M. (1995) Hyphenated flow injection systems and high discrimination instruments. *Talanta*, 42, 151-169.

[3] Bier, M. E. and Cooks, R. G. (1987) Membrane interface for selective introduction of volatile compounds directly into the ionization chamber of a mass spectrometer. *Analytical Chemistry*, 59, 597-601.

[4] Wang, X., Viczian, M., Lasztity, A. and Barnes, R. (1988) Lead hydride generation for isotope analysis by inductively coupled plasma mass spectrometry. *Journal of Analytical Atomic Spectrometry*, 3, 821-827.

[5] Hansen, E. H. (2005) Use of flow injection and sequential injection analysis schemes for the determination of trace-level concentrations of metals in complex matrices by ETAAS and ICPMS. *Journal of Environmental Science and Health Part A-Toxic/Hazardous Substances & Environmental Engineering*, 40, 1507-1524.

[6] Holmes, J. C. and Morrell, F. A. (1957) Oscillographic mass spectrometric monitoring of gas chromatography. *Applied Spectroscopy*, 2, 86-88.

[7] Jones, P. R. and Yang, S. K. (1975) A liquid chromatography/mass spectrometer interface. *Analytical Chemistry*, 47, 1000-1003.

[8] Smith, R. D., Olivares, J. A., Nguyen, N. T. and Udseth, H. R. (1988) Capillary zone electrophoresis mass spectrometry using an electrospray ionization interface. *Analytical Chemistry*, 60, 436-441.

[9] Beckett, R. (1991) Field flow fractionation ICP-MS a powerful new analytical tool for characterizing macromolecules and particles. *Atomic Spectroscopy*, 12, 228-232.

[10] Centineo, G., Montes Bayon, M., De la Campa, R. F. and Sanz-Medel, A. (2000) Flow injection analysis with inductively coupled plasma time-of-flight mass spectrometry for the simultaneous determination of elements forming hydrides and its application to urine. *Journal of Analytical Atomic Spectrometry*, 15, 1357-1362.

[11] Schröder, H. Fr. (1997) Mass spectrometric detection and identification of polar pesticides and their degradation products a comparison of different ionization methods. *Environmental Monitoring and Assessment*, 44, 503-513.

[12] Cole, R. B. (1997) *Electrospray Ionization Mass Spectrometry Fundamentals, Instrumentation and Applications*, John Wiley & Sons, New York.

[13] Vestal, M. L. (1983) Studies of ionization mechanisms involved in thermospray LC-MS. *International Journal of Mass Spectrometry*, 46, 193-196.

[14] Takáts, Z., Wiseman, J. M., Gologan, B. and Cooks, R. G. (2004) Mass spectrometry sampling under ambient conditions with desorption electrospray ionization. *Science*, 306, 471-473.

[15] Cody, R. B., Laramée, J. A. and Durst, D. (2005) Versatile new ion source for the analysis of materials in open air under ambient conditions. *Analytical Chemistry*, 77, 2297-2302.

[16] Yinon, J. and Klein, F. S. (1971) The quadrupole and its applications in vacuum technology and mass spectrometry. *Vacuum*, 21, 379-383.

[17] Wollnik, H. (1999) Ion optics in mass spectrometry. *Journal of Mass Spectrometry*, 34, 991-1006.

[18] Jarvis, K. E., Gray, A. L. and Houk, R. S. (1992) *Handbook of Inductively Coupled Plasma Mass Spectrometry*, Blackie, Glasgow.

[19] Montaser, A. (1998) *Inductively Coupled Plasma Mass Spectrometry*, Wiley-VCH, New York.

[20] Günther-Leopold, I., Wernli, B., Kopajtic, Z. and Günther, D. (2004) Measurement of isotope ratios on transient signals by MC-ICP-MS. *Analytical and Bioanalytical Chemistry*, 378, 241-249.

[21] Rehkämper, M., Schönbächler, M. and Stirling, C. H., (2001) Multicollector ICP-MS: introduction to instrumentation, measurement techniques and analytical capabilities. *Geostandards Newsletter*, 25, 23-40.

[22] Price, D. and Milnes, G. J. (1990) The renaissance of time-of-flight mass spectrometry. *International Journal of Mass Spectrometry and Ion Processes*, 99, 1-39.

[23] Ray, S. J. and Hieftje, G. M. (2001) Mass analyzers for inductively coupled plasma time-of-flight mass spectrometry. *Journal of Analytical Atomic Spectrometry*, 16, 1206-1216.

[24] Ray, S. J., Andrade, F., Gamez, G., McClenathan, D. M., Rogers, D., Schilling, G., Wetzel, W. and Hieftje, G. M. (2004) Plasma-source mass spectrometry for speciation analysis: state-of-the-art. *Journal of Chromatography A*, 1050, 3-34.

[25] Carrión, M. C., Andrés, J. R., Rubí, J. A. M. and Emteborg, H. (2003) Performance optimization of isotope ratio measurements in transient signals by FI-ICP-TOFMS. *Journal of Analytical Atomic Spectrometry*, 18, 437-443.

[26] Bier, M. E. and Cooks, R. G. (1987) Membrane interface for selective introduction of volatile compounds directly into the ionization chamber of a mass spectrometer. *Analytical Chemistry*, 59, 597-601.

[27] Hayward, M. J., Kotiaho, T., Lister, A. K., Cooks, R. G., Austin, G. D., Narayan, R. and Tsao, G. T. (1990) On-line monitoring of bioreactions of bacillus *Polymyxa* and *Klebsiella oxytoca* by membrane introduction tandem mass spectrometry with flow injection analysis sampling. *Analytical Chemistry*, 62, 1798-1804.

[28] Srinivasan, N., Johnson, R. C., Kasthurikrishnan, N., Wong, P. and Cooks, R. G. (1997) Membrane introduction mass spectrometry. *Analytica Chimica Acta*, 350, 257-271.

[29] Tsai, G. -J., Austin, G. D., Syu, M. J., Tsao, G. T., Hayward, M. J., Kotiaho, T. and Cooks, R. G. (1991) Theoretical analysis of probe dynamics in flow injection/ membrane introduction mass spectrometry. *Analytical Chemistry*, 63, 2460-2465.

[30] Bier, M. E., Kotiaho, T. and Cooks, R. G. (1990) Direct insertion membrane probe for selective introduction of organic compounds into a mass spectrometer. *Analytica Chimica Acta*, 231, 175-190.

[31] Srinivasan, N., Kasthurikrishnan, N., Cooks, R. G., Krishnan, M. S. and Tsao, G. T. (1995) On-line monitoring with feedback control of bioreactors using a high ethanol tolerance yeast by membrane introduction mass spectrometry. *Analytica Chimica Acta*, 316, 269-276.

[32] Thompson, J. J. and Houk, R. S. (1986) Inductively coupled plasma mass spectrometry detection for multielement flow injection analysis and elemental speciation by reversed-phas liquid chromatography. *Analytical Chemistry*, 58, 2541-2548.

[33] Schröder, H. Fr. (1993) Surfactants: non-biodegradabls, significant pollutants in seawage treatment plant effluents Separation, identification and quantification by liquid chromatography, flow injection analysis-mass spectrometry and tandem mass spectrometry. *Journal of Chromatography*, 647, 219-234.

[34] Barco, M., Planas, C., Palacios, O., Ventura, F., Rivera, J. and Caixach, J. (2003) Simultaneous quantitative analysis of anionic, cationic and nonionic surfactants in water by electrospray ionization mass spectrometry with flow injection analysis. *Analytical Chemistry*, 75, 5129–5136.

[35] Kristiansen, G. K., Brock, R. and Bojesen, G. (1994) Comparison of Flow injection/thermospray MS/MS and LC/Thermospray MS/MS methods for determination of sulfonamides in meat and blood. *Analytical Chemistry*, 66, 3253–3258.

[36] Schröder, H. Fr. and Meesters, R. J. W. (2005) Stability of fluorinated surfactants in advanced oxidation processes A follow up of degradation products using flow injection mass spectrometry, liquid chromatography mass spectrometry and liquid chromatography multiple stage mass spectrometry. *Journal of Chromatography A*, 1082, 110–119.

[37] Gómez-Ariza, J. L., Arias-Borrego, A. and García-Barrera, T. (2006) Use of flow injection atmospheric pressure photoionization quadrupole time-of-flight mass spectrometry for fast olive oil fingerprinting. *Rapid Communications in Mass Spectrometry*, 20, 1181–1186.

[38] Wang, A. P. L. and Li, L. (1992) Pulsed sample introduction interface for combining flow injection analysis with multiphoton ionization time-of-flight mass spectrometry. *Analytical Chemistry*, 64, 769–775.

[39] Koeber, R., Niesser, R. and Bayona, J. M. (1997) Comparison of liquid chromatography-mass spectrometry interfaces for the analysis of polar metabolites of benzopyrene. *Fresenius' Journal of Analytical Chemistry*, 359, 267–273.

[40] Wiederin, D. R. and Houk, R. S. (1991) Measurements of aerosol-particle sizes from a direct injection nebulizer. *Applied Spectroscopy*, 45, 1408–1412.

[41] Bouyssiere, B., Ordóñez, Y. N., Lienemann, C. P., Schaumlöffel, D. and Łobiński, R. (2006) Determination of mercury in organic solvents and gas condensates by (flow injection-inductively coupled plasma mass spectrometry using a modified total consumption micronebulizer fitted with single pass spray chamber. *Spectrochimica Acta Part B–Atomic Spectroscopy*, 61, 1063–1068.

[42] Koropchak, J. A. and Veber, M. (1992) Thermospray sample introduction to atomic spectrometry. *Critical Reviews in Analytical Chemistry*, 23, 113–141.

[43] Machado, L. F. R., Jacintho, A. O., Menegário, A. A., Zagatto, E. A. G. and Giné, M. F. (1998) Electrochemical and chemical processes for hydride generation in flow injection ICP-MS: determination of arsenic in natural waters. *Journal of Analytical Atomic Spectrometry*, 13, 1343–1346.

[44] Al-Ammar, A. S. and Barnes, R. M. (2001) Improving isotope ratio precision in inductively coupled plasma quadrupole mass spectrometry by common analyte internal standardization. *Journal of Analytical Atomic Spectrometiy*, 16, 327–332.

[45] Houk, R. S. and Thompson, J. J. (1983) Trace-metal isotopic analysis of microliter solution volumes by inductively coupled plasma mass-spectrometry. *Biomedical Mass Spectrometry*, 10, 107–112.

[46] Schaumlöffel, D., Giusti, P., Zority, M. V., Pickhardt, C., Szpunar, J., Łobiński, R. and Becker, J. S. (2005) Ultratrace determination of uranium and plutonium by nano-volume flow injection doublefocusing sector field inductively coupled plasma mass spectrometry (nFI-ICP-SFMS). *Journal of Analytical Atomic Spectrometry*, 20, 17–21.

[47] Menegário, A. A., Packer, A. P. and Giné, M. F. (2001) Determination of Ba, Cd, Cu and Zn in saliva by isotope dilution direct injection inductively coupled plasma mass spectrometry. *Analyst*, 126, 1363–1366.

[48] Willie, S. N., Lam, J. W. H., Yang, L. and Tao, G. (2001) On-line removal of Ca, Na and Mg from imidioacetate resin for the determination of trace elements in seawater and fish otoliths by flow injection ICP-MS. *Analytica Chimica Acta*, 447, 143-152.

[49] Menegário, A. A. and Giné, M. F. (1997) On-line removal of anions for plant analysis by inductively coupled plasma mass spectrometry. *Journal of Analytical Atomic Spectrometiy*, 12, 671-674.

[50] Sun, Y. C., Lin, C. Y., Wu, S. F. and Chung, Y. T. (2006) Evaluation of on-line desalter-inductively coupled plasma-mass spectrometry system for determination of Cr (III), Cr (VI), and total chromium concentrations in natural water and urine samples. *Spectrochimica Acta Part B-Atomic Spectroscopy*, 61, 230-234.

[51] Wang, J. H. and Hansen, E. H. (2002) FI/SI on-line solvent extraction/back extraction preconcentration coupled to direct injection nebulization inductively coupled plasma mass spectrometry. *Journal of Analytical Atomic Spectrometry*, 17, 1284-1289.

[52] Klinkenberg, H., Beeren, T. and Borm, W. V. (1993) The use of an enriched isotope as an on-line internal standard in inductively coupled plasma mass spectrometry: a reference method for a proposed determination of tellurium in industrial waste water by means of graphite furnace atomic absorption spectrometry. *Spectrochimica Acta Part B-Atomic Spectroscopy*, 48, 649-661.

[53] Sperling, M., Fang, Z. and Welz, B. (1991) Expansion of dynamic working range and correction for interferences in flame atomic absorption spectrometers using flow injection gradient ratio calibration with a single standard. *Analytical Chemistry*, 63, 151-159.

[54] McClenathan, D. M., Ray, S. J. and Hieftje, G. M. (2001) Novel flow injection strategies for study and control of matrix interferences by inductively coupled plasma time-of-flight mass spectrometry. *Journal of Analytical Atomic Spectrometry*, 16, 987-990.

[55] Giné, M. F., Reis, B. F., Krug, F. J., Jacintho, A. O., Bergamin Filho, H. and Zagatto, E. A. G. (1994) Flow injection plasma spectrometry. *ICP Information Newsletter*, 20, 413-417.

[56] Araújo, M. C. U., Pasquini, C., Bruns, R. E. and Zagatto, E. A. G. (1985) A fast procedure for standard additions in flow injection analysis. *Analytica Chimica Acta*, 171, 337-343.

[57] Antler, M., Mazwell, E. J., Duford, D. A. and Salin, E. D. (2007) On-line standard additions calibration of transient signals for inductively coupled plasma mass spectrometry. *Analytical Chemistry*, 79, 688-694.

[58] Goossens, J., Moens, L. and Dams, R. (1994) Determination of lead by flow injection inductively coupled plasma mass spectrometry comparing several calibration techniques. *Analytica Chimica Acta*, 293, 171-181.

[59] Parent, P. M., Vanhoe, H., Moens, L. and Dams, R. (1996) Evaluation of a flow injection system combined with an inductively coupled plasma mass spectrometry with a thermospray nebulization for the determination of trace level of platinum. *Analytica Chimica Acta*, 320, 1-10.

[60] Magnusson, B., Trešl, I. and Haraldsson, C. (2005) Isotope dilution with ICP-MS simplified uncertainty estimation using a robust procedure based on a higher target value of uncertainty. *Journal of Analytical Atomic Spectrometry*, 20, 1024-1029.

[61] Lasztity, A., Viczián, M., Wang, X. and Barnes, R. M. (1989) Sample analysis by on-line isotope dilution inductively coupled plasma mass spectrometry. *Journal of Analytical Atomic Spectrometry*, 4, 761-766.

[62] Viczián, M., Lasztity, A., Wang, X. and Barnes, R. M. (1990) On-line isotope dilution and

sample dilution by flow injection and inductively coupled plasma mass spectrometry. *Journal of Analytical Atomic Spectrometry*, 5, 125-133.

[63] Beauchemin, D. and Specht, A. A. (1997) On-line isotope dilution analysis with ICPMS using reverse flow injection. *Analytical Chemistry*, 69, 3183-3187.

[64] Specht, A. A. and Beauchemin, D. (1998) Automated on-line isotope dilution analysis with ICP-MS using sandwich flow injection. *Analytical Chemistry*, 70, 1036-1040.

[65] Furuta, N. (1991) Optimization of the mass scanning rate for the determination of lead isotope ratio using an inductively coupled plasma mass spectrometry. *Journal of Analytical Atomic Spectrometry*, 6, 199-203.

[66] Bellato, A. C. S., Menegário, A. A. and Giné, M. F. (2003) Boron Isotope Dilution in cellular fractions of coffee leaves evaluated by Inductively Coupled Plasma Mass Spectrometry with Direct Injection nebulization (DIN-ICP-MS). *Journal of the Brazilian Chemical Society*, 14, 269-273.

[67] Benkhedda, K., Infante, H. G. and Addams, F. C. (2002) Inductively coupled plasma mass spectrometry for trace analysis using flow injection on-line preconcentration and time-of-flight mass analyser. *Trends in Analytical Chemistry*, 21, 332-342.

[68] Packer, A. P. C., Giné, M. F., Reis, B. F. and Menegário, A. A. (2001) Micro flow system to perform programmable isotope dilution for inductively coupled plasmamass spectrometry. *Analytica Chimica Acta*, 438, 267-272.

[69] Giné, M. F., Bellato, A. C. S. and Menegário, A. A. (2004) Determination of trace elements in serum samples by isotope dilution inductively coupled plasma mass spectrometry using on-line dyalisis. *Journal of Analytical Atomic Spectrometry*, 19, 1252-1256.

[70] Bellato, A. C. S., Giné, M. F. and Menegário, A. A. (2004) Determination of B in body fluids by isotope dilution inductively coupled plasma mass spectrometry with direct injection nebulization. *Microchemical Journal*, 77, 119-123.

[71] Heumann, K. G. (2004) Isotope dilution ICP-MS for trace element determination and speciation: from a reference method to a routine method? *Analytical and Bioanalytical Chemistry*, 378, 318-329.

[72] Schaumlöffel, D., Giusti, P., Preud' Homme, H., Szpunar, J. and Łobiński, R. (2007) Pre-column isotope dilution analysis in nano HPLC-ICPMS for absolute quantification of sulfur containing peptides. *Analytical Chemistry*, 79, 2859-2868.

[73] Huang, Z. Y., Chen, F. R., Zhuang, Z. X., Wang, X. R. and Lee, F. S. C. (2004) Trace lead measurement and on-line removal of matrix interferences in seawater by isotope dilution coupled with flow injection and ICP-MS. *Analytica Chimica Acta*, 508, 239-245.

[74] Dimitrova-Koleva, B., Benkhedda, K., Ivanova, E. and Adams, F. (2007) Determination of trace elements in natural waters by inductively coupled plasma time of flight mass spectrometry after flow injection preconcentration in a knotted reactor. *Talanta*, 71, 44-50.

[75] DeNicola Cafferky, K., Thompson, R. L., Richardson, D. D. and Caruso, J. A. (2007) Determination, by inductively coupled plasma mass spectrometry, of changes in cellular metal content resulting from herpes simplex virus (HSV-1) infection. *Analytical and Bioanalytical Chemistry*, 387, 2037-2043.

[76] Dressier, V. L., Pozebon, D., Matusch, A. and Becker, J. S. (2007) Micronebulization for trace analysis of lanthanides in small biological specimens by ICP-MS. *International Journal of Mass Spectrometry*, 266, 25-33.

[77] Giné, M. F., Gervasio, A. P. G., Lavorante. A. F., Miranda, C. E. S. and Carrilho, E. (2002)

Interfacing flow injection with capillary electrophoresis and inductively coupled plasma mass spectrometry for Cr speciation in water samples. *Journal of Analytical Atomic Spectrometry*, 17, 736-738.

[78] Packer, A. P., Gervasio, A. P. G., Miranda, C. E. S., Reis, B. F., Menegário, A. A. and Giné, M. F. (2003) On-line electrolytic dissolution for lead determination in high purity copper by isotope dilution inductively coupled plasma mass spectrometry. *Analytica Chimica Acta*, 485, 145-153.

[79] Yang, L., Willie, S. and Sturgeon, R. (2005) Ultra-trace determination of mercury in water by cold-vapor generation isotope dilution mass spectrometry. *Journal of Analytical Atomic Spectrometry*, 20, 1226-1231.

[80] Falta, T., Limbeck, A., Stingeder, G. and Hann, S. (2007) Ultra-trace determination of palladium in human urine samples via flow injection coupled with ICP-MS. *Atomic Spectroscopy*, 28 (3), 81-89.

[81] Long, S. E. and Vetter, T. W. (2002) Determination of sodium in blood serum by inductively coupled plasma mass spectrometry. *Journal of Analytical Atomic Spectrometry*, 17 (12), 1589-1594.

[82] Saint' Pierre, T. D., Tormen, L., Frescura, V. L. A. and Curtius, A. J. (2006) The direct analysis of fuel ethanol by ICP-MS using a flow injection system coupled to an ultrasonic nebulizer for sample introduction. *Journal of Analytical Atomic Spectrometry*, 21, 1340-1344.

[83] Lu, P.-L., Huang, K.-S. and Jiang, S.-J. (1993) Determination of traces of copper cadmium and lead in biological and environmental samples by flow injection isotope dilution inductively coupled plasma mass spectrometry. *Analytica Chimica Acta*, 284, 181-188.

[84] Coedo, A. G., Dorado, M. T. and Padilla, I. (2005) Evaluation of different sample introduction approaches for the determination of boron in unalloyed steels by inductively coupled plasma mass spectrometry. *Spectrochimica Acta Part B-Atomic Spectroscopy*, 60, 73-79.

[85] Koellensperger, G., Hann, S., Nurmi, J., Prohaska, T. and Stingeder, G. (2003) Uncertainty of species unspecific quantification strategies in hyphenated ICP-MS analysis. *Journal of Analytical Atomic Spectrometry*, 18, 1047-1055.

[86] Infante, H. G., Van Campenhout, K., Schaumloffel, D., Blust, R. and Adams, F. C. (2003) Multi-element speciation of metalloproteins in fish tissue using size-exclusion chromatography coupled "online" with ICP-isotope dilution-time-of-flight-mass spectrometry. *Analyst*, 128, 651-657.

[87] Chen, S., Xiao, M., Lu, D. and Zhan, X. (2007) Use of a microcolumn packed with modified carbon nanofibers coupled with inductively coupled plasma mass spectrometry for simultaneous on-line preconcentration and determination of trace rare earth elements in biological samples. *Rapid Communications in Mass Spectrometry*, 21, 2524-2528.

[88] Epov, V. N., Benkhedda, K. and Evans, R. D. (2005) Determination of Pu isotopes in vegetation using a new on-line FI-ICP-DRC-MS protocol after microwave digestion. *Journal of Analytical Atomic Spectrometry*, 20, 990-992.

[89] Fang, J., Liu, L.-W. and Yang, X.-P. (2006) Minimization of matrix interferences in quadrupole inductively coupled plasma mass spectrometric (ICP-MS) determination of Palladium using a flow injection mass displacement solid-phase extraction protocol. *Spectrochimica Acta Part B-Atomic Spectroscopy*, 61, 864-869.

[90] Tolmachyov, S. Y., Kuwahara, J. and Noguchi, H. (2004) Flow injections extraction with ICP-MS for thorium and uranium determination in human body fluids. *Journal of Radioanalytical and Nuclear Chemistry*, 261, 125-131.

[91] Sun, Y., Lu, Y. and Chung, Y. (2007) Online in-tube solid-phase extraction coupled to ICP-MS for *in vivo* determination of the transfer kinetics of trace elements in the brain extracellular fluid of anesthetized rats. *Journal of Analytical Atomic Spectrometry*, 22, 77-83.

[92] Benkhedda, K., Dimitrova, B., Infante, H. G., Ivanova, E. and Adams, F. C. (2003) Simultaneous on-line preconcentration and determination of Pt, Rh and Pd in urine, serum and road dust by flow injection combined with inductively coupled plasma time-of-flight mass spectrometry. *Journal of Analytical Atomic Spectrometry*, 18, 1019-1025.

[93] Lopes, C. M. P. V., Almeida, A. A., Saraiva, M. L. M. F. S. and Lima, J. L. F. C. (2007) Determination of Rh, Pd and Pt in urine samples using a preconcentration sequential injection analysis system coupled to a quadrupole-inductively coupled plasma-mass spectrometer. *Analytica Chimica Acta*, 600, 226-232.

[94] Hirata, S., Kajiya, T., Takano, N., Aihara, M., Honda, K., Shikino, O. and Nakayama, E. (2003) Determination of trace metals in seawater by on-line column preconcentration inductively coupled plasma mass spectrometry using metal alkoxide glass immobilized 8-quinolinol. *Analytica Chimica Acta*, 499, 157-165.

[95] Hirata, S., Kajiya, T., Aihara, M., Honda, K., Shikino, O. and Nakayama, E. (2002) Determination of rare earth elements in seawater by on-line column preconcentration inductively coupled plasma mass spectrometry. *Talanta*, 58, 1185-1194.

[96] Epov, V. N., Benkhedda, K., Brownell, D., Cornett, R. J. and Evans, R. D. (2005) Comparative study of three sample preparation approaches for the fast determination of americium in urine by flow injection ICP-MS. *Canadian Journal of Analytical Sciences and Spectroscopy*, 50, 14-22.

[97] Epov, V. N., Benkhedda, K., Cornett, R. J. and Evans, R. D. (2005) Rapid determination of plutonium in urine using flow injection on-line preconcentration and inductively coupled plasma mass spectrometry. *Journal of Analytical Atomic Spectrometry*, 20, 424-430.

[98] Benkhedda, K., Lariviere, D., Scott, S. and Evans, D. (2005) Hyphenation of flow injection on-line preconcentration and ICP-MS for the rapid determination of Ra-226 in natural waters. *Journal of Analytical Atomic Spectrometry*, 20, 523-528.

[99] Geerdink, R. B., Berg, P. J., Kienhuis, P. G. M. and Brinkman, U. A. T. (1996) Flow-injection analysis thermospray tandem mass spectrometry of triazine herbicides and some of their degradation products in surface water. *International Journal of Environmental Analytical Chemistry*, 64, 265-278.

[100] Kristiansen, G. K., Brock, R. and Bojesen, G. (1994) Comparison of flow injection/thermospray MS/MS and LC/thermospray MS/MS methods for determination of sulfonamides in meat and blood. *Analytical Chemistry*, 66, 3253-3258.

[101] Schröder, H. Fr. (1995) Polar organic pollutants in the Elbe river - liquid-chromatographic mass-spectrometric and flow injection analysis mass-spectrometric analyses demonstrating changes in quality and concentration during the unification process in germany. *Journal of Chromatography A*, 712, 123-140.

[102] Ito, Y., Goto, T., Oka, H., Matsumoto, H., Miyasaki, Y., Takahashi, N. and Nakazawa, H. (2003) Simple and rapid determination of thiabendazole, imazalil and *o*-phenylphenol in citrus fruit using flow injection electrospray ionization tandem mass spectrometry. *Journal of Agricultural and Food Chemistry*, 51, 861-866.

[103] Basilico, F., Sauerwein, W., Pozzi, F., Wittig, A., Moss, R. and Mauri, P. L. (2005) Analysis of B-10 antitumoral compounds by means of flow injection into ESI-MS/MS. *Journal of Mass*

Spectrometry, 40, 1546-1549.

［104］Santos, B., Simonet, B. M., Ríos, A. and Valcárcel, M. (2004) Direct automatic determination of biogenic amines in wine by flow injection–capillary electrophoresis–mass spectrometry. *Electrophoresis*, 25, 3427-3433.

［105］Santos, B., Simonet, B. M., Lendl, B., Ríos, A. and Valcárcel, M. (2006) Alternatives for coupling sequential injection systems to commercial capillary electrophoresis–mass spectrometry equipment. *Journal of Chromatography A*, 1127, 278-285.

［106］Carducci, C., Santagata, S., Leuzzi, V., Carducci, C., Artiola, C., Giovanniello, T., Battini, R. and Antonozzi, I. (2006) Quantitative determination of guanidinoacetate and creatine in dried blood spot by flow injection analysis–electrospray tandem mass spectrometry. *Clinica Chimica Acta*, 364, 180-187.

［107］Johnson, D. W., Gerace, R., Rainieri, E., Trihn, M. and Fingerhut, R. (2007) Analysis of succinylacetone, as a Girard T derivative, in urine and dried bloodspots by flow injection electrospray ionization tandem mass spectrometry. *Rapid Communications in Mass Spectrometry*, 21, 59-63.

［108］Charles, L. (2003) Flow injection of the lock mass standard for accurate mass measurement in electrospray ionization time–of–flight mass spectrometry coupled with liquid chromatography. *Rapid Communications in Mass Spectrometry*, 17, 1383-1388.

［109］McCooeye, M. and Mester, Z. (2006) Comparison of flow injection analysis electrospray mass spectrometry and tandem mass spectrometry and electrospray high-field asymmetric waveform ion mobility mass spectrometry and tandem mass spectrometry for the determination of underivatized amino acids. *Rapid Communications in Mass Spectrometry*, 20, 1801-1808.

［110］Mauri, P., Minoggio, M., Simonetti, P., Gardana, C. and Pietta, P. (2002) Analysis of saccharides in beer samples by flow injection with electrospray mass spectrometry. *Rapid Communications in Mass Spectrometry*, 16, 743-748.

［111］McCooeye, M., Ding, R., Gardner, G. J., Fraser, C. A., Lam, J., Sturgeon, R. E. and Mester, Z. (2003) Separation and quantitation of the stereoisomers of ephedra alkaloids in natural health products using flow injection–electrospray ionization high field asymetric waveform ion mobility spectrometry–mass spectrometry. *Analytical Chemistry*, 75, 2538-2542.

［112］Wade, N. and Miller, K. (2005) Determination of active ingredient within pharmaceutical preparations using flow injection mass spectrometry. *Journal of Pharmaceutical and Biomedical Analysis*, 37, 669-678.

［113］Numan, A. and Danielson, N. D. (2002) On-line photo derivatization with flow injection and liquid chromatographic–atmospheric pressure electrospray mass spectrometry for the identification of indoles. *Analytica Chimica Acta*, 460, 49-60.

［114］Vaidyanathan, S., Kell, D. B. and Goodacre, R. (2002) Flow-injection electrospray ionization mass spectrometry of crude cell extracts for high throughput bacterial identification. *Journal of the American Society for Mass Spectrometry*, 13, 118-128.

［115］Keith-Roach, M. J., Buratti, M. V. and Worsfold, P. J. (2005) Thorium complexation by hydroxamate siderophores in perturbed multicomponent systems using Flow Injection Electrospray Ionization Mass Spectrometry. *Analytical Chemistry*, 77, 7335-7341.

8 流动分析的环境应用

Shoji Motomizu

8.1 引言

环境科学是需要化学分析的最广泛的领域之一。在环境分析中，样品数量非常大，分析必须尽快进行，或者在现场进行。此外，样品非常复杂，测量前必须去除基质；分析物浓度通常很低，因此需要一些富集程序。

为了克服这些问题，必须开发一种与带有预处理功能的传统流动注射分析（FIA）和顺序注射分析（SIA）相结合的计算机控制流动方法。现在可以使用计算机控制的泵、阀门和在线预处理设备，因此，可以以合理的成本轻松构建复杂的分析系统。

8.2 通过流动方法分析水环境

8.2.1 与富营养化有关的物质

由于生活污水和工业废水的流入，湖泊、沿海和内海可能发生富营养化，有时会发生赤潮，导致鱼类和贝类无法生存。参与富营养化的主要物质是无机和有机的氮和磷化合物。无机氮化合物包括亚硝酸盐、硝酸盐和氨，无机磷化合物包括正磷酸盐和缩合磷化合物，例如焦磷酸盐和三磷酸盐。一般来说，关于富营养化，必须测定废水、河湖水和海水中的总氮和总磷含量。该测定基于氮和磷化合物分别在高温和高压下与过二硫酸盐氧化分解为硝酸盐和正磷酸盐。因此，硝酸盐和正磷酸盐的测定方法对于富营养化分析很重要。

8.2.1.1 含氮化合物

（1）亚硝酸盐和硝酸盐的测定　一般情况下，水生环境中亚硝酸盐和硝酸盐的总浓度水平约为 $10^{-5}\sim 10^{-4}$ mol/L，亚硝酸盐与硝酸盐的浓度比约为 1∶10。因此，亚硝酸盐的可测定范围和检测限（LOD）必须分别处于 10^{-6} mol/L 和 10^{-7} mol/L 的浓度水平。

光度检测可用于通过流动进样技术测定亚硝酸盐。所用的检测反应是亚硝酸盐与磺胺（SA）的重氮化，然后将重氮离子与 N-（1-萘基）-乙二胺（NEDA）[1]偶联。可以在 540nm 处测量产物。

这种重氮化偶联反应可用于通过适当的还原剂将硝酸盐还原为亚硝酸盐后测定硝酸盐。在将硝酸盐还原为亚硝酸盐的便捷方法中，铜化镉（Cd/Cu）颗粒（0.5~1.0mm）填充到玻璃柱 [2mm（i.d.）×10cm] 中并安装在载体管线中，其中样品溶液由六通切换阀引入，如图 8.1（1）所示。试剂溶液制备如下：在 0.1mol/L HCl 中含有 0.1% SA 和 0.01% NEDA（二盐酸盐），载体包含 EDTA 用于更新 Cd/Cu 颗粒的表面[2]。

对于亚硝酸盐的测定，可以通过修改系统并使用相同的流程图：用水代替载体并卸载还原柱。

在图 8.1（2）中，显示了亚硝酸盐和硝酸盐的流动信号。从中可以看出，硝酸盐的峰高与亚硝酸盐的峰高几乎相同，说明硝酸盐还原为亚硝酸盐的比率几乎为 100%。

该方法可以很好地应用于河流和海水的测定；由所提出方法获得的结果与间歇标准方法（日本工业标准）获得的结果之间具有非常好的相关性[2]。

要将硝酸盐还原为亚硝酸盐，可以使用低压汞灯[3]；反应管［PTFE：0.8mm（i.d.），1.6mm（o.d.）×3m］缠绕在灯［14mm（o.d.）×134mm，4W］周围，并用铝箔包裹。作为载体，使用了0.1mol/L KH_2PO_4 和1mmol/L EDTA。灯安装在进样阀之后；流程图类似于图8.1（1）。

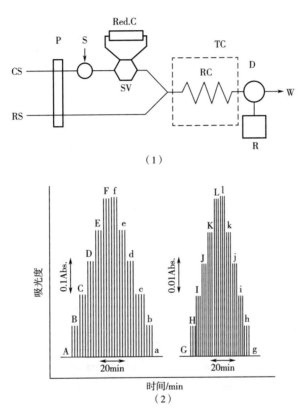

图8.1 （1）使用还原柱的流程图[2] 测定硝酸盐的流程图 （2）亚硝酸盐和硝酸盐的流动分布
CS—载体（EDTA，pH 8.2） RS—试剂溶液（SA+NEDA） P—双柱塞泵 S—进样器（100pL）
SV—六通切换阀 Red. C—还原柱 TC—温控浴（40℃） RC—反应柱 ［0.5mm（i.d.）×2m］
D—检测器（540nm） R—记录仪 W—废弃物

[$N-NO_2^-$]/ppm：A，0；B，0.2；C，0.4；D，0.6；E，0.8；F，1.0。[$N-NO_3^-$]/ppm：a，0；b，0.2；c，0.4；d，0.6；e，0.8；f，1.0。[$N-NO_2^-$]/ppb：G，0；H，20；I，40；J，60；K，80；L，100。[$N-NO_3^-$]/ppb：g，0；h，20；i，40；j，60；k，80；l，100。

作为简单的流动进样系统，间苯三酚（1,3,5-三羟基苯）可用于检测亚硝酸盐；亚硝化反应的产物可以在312nm处进行分光光度法检测[4]。同样可以使用 N,N-二-(2-羟丙基)苯胺，检测波长为500nm[5]。

可以通过基于C酸（3-氨基-1,5-萘二磺酸）在酸性介质中重氮化的荧光光度法对硝酸盐和硝酸盐进行高灵敏度检测，然后向反应溶液中加入NaOH以产生重氮酸的去质子化物质：λ_{ex} = 365nm，λ_{em} = 470nm。该方法可应用于测定河水和海水[6,7]。

顺序进样/分光光度法可用于测定亚硝酸盐，其试剂与流动进样技术中的试剂相同。然而，其灵敏度不足以测定浓度为 10^{-6} mol/L 或更低的亚硝酸盐；它足以测定 10^{-5} mol/L 或更高浓度的硝酸盐，但 Cd/Cu 柱的更新是单调且不可重现的。

多注射器和电磁泵系统可用于通过分光光度检测测定亚硝酸盐/硝酸盐[8]。

（2）氨的测定　水溶液中的氨可以通过基于靛酚蓝反应的分光光度法测定，其中使用苯酚和次氯酸盐进行着色。在硝普钠作为催化剂的情况下，氨可以与次氯酸盐反应生成一氯胺（NH_2Cl），后者又与酚类反应生成吲哚酚蓝。各种苯酚已被建议用于氨的测定，其中推荐使用水杨酸，因为苯酚（水杨酸）具有稳定性和高选择性[9]。但硝普钠是一种有毒物质，因此必须高度重视对实验废物的处理。

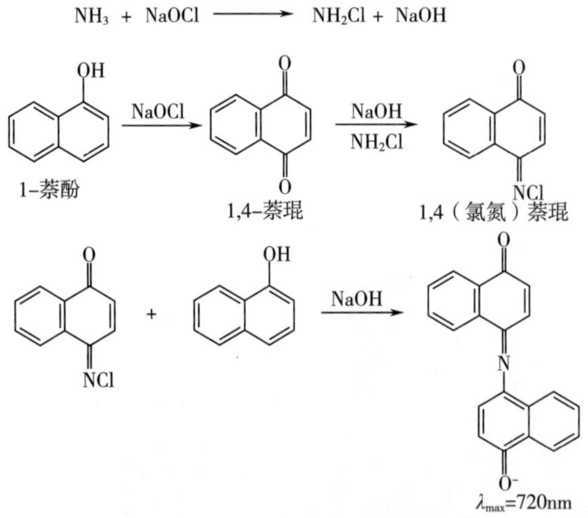

图 8.2　通过 1-萘酚法测定氨的检测反应

最有趣和最灵敏的靛酚蓝反应之一是基于 1-萘酚和次氯酸盐在碱性溶液中的反应，其中不含作为催化剂的硝普钠[10]。该方法使用丙酮来提高灵敏度[11]，异氰尿酸钠溶液的使用方法与次氯酸钠方法类似：异氰尿酸盐在碱性溶液中可转化为次氯酸盐。反应可如图 8.2 所示发生。

在图 8.3（1）和（2）中，所示分别为氨测定的流程图和信号曲线。通过所提出的方法，可以确定浓度约为 10^{-7} mol/L 的氨。

一种替代方法基于氨在碱性溶液中通过气体扩散穿过疏水膜进入含有显色试剂的吸收溶液。水样中的铵离子在碱性载流中转化为气态氨，气态氨可通过透气膜被吸收到含有酸碱指示剂的试剂溶液中。作为透气膜，可以使用微孔 PTFE 管[12-14]。在图 8.4（1）和（2）中，显示了双管式气体扩散装置和带有气体扩散装置的流程图[15]。

在气体扩散装置中，在水溶液长时间流动之后，疏水膜可能变得不那么疏水并且渗透效率降低。为了避免膜疏水性的降低，内管和外管中的水溶液可以在实验后用空气代替，具体操作方法是打开额外的用于吸入空气的开关阀，并用吸水器或真空泵将空气吸入管中几分钟[15]。

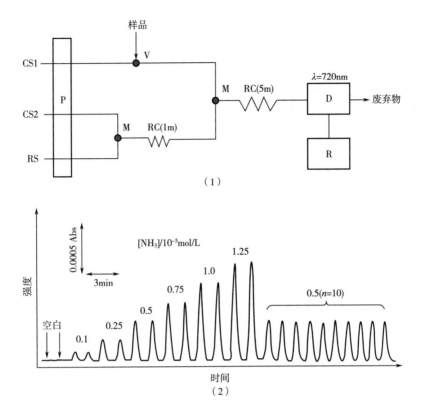

图 8.3 （1）氨测定流程图 （2）氨测定校准图的流动曲线

CS1—纯水　CS2—0.01mol/L $C_3Cl_2N_3NaO_3$、0.1 mol/L NaOH　RS—35%丙酮中加入 0.5% 1-萘酚
V—注射阀　M—混合接头　RC—反应盘管　D—探测器　R—记录仪器
RSD：0.81%（n = 10）；LOD：0.79×10^{-7} mol/L；样品吞吐量：45/h。

一般情况下，氨、二氧化碳等气态物质的渗透效率约为 10%，因此 LOD 可能会比用吲哚酚蓝法直接检测更差。但是，几乎可以消除实际样品中存在的干扰物质的所有影响。NH_3-N 的可测定范围和 LOD 分别为 0~1.0ppm 和 0.01ppm[15]。

（3）有机氮化合物　水样中的总氮含量可以通过分光光度法测量。样品中的含氮化合物在 120℃ 高压釜中用过二硫酸钾氧化分解后，硝酸盐可直接用 220nm 分光光度法测定[16]，也可采用基于硝酸盐还原为亚硝酸盐的分光光度法和重氮化偶联反应测定。

8.2.1.2 含磷化合物

磷是所有生命（包括水生系统中的浮游植物）的必需营养素，因此，来自生活和工业废物以及陆地土壤中过量的磷化合物会导致湖泊、沿海和内海富营养化，这可能会导致赤潮。赤潮往往伴随着有毒藻类的异常生长，有时会对鱼类造成损害。

正磷酸盐（PO_4^{3-} 及其质子化形式）是所有磷形式中生物利用度最高的。非富营养化水域中磷浓度的标准被认为小于 20ppb[17]，这意味着水生环境中磷的最低可测定浓度可能<5ppb。

（1）正磷酸盐　测定正磷酸盐（PO_4^{3-}）的传统分光光度法之一是基于在还原剂存

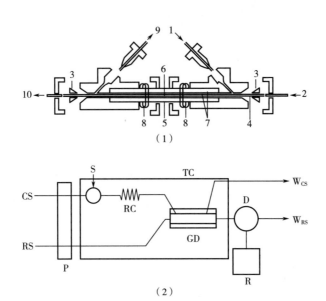

图8.4 （1）双管气体扩散装置[15] （2）气体扩散法测定氨的流程图
1—载体进出 2—试剂溶液的进出 3—套圈 4—多孔聚四氟乙烯管 5—玻璃管
6—试剂流 7—载流 8—O型圈 9—载体废弃物 10—试剂进出到检测器
CS—载流（0.02mol/LNaOH） RS—试剂流（4×10^{-4} mol/L HEPES，1.25×10^{-4} mol/L 甲酚红，pH7.0）
P—泵（1.0mL/min） S—样品（200μL） RC—反应盘管[0.5mm（内径）×1m] GD—气体扩散装置（气体扩散：5cm）
TC—温控浴（40℃） D—检测器（550nm） R—记录仪 W$_{CS}$—载体废物 W$_{RS}$—试剂废物

在下在酸性溶液中与钼酸盐形成钼蓝的经典反应。正磷酸盐能与钼酸盐反应生成黄色杂多酸、钼磷酸盐或 $H_3PMo_{12}O_{40}$，再用抗坏血酸、氯化亚锡等还原剂还原生成钼蓝，摩尔吸光度为 $(1\sim2)\times10^4$ L/(mol·cm)。钼蓝法在间歇法中的缺陷之一是缺乏反应的重现性，尤其是还原反应。然而，这个缺陷可以通过使用流动进样技术完全克服[18]。流程图是一个传统的双管线系统：一个用于载流，另一个用于试剂流。例如，载体溶液可能只是水，而试剂溶液是通过将等体积的溶液（A）和（B）混合来制备的：（A）含有5.5g钼酸铵（四水合物），0.25g 酒石酸锑钾和浓硫酸的 1000mL水溶液；（B）在 1000mL 水溶液中含有 3.0g 抗坏血酸和 1.0g 十二烷基硫酸钠[19]。

孔雀石绿（MG）是一种阳离子三苯甲烷染料，在酸性介质中与磷钼酸反应形成有色离子缔合物[20]，通过使用该方法，可以开发更灵敏的分光光度法测定正磷酸盐。这种着色反应基于磷钼酸与黄色质子化孔雀石绿 HMG^{2+} 的离子缔合反应。这种着色反应可以在酸性介质中进行，如下所示：

$$H_3PMo_{12}O_{40} + HMG^{2+} \longrightarrow (MG^+)(H_2PMo_{12}O_{40}^-) + 2H^+ \quad (8.1)$$
（黄） （黄，$\lambda_{max}=446nm$） （绿，$\lambda_{max}=650nm$）

$$(MG^+)(H_2PMo_{12}O_{40}^-) + HMG^{2+} \longrightarrow (MG^+)_2(HPMo_{12}O_{40}^{2-}) + 2H^+ \quad (8.2)$$

$$(MG^+)_2(HPMo_{12}O_{40}^{2-}) + HMG^{2+} \longrightarrow (MG^+)_3(PMo_{12}O_{40}^{3-}) + 2H^+ \quad (8.3)$$

如式（8.3）中形成的（3:1）离子缔合物的最终产物可以在水溶液中沉淀，该产物可以萃取到有机相中[21]，或者吸附在疏水固相上，例如硝酸纤维素膜过滤器[22,23]，通过在反应溶液中加入聚乙烯醇（PVA），离子缔合反应（8.2）和（8.3）终止，离子缔合体可以溶解在溶液中。在650nm处，水溶液中的摩尔吸收率约为$8\times10^4/$（mol·cm）：这种着色反应称为孔雀石绿方法。孔雀石绿方法的优点：①无需加热即可进行快速显色反应；②比钼蓝方法灵敏度更高；③最大波长比钼黄方法更长。

孔雀石绿方法与流动注射分析联用可用于测定水样中的磷酸盐：在650nm处磷的LOD为1ppb。孔雀石绿方法可用于通过在线消解/分光光度法流动注射分析测定工业废水中的总磷[24]。

通过使用双管线流动注射分析系统，荧光光度法可用于测定正磷酸盐的灵敏度：载体是水，试剂流包含钼酸盐、罗丹明6G和盐酸[25]。该方法基于由磷钼酸与罗丹明6G形成离子缔合物而导致的阳离子染料的荧光猝灭：磷的LOD为0.1ppb，0.5mol/L NaCl不会干扰测定。该方法可用于河流、海水等实际水域中正磷酸盐的测定。可以在聚乙烯醇存在下使用罗丹明B（RB）代替罗丹明6G[26]：LOD约为1×10^{-8} mol/L。在通过重新检查实验条件改进的RB/荧光猝灭方法中，LOD改善到5×10^{-9} mol/L[27]。

（2）总磷 一般而言，在有机和无机磷化合物、有机和无机缩合磷化合物、固体和溶解的磷化合物分解为正磷酸盐之后，钼蓝法可用于测定实际水样中的总磷。而磷化合物的分解可以通过以下方式进行，如在过二硫酸钾和硫酸存在下的高温条件（120~160°C）[28,29]，在反应盘管中用铂丝高温加热[24]，或辐照含有过氧化物的水样——用紫外线照射二硫酸盐[19,30]，通过将反应盘管缠绕在两个低压汞灯（外径14mm，长度134mm，杀菌用途4W）上，大多数有机磷化合物在70°C下分解为正磷酸盐：磷在830nm处的LOD为1ppb[19]。

硫胺素与磷酸钼酸的氧化反应可用于测定水样中的总溶解磷，其中在过二硫酸盐存在下使用简单的紫外光反应器可以分解磷化合物：LOD为1×10^{-8} mol/L[31]。虽然硫胺法非常灵敏，但该反应对正磷酸盐的选择性较低，氧化剂可能会干扰测定并导致阳性误差。

可以使用鲁米诺化学发光检测系统：该方法基于鲁米诺与磷酸钼酸的氧化反应，LOD为磷的30ppt[32]。该方法非常灵敏实用，可应用于测定水环境样品中的痕量磷。

8.2.2 与水污染有关的有机化合物

8.2.2.1 *表面活性剂*

在全球已经大量生产了各种表面活性剂，如今它们是工业和家庭中的必需品。部分表面活性剂，特别是家庭使用的表面活性剂，会作为生活垃圾排放到河流、湖泊和海洋中，并可能污染这些水源。在表面活性剂中，经常用于家庭洗涤的阴离子表面活性剂是造成水污染的重要原因。

通常，可以通过用阳离子染料作为离子缔合物进行溶剂萃取来测定阴离子表面活性剂。在流动注射分析中，可渗透疏水性有机溶剂的PTFE多孔膜用于相分离[33]。

在图8.5（1）和（2）中，显示了用于测定阴离子表面活性剂的流程图和信号曲线。在该系统中，亚甲蓝用作阴离子表面活性剂的配对离子，邻二氯苯因其较低的挥发性和毒性而取代氯仿[34]。在系统中安装了一种简单的相分离器，该分离器由两个模块组成，而这两个模块分别由聚（三氟氯乙烯）（CTFE）和 PTFE 膜（孔径：0.8μm）[34]制备。圆柱形空腔型相分离器[36]可用于在溶剂萃取中更好地进行相分离。

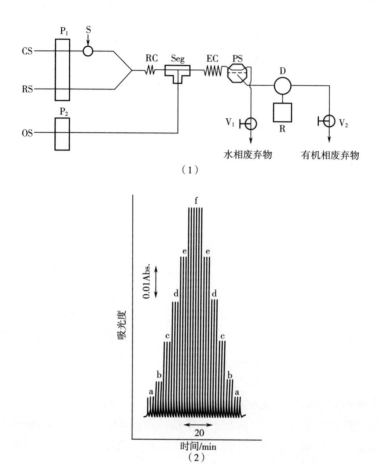

图 8.5 （1）用于测定阴离子表面活性剂的溶剂萃取/流动注射分析流程图
（2）用于测定阴离子表面活性剂（十二烷基硫酸酯）/10^{-6}mol/L 的流动曲线
CS—载体 RS—试剂溶液 OS—萃取溶剂 P_1、P_2—泵 S—样品 RC—反应盘管 Seg—分段器
EC—提取线圈 PS—相分离器 D—探测器 R—记录仪 V_1、V_2—针阀
(a) 0.5；(b) 1；(c) 2；(d) 3；(e) 4；(f) 5

阴离子和阳离子表面活性剂可以在水相中根据阴离子染料和阳离子表面活性剂的离子缔合反应/颜色变化来测定，无需溶剂萃取[37]。与溶剂萃取系统相比，流动系统和试剂组成非常简单。然而，在环境样品中必须考虑来自天然物质（例如延胡索酸）的干扰，它们可能会以正误差的形式干扰阴离子表面活性剂的测定。

8.2.2.2 化学需氧量（COD）

化学需氧量（COD）是有机物污染的指标之一，定义为用一定的氧化剂氧化有机物所需要的氧气量，例如高锰酸盐、铬酸盐或 Ce^{4+}。用于测量水生环境样品中 COD 的氧化剂有不同的标准方法。一般来说，要改善有机物的氧化，需要较长的反应盘管和较高的温度。因此，需要一个双柱塞泵来在高压下推动两种溶液。

在用于铬酸盐 COD 的 FIA 系统中[38]，可以使用两管线流动注射分析系统：一个用于载流（H_2O），另一个用于试剂蒸气（H_2SO_4 中的 $K_2Cr_2O_7$）。其他条件如下。反应盘管：50m（PTFE，0.5mm）在120℃浴槽中；背压盘管：0.25mm（i.d）×3m 聚四氟乙烯；样品量：100μL；波长：445nm，使用 10mm 光程石英池；每次流速：0.3mL/min；样品通量：15/h；LOD：5mg/L；以 D-葡萄糖计。

在具有高锰酸盐的流动注射分析系统中[39]，铂管[0.5mm（i.d）×1m 或 3m]可用作热浴（95℃）中的反应盘管。在 0.9mol/L 硫酸溶液和 0.1% H_3PO_4 的酸溶液中，铂可以作为高锰酸盐氧化的催化剂。

可以使用循环流动注射分析，以 $KMnO_4$ 作为氧化剂重复测定 COD；样品（20μL）注入试剂溶液（0.8mol/L H_2SO_4 和 0.001mol/L HIO_4 溶液中的 0.2mol/L $KMnO_4$），该溶液在 PTFE 管中从 50mL 储液器循环到流动池（5mm 路径，9μL）通过热浴（70℃）中的反应盘管（0.5mm × 2m）、冷却盘管[0.5mm（i.d）× 1.5m]和背压盘管（0.5mm × 10m）连接到储液器；流速：0.75mL/min；波长：525nm[40]。

8.2.3　与有毒/危险问题相关的有机化合物

水环境中各种杀虫剂、除草剂和内分泌干扰化合物的分析是环境和分析科学中最重要的问题之一。这种分析的难点是：①方法的检测浓度必须非常低，因为分析物的浓度通常很低；②检测方法的选择性必须很好，因为各种干扰物质可能存在于高浓度样品中。通常，在测量此类化合物时，与间歇测量方法类似，在通过流动注射分析测量之前需要进行用于提高灵敏度和选择性的预处理。

通过带有柱和筒的固相萃取（SPE）进行的常规预处理程序可用于富集分析物以及从分析物中去除或分离干扰物/基质物质，以便在流动注射分析测量或在线预处理之前进行间歇预处理。对于有机分析物或干扰物质的收集/富集，疏水性吸附剂（例如 ODS）用于非离子有机物质，而离子交换剂用于离子物质。

可以在安装在流动注射分析系统中的 Amberlite XAD-4 色谱柱（pH2）上预浓缩水样中的痕量酚类，并与样品基质分离。在色谱柱上富集的酚可以用碱性溶液（pH13）洗脱，并通过流动进样技术使用 4-氨基安替比林（4-AAP）方法在 510nm[41]处进行测量。

对于荧光测定 17β-雌二醇（内分泌干扰物之一），可以在微柱中填充的分子印迹聚合物（MIP）上进行富集[42]，分子印迹聚合物/SPE（MIP/SPE）与流动进样化学发光法的组合可用于测定四环素[43]。

8.2.4　与水污染和有毒/危险问题相关的金属和金属化合物

8.2.4.1　光谱检测

流动注射分析与光谱检测方法（如原子吸收、ICP-AES 和 ICP-MS）相结合，是比分光光度法或荧光光度法检测方法更方便、更通用的水生环境样品金属分析系统。

在结合了原子吸收光谱法的经典流动注射分析方法外，使用单管线系统（载体流），并将样品注入载体中。这种系统的主要目的是使测量过程自动化并提高样品吞吐量。然而，在水污染物分析中，分析物浓度通常非常低，因此，需要一些预浓缩和/或净化程序来提高原子吸收光谱和 ICP-AES 的检测灵敏度并去除 ICP-MS 中的样品基质。

在表 8.1 中，总结了一个使用 ICP-AES 提高灵敏度的微型柱前处理的例子，可以检测到几个或几十个 ppt 浓度的痕量和超痕量金属离子[44]，在线预处理系统如图 8.6 所示，其中在线预处理是通过使用顺序进样系统进行的。

表 8.1　通过微型柱并使用在线预处理系统进行收集/浓缩金属离子的分析特征

元素	波长/nm①	线性区间/(ng/mL)	线性度	富集因子②	相对标准差/%③	检出限/(ng/mL) 测量值②,④	检出限/(ng/mL) 直接 ICP-AES⑤
Ba	493.408	0.1~10	0.9973	5	4.6	0.02	0.4
Be	313.042	0.001~10	0.9946	10	5.1	0.001	0.4
Cd	226.502	0.01~10	0.9991	16	6.7	0.018	0.6
Co	228.615	0.1~10	0.9987	9	9.6	0.10	1.5
Cr	205.560	0.1~10	0.9998	14	4.6	0.09	1.3
Cu	324.754	0.1~10	0.9909	15	2.1	0.08	1.5
Fe	259.940	0.1~10	0.9968	13	2.7	0.05	1.3
Mn	257.610	0.01~10	0.9997	10	4.4	0.008	0.4
Ni	231.604	0.05~10	0.9971	12	6.7	0.16	2.4
Pb	220.353	0.1~10	0.9980	16	5.5	0.18	4.4
Sc	361.383	0.01~10	0.9916	19	4.5	0.01	0.3
V	292.401	0.1~10	0.9986	17	2.9	0.09	1.4
Zn	213.856	0.1~10	0.9955	12	8.7	0.02	1.3

注：①使用的发射波长基于 200.7 US-EPA 方法。
②使用了 5mL 的样品溶液。
③通过使用 5mL 混合标准溶液获得，所使用的每种金属离子的浓度为 0.5ppb（$n = 7$）。
④检测限，对应为 3（S/N）。
⑤仪器检测限，对应 3 SD（标准偏差）的 0.01mol/L HNO_3（$n = 10$）。

表 8.2 给出了在线系统在标准参考物质分析中的应用示例：河水样品中的大多数痕量和超痕量金属离子可以通过 ICP-AES 进行测定。

表 8.2　标准河水的分析结果

元素	SLRS-4① 认证值/(ng/mL)	SLRS-4① 测量值/(ng/mL)	JSAC 030 1-1② 认证值/(ng/mL)	JSAC 030 1-1② 测量值/(ng/mL)	JSAC 0302③ 认证值/(ng/mL)	JSAC 0302③ 测量值/(ng/mL)
Ba	12.2±0.6	OR④	0.60±0.02	0.55±0.03	0.60±0.01	0.58±0.01
Be	0.007±0.002	0.009±0.001	—	(0.006±0.001)⑥	0.99±0.04	1.02±0.01

续表

元素	SLRS-4[①]		JSAC 030 1-1[②]		JSAC 0302[③]	
	认证值/(ng/mL)	测量值/(ng/mL)	认证值/(ng/mL)	测量值/(ng/mL)	认证值/(ng/mL)	测量值/(ng/mL)
Cd	0.012±0.002	0.014±0.005[⑤]	0.0023±0.0007	0.005±0.002[⑤]	1.01±0.01	0.92±0.06
Co	0.033±0.006	0.040±0.012[⑤]	—	(0.040±0.010)[⑤⑥]	—	(0.032±0.013)[⑤⑥]
Cr	0.33±0.02	0.33+0.02	0.15±0.01	0.18±0.02	10.1±0.2	9.8±0.2
Cu	1.81±0.08	1.92±0.13	0.57±0.07	0.61±0.05	10.3±0.2	10.7±0.3
Fe	103±5	OR[④]	4.7±0.3	4.6±0.5	56±1	OR[④]
Mn	3.37±0.18	3.33±0.16	0.125±0.007	0.13±0.03	5.0±0.1	4.9±0.1
Ni	0.67±0.08	0.65±0.02	—	(0.082±0.013)[⑤⑥]	9.9±0.2	9.4±0.3
Pb	0.086±0.007	0.092±0.002	(0.005)[⑥]	ND[④]	10.1±0.2	9.8±0.6
Sc	—	(0.026±0.003)[⑥]	—	(0.036±0.008)[⑥]	—	(0.015±0.004)[⑥]
V	0.32±0.03	0.34±0.04	—	(0.47±0.04)[⑥]	—	(0.49±0.04)[⑥]
Zn	0.93±0.10	0.92±0.08	0.19±0.03	0.17±0.03	10.2±0.3	9.8±0.3

注：①加拿大国家研究委员会发布的用于痕量金属的河水参考材料。
②日本分析化学学会发布的用于痕量金属的河水参考材料（未加标）。
③日本分析化学学会（加标）发布的用于痕量金属的河水参考材料，用 0.01mol/L HNO_3 稀释 2 倍。
④OR，超出范围；ND，无法检测。
⑤样品体积 15mL。在使用 15mL 时 Cd、Co、Ni 和 Pb 的 LODs 分别为 0.005ng/mL、0.03ng/mL、0.05ng/mL 和 0.06ng/mL。
⑥括号中的数字为信息值。

在与多检测系统（例如 ICP-AES 和 ICP-MS）联用的在线系统中，拥有亚氨基二乙酸酯官能团（用于多元素收集）的多功能螯合树脂（例如 Chelex 100 和 Muromac A-l）非常有用。

在线富集/原子吸收光谱与流动注射分析相结合可用于痕量和超痕量金属。例如，铅可以通过火焰原子吸收光谱法测定[45,46]，而铋、镉和铅可以通过电热（石墨炉）原子化原子吸收光谱法（ET-AAS）[47]测定。

在 ET-AAS 中，分析物浓度最高的洗脱液区必须保持在 ET-AAS 自动进样器的注射喷嘴尖端，这可以轻松且可重现地重复顺序注射分析的计算机控制的注射泵系统。

ICP-AES 与在线预处理系统相结合，如图 8.6 所示，也可用于单一金属分析。例如，可以测定痕量和超痕量的铅[48,49]。

对于金属的形态分析，可以在流动注射分析中安装两个或多个微型色谱柱，并与原子吸收光谱或 ICP-AES 结合使用。可以使用小型薄固相（STSP）柱对 Cr^{3+} 和 Cr^{6+} 进行形态分析：一个装有用于收集 Cr^{3+} 的阳离子交换剂，另一个装有用于收集 CrO_4^{2-} 的阴

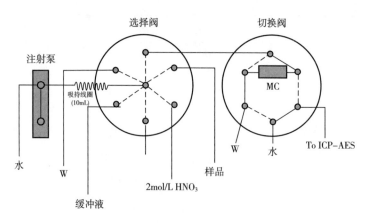

图 8.6 带有用于收集/浓缩金属离子的色谱柱的在线预处理系统
柱尺寸：40mm×2mm（i.d.）。

离子交换剂[50]，均通过 ICP-AES 对两种物质的进行检测。亚 μg/L 水平的亚硒酸盐和硒酸盐的形态可以通过使用 ICP-MS 并使用填充有阴离子交换剂的微型柱进行分析。在 pH 为 1.5 时，硒酸盐以 $HSeO_4^-$ 形式存在，亚硒酸盐以不带电的 H_2SeO_3 形式存在，因此，在较高 pH 时，只能在第一根色谱柱上收集阴离子硒酸盐，而在第二根色谱柱上收集亚硒酸盐[51]。

8.2.4.2 其他检测方法

通常，水环境中金属的痕量和超痕量分析是通过光谱方法进行的。然而，在特殊情况下，分光光度法、荧光光度法和化学发光方法比波谱分析法灵敏得多。河水样品中的痕量铁和铜可以通过流动注射分析结合使用催化反应的分光光度检测进行测定[52-54]；在 N,N-二甲基-1,4-戊二胺（DPA）和过氧化氢存在下，铁和铜可以作为 DPA 氧化的催化剂，Fe 的检测限为 0.02ppb[52]，铁和铜的检测限分别为 0.01ppb 和 0.07ppb。

鲁米诺与流动注射分析结合的化学发光检测可用于海水中铁的超痕量测定：总铁（Fe^{2+} 和 Fe^{3+}）的 LOD 为 40ppm。在系统中，安装了一个 8-羟基喹啉螯合柱，用于在线基质消除/预浓缩[55]。

8.3 用流动方法分析大气环境

空气样品中的痕量和超痕量分析物可以通过间歇或在线方法以某种吸收溶液收集/浓缩分析物后由流动注射分析测定。在经典方法中，分析物从气相到液相的传质可以通过将样品气体鼓泡到吸收溶液中来完成。这种方法原理非常简单，收集效率几乎是 100%。然而，收集过程需要很长时间才能达到高富集因子，并且难以小型化或安装在流动注射分析系统中。为了克服传统鼓泡法的缺点，人们提出了几种安装在流动注射分析中的用于在线收集/浓缩的气液萃取装置，包括扩散管（DN）、气体扩散洗涤器（GDS）和色谱膜流动池（CMC）。

8.3.1 扩散管（DN)和气体扩散洗涤器（GDS)

在通常的扩散管装置中，黏性液体或固体等吸收材料被涂覆在圆柱管的内壁上，气体样品流经该圆柱管。分析物通过气体扩散从气相转移到液相或固相。采样结束后，用合适的洗脱液冲洗管内壁，从洗脱液中回收分析物。扩散管装置的收集方法不适合流动分析系统。

在气体扩散洗涤器装置中，吸收溶液保留在透气膜的一侧，而气体样品则流过膜的另一侧。气体样品可以连续流过气体扩散洗涤器装置中的吸收层，其可以回收并用于分析物的连续检测。因此，气体扩散洗涤器装置可用于连续流动测量方法。

气体扩散洗涤器是在线气液萃取最有用的设备之一，它可以小型化并安装在各种流动分析系统中。自 1984 年以来，Dasgupta 及其同事开发了许多基于气体扩散洗涤器的分析系统并将其应用于空气分析[56-62]。

在将气体样品中的分析物收集到吸收液中时，扩散管和气体扩散洗涤器装置的收集效率通常为 10%~20%，因此，标准气体样品是绘制校准图所必需的。

8.3.2 色谱膜流动池（CMC）

CMC 装置是一种不同于扩散管和气体扩散洗涤器装置的气体收集器[62]。自 1994 年以来，CMC 设备已被用于将水溶液中的分析物浓缩为有机溶液[63,64]以及气体环境中收集/浓缩气体分析物[65-68]。

CMC 的膜块由双孔聚四氟乙烯制成，有微孔（0.1~0.5μm）和大孔（250~500μm）两种。CMC 装置及其收集机制分别如图 8.7 和图 8.8 所示。比界面面积约为 $65cm^2/cm^3$[69]，在膜块中，极性液体填充到大孔中，而微孔只能用于非极性相，如有机溶剂或气体，因为极性液体的毛细压力阻止它们渗透到微孔中。因此，即使极性相，如水溶液，充满了大孔，非极性相，如空气或有机溶剂，也可以通过膜块。当非极性相通过 CMC 膜块时，其中的大孔中充满了极性相，非极性相中的分析物可以分布在两相之间，并通过传质从非极性相中转移到极性相中，其类似于在填充有一些填料的色谱柱中的分布；因此，在膜中，峰形区域如图 8.8（2）所示[70]。

在实际应用中，通过吸收性水溶液收集/浓缩空气样品中的分析物的具体过程如下，首先将吸收溶液作为固定相填充到大孔中，然后将空气样品连续流入 CMC 膜中，其中气相通过微孔并离开 CMC 膜，如图 8.7 所示。气态样品中分析物在流过双孔 PTFE 块时发生传质。在 CMC 膜中，可以轻松快速地实现分析物的完全转移，因为两相之间的界面非常大。此外，可以实现分析物的高效富集，因为只有小体积的吸收液作为固定相填充在大孔中，而大量的气体样品流过 CMC 膜；气体样品与吸收液的体积比很大，通常可以达到几百倍的富集。此外，通过使样品以合理的流速在膜中流动，空气样品中分析物的收集效率几乎为 100%。因此，不需要气体标准来制备校准图，通常，标准水溶液可用于校准[70]。

可以通过将 CMC 设备与传统或微型流动注射分析耦合来在线收集/浓缩大气样品中的痕量和超痕量分析物。一个自动化的 CMC 系统，它由一个六通切换阀、一个用于

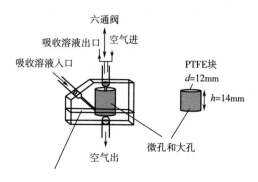

图 8.7 带有双孔 PTFE 膜的色谱膜流动池（CMC）

CMC：三孔色谱流动池。

将吸收溶液填充到 CMC 膜中的液体泵、一个用于吸入空气样品并将其输送到 CMC 膜的气泵、一个脱气装置、一个三通切换阀和一个装有双孔 PTFE 块［12mm（o.d.）× 14mm］的 CMC 装置构成，如图 8.9 所示。在该系统中，CMC 装置连接到切换阀，阀的两个端口连接到流动注射分析系统的载流。六通切换阀可用于引入标准溶液以制备校准图。CMC 系统与流动注射分析联用可用于测定空气中的二氧化氮（LOD：0.9ppb）[70-72]、空气中的二氧化硫（0.5ppb）[73]和空气中的甲醛（0.03ppb）[74]，表 8.3 总结了测定二氧化氮、二氧化硫和甲醛的实验条件。

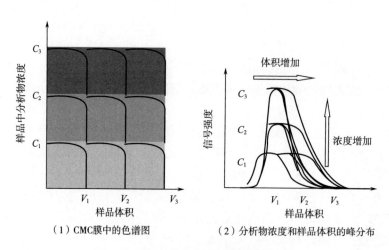

（1）CMC 膜中的色谱图　　　　（2）分析物浓度和样品体积的峰分布

图 8.8　CMC 膜中分析物的分布示意图

流动进样法与 CMC 装置结合可用于测定大气中的其他气态分析物，如氨、臭氧、过氧化氢和硫化氢。

8.3.3　空气中物质的简单间歇收集/浓缩方法

在分析气体样品（包括大气）中，最重要的程序之一是在吸收溶液中收集分析物。GDS 和 CMC 设备对此非常有用。然而，有时由于各种原因，分析仪器不能在现场使

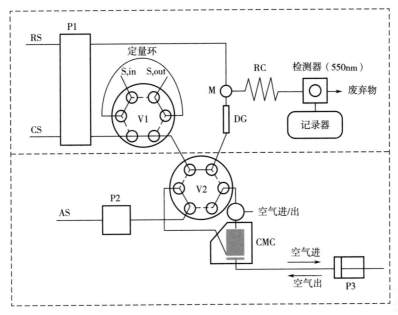

图 8.9 流动注射分析系统与三孔 CMC 耦合

RS—$5×10^{-5}$ mol/L 碱性副品红 + $4.5×10^{-2}$ mol/L 甲醛，pH=1.4　CS/AS—吸收溶液 [2g/L 三乙醇胺（TEA）]
P1—双柱塞泵（每个流速 0.2mL/min）　P2—蠕动泵（流速：0.6mL/min）　P3—注射式泵（5mL/min）
DG—脱气装置　V1 和 V2—六通阀　RC—反应盘管 [0.5mm（i.d.）×200cm]

用。在这种情况下，间歇方法是需要的，该方法必须具备程序简单、设备简单、成本低的特点。此外，收集效率应该是可以定量的，可以进行合理的富集。

表 8.3　测定空气中二氧化硫、二氧化氮和甲醛的实验条件

条件	SO_2 测定	NO_2 的测定	HCHO 测定
载体溶液	2g/L 三乙醇胺溶液		纯水
试剂溶液	$5×10^{-5}$ mol/L 碱性副品红+$4×10^{-2}$ mol/L CH_2O，pH1.25	20g/L 磺胺 + 0.5 g NED + 25mL/L HCl	0.03mol/L 乙酰丙酮 + 2mol/L 醋酸铵缓冲液 pH5.8~6.0
波长	550nm	525nm	412nm
每个流速	0.20mL/min		0.40mL/min
标样	300μL		
混合盘管长度	200cm		500cm
采集的气体样品	20mL		
空气样品的流速	5~6mL/min		

体积为50mL的塑料注射器可用于收集/浓缩空气中的物质。程序非常简单,不需要特殊技术。空气中气态物质的简单分批收集/浓缩方法如图8.10所示。该方法可应用于收集/浓缩空气中的甲醛[75]和氨气[11],然后通过常规流动注射分析进行测定。当充气至活塞停止时,注射器的总体积为(62.2±0.2)mL,如图8.10[2]所示。

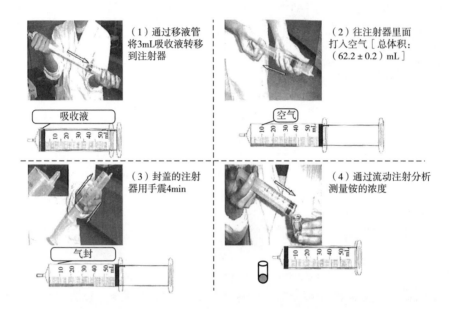

图8.10 空气中气态物质的简单分批收集/浓缩方法

注射器收集法对现场收集/浓缩是有用且方便的,并且通常可以在现场或在实验室中进行测量。

8.4 用流动方法分析地圈环境

在地圈环境分析中,固体样品必须首先通过常规方法转化为水溶液,例如从样品中溶解分析物,在加热或不加热和微波的情况下对样品进行酸消解,然后通过预处理,使用或不使用色谱柱和小柱来进行净化和富集。如此制备的水性样品溶液通过常规流动方法测量目标分析物(以与水性样品中类似的方式)。

地圈样品分析中最重要的程序与大气样品分析相同;预处理水样的制备是可靠分析的必要条件。从标准方法、推荐方法等发布的手册中可以找到合适的方法制备可靠的样品。

8.5 环境分析的未来展望

通常,在环境分析中,必须在短时间内分析大量样品,因此,需要基于流动的方法。此外,环境样品有时非常复杂,必须始终考虑干扰效应。与二噁英分析一样,分析物浓度通常非常低,因此可能需要有效的富集方法或复杂且成本非常高的仪器。此外，还必须降低样品、试剂、耗材、能源、劳动力和时间的消耗。

为了解决这些问题，环境样品分析需要一个计算机辅助流动化学分析系统（CAFCA 系统），它可以根据十分精细的程序完成复杂的过程。CAFCA 系统可以通过计算机控制的泵、选择阀、切换阀、电磁泵和阀门，以及在线预处理的专用装置和计算机控制系统进行组装，包括自动采样和数据采集/反馈。然而，预处理样品的可靠采样和制备对于分析来说是基本要求并且是必不可少的。分析师必须始终对所获分析数据的可靠性负责。

参考文献

[1] Anderson, L. (1979) Simultaneous spectrophotometric determination of nitrite and nitrate by flow injection analysis. *Analytica Chimica Acta*, 110, 123-128.

[2] Inoue, A., Higuchi, K. and Tamanouchi, H. (2000) Spectrophotometric determination of nitrite and nitrate in water samples by flow injection techniques. *Journal of Flow Injection Analysis*, 16 (Supplement), 57.

[3] Motomizu, S. and Sanada, M. (1995) Photo-induced reduction of nitrate to nitrite and its application to the sensitive determination of nitrate in natural waters. *Analytica Chimica Acta*, 308, 406-412.

[4] Burakham, R., Oshima, M., Grudpan, K. and Motomizu, S. (2004) Simple flow-injection system for the simultaneous determination of nitrite and nitrate in water samples. *Talanta*, 64, 1259-1265.

[5] Motomizu, S., Rui, S. C., Oshima, M. and Toei, K. (1987) Spectrophotometric determination of trace amounts of nitrite based on the nitrosation reaction with N,N-bis (2-hydroxypropyl) aniline and its application to flow injection analysis. *Analyst*, 112, 1261-1263.

[6] Motomizu, S., Mikasa, H. and Toei, K. (1987) Fluorimetric determination of nitrate in natural waters with 3-amino-1, 5-naphthalenedisulphonic acid in a flow injection system. *Analytica Chimica Acta*, 193, 343-347.

[7] Motomizu, S., Mikasa, H. and Toei, K. (1987) Fluorimetric determination of nitrite in natural waters with 3-aminonaphthalene-1, 5-disulphonic acid by flow injection analysis. *Talanta*, 33, 729-732.

[8] Joichi, Y., Lenghor, N., Takayanagi, T., Oshima, M. and Motomizu, S. (2006) Development of computer-controlled flow injection instruments and its application to determination of nitrate, nitrite, and ammonium ions in environmental samples. *Bunseki Kagaku*, 55, 707-713.

[9] Muraki, H., Higuchi, K., Sasaki, M., Korenaga, T. and Tôei, K. (1992) Fully automated system for the continuous monitoring of ammonium ion in fish farming plant sea water by flow injection analysis. *Analytica Chimica Acta*, 261, 345-349.

[10] Morita, Y. and Kogure, Y. (1963) Spectrophotometric determination of nitrogen with hypochlorite and a-naphthol. *Nippon Kagaku Zasshi*, 84, 816-823.

[11] Suekane, T., Oshima, M. and Motomizu, S. (2005) Determination of trace amounts of ammonia in air using batchwise collection/concentration method by spectrophotometry. *Bunseki Kagaku*, 54, 953-957.

[12] Aoki, T., Uemura, S. and Munemori, M. (1983) Continuous flow fluorometric determination of ammonia in water. *Analytical Chemistry*, 55, 1620-1622.

[13] Motomizu, S., Toei, K., Kuwaki, T. and Oshima, M. (1987) Gas-diffusion unit with tubular microporous poly (tetrafluoroethylene) membrane for flow-injection determination of carbon dioxide. *Analytical*

Chemistry, 59, 2930-2932.

［14］ Sanada, M., Oshima, M. and Motomizu, S. (1993) Assembly of a new gas-diffusion unit and its application to the determination of total carbonate and ammoniacal nitrogen by FIA. *Bunseki Kagaku*, 42, T123-T128.

［15］ Higuchi, K., Inoue, A., Tsuboi, T. and Motomizu, S. (1999) Development of a new gas-permeation system and its application to the spectrophotometric determination of ammonium ion by FIA. *Bunseki Kagaku*, 48, 253-259.

［16］ Goto, M., Murobushi, S. and Ishii, D. (1988) Continuous monitoring method of total nitrogen in wastewater using continuous microflow analysis. *Bunseki Kagaku*, 37, 47-51.

［17］ Hori, T., Kanada, Y. and Fujinaga, T. (1983) Fractionation of total phosphorus occurring in Lake Biwa by sensitive indirect phosphate determination. *Bunseki Kagaku*, 31, 592-597.

［18］ Ruzicka, J. and Hansen, E. H. (1975) Flow injection analyses. Part I. A new concept of fast continuous flow analysis. *Analytica Chimica Acta*, 78, 145-157.

［19］ Higuchi, K., Tamanouchi, H. and Motomizu, S. (1998) On-line photo-oxidative decomposition of phosphorus compounds to orthophosphate and its application to flow injection spectrophotometric determination of total phosphorus in river water and waste waters. *Analytical Sciences*, 14, 941-946.

［20］ Itaya, K. and Ui, M. (1966) A new micromethod for the colorimetric determination of inorganic phosphate. *Clinica Chimica Acta*, 14, 361-366.

［21］ Motomizu, S., Wakimoto, T. and Toei, K. (1984) Solvent extraction - spectrophoto - metric determination of phosphate with molybdate and malachite green in river water and sea water. *Talanta*, 31, 235-240.

［22］ Matsubara, C., Yamamoto, Y. and Takamura, K. (1987) Rapid determination of trace amounts of phosphate and arsenate in water by spectrophotometric detection of their heteropoly acid-Malachite Green aggregates following preconcentration by membrane filtration. *Analyst*, 112, 1257-1260.

［23］ Susanto, J. P., Oshima, M. and Motomizu, S. (1995) Determination of micro amounts of phosphorus with Malachite Green using a filtration-dissolution preconcentration method and flow injection-spectrophoto-metric detection. *Analyst*, 120, 187-191.

［24］ Aoyagi, M., Yasumasa, Y. and Nishida, A. (1988) Rapid spectrophotometric determination of total phosphorus in industrial wastewaters by flow injection analysis including a capillary digestor. *Analytica Chimica Acta*, 214, 229-237.

［25］ Motomizu, S., Mikasa, H., Oshima, M. and Toei, K. (1984) Continuous flow method for the determination of phosphorus using the fluorescence quenching of rhodamine 6G with molybdophosphate. *Bunseki Kagaku*, 33, 116-119.

［26］ Motomizu, S., Oshima, M. and Katsumura, N. (1995) Fluorimetric determination of phosphate in sea water by flow injection analysis. *Analytical Science & Technology*, 8, 843-848.

［27］ Li, Z., Oshima, M., Sabarudin, A. and Motomizu, S. (2005) Trace and ultratrace analysis of purified water samples and hydrogen peroxide solutions for phosphorus by flow injection method. *Analytical Sciences*, 21, 263-268.

［28］ Korenaga, T. and Okada, K. (1984) Automated system for total phosphorus in wastewaters by flow injection analysis. *Bunseki Kagaku*, 33, 683-686.

［29］ Goto, M., Nishimura, M., Tominaga, T. and Ishii, D. (1988) Continuous monitoring method of total phosphorus in wastewater using continuous microflow analysis. *Bunseki Kagaku*, 37, 52-55.

[30] Vlessidis, A. G., Kotti, M. E. and Evmiridis, N. P. (2004) A Study for the validation of spectrophotometric methods for detection, and of digestion methods using a flow injection manifold, for the determination of total phosphorus in wastewaters. *Journal of Analytical Chemistry*, 59, 77-85.

[31] Perez-Ruiz, T., Martinez-Lozano, C., Tomas, V. and Martin, J. (2001) Flowinjection spectrofluorimetric determination of dissolved inorganic and organic phosphorus in waters using online photo-oxidation. *Analytica Chimica Acta*, 442, 147-153.

[32] Yaqoob, M., Nabi, A. and Worsfold, P. J. (2004) Determination of nanomolar concentrations of phosphate in freshwaters using flow injection with luminol chemiluminescence detection. *Analytica Chimica Acta*, 510, 213-218.

[33] Kawase, J., Nakae, A. and Yamanaka, M. (1979) Determination of anionic surfactants by flow injection analysis based on ion-pair extraction. *Analytical Chemistry*, 51, 1640-1643.

[34] Motomizu, S., Oshima, M. and Kuroda, T. (1988) Spectrophotometric determination of anionic surfactants in water after solvent extraction coupled with flow injection. *Analyst*, 113, 747-753.

[35] Sakai, T., Chung, Y. S., Ohno, N. and Motomizu, S. (1993) Double-membrane phase separator for liquid-liquid extraction in flow injection analysis. *Analytica Chimica Acta*, 276, 127-131.

[36] Motomizu, S. and Korechika, K. (1989) Modified cylindrical cavity-type phase separator for liquid/liquid extraction in flow injection system. *Analytica Chimica Acta*, 220, 275-280.

[37] Motomizu, S., Oshima, M. and Hosoi, Y. (1992) Spectrophotometric determination of cationic and anionic surfactants with anionic dyes in the presence of nonionic surfactants, Part II: Development of batch and flow injection methods. *Microchimica Acta*, 106, 67-74.

[38] Korenaga, T. and Ikatsu, H. (1982) The determination of chemical oxygen demand in wastewaters with dichromate by flow injection analysis. *Analytica Chimica Acta*, 141, 301-309.

[39] Tsuboi, T., Hirano, Y., Kinoshita, K., Oshima, M. and Motomizu, S. (2004) Rapid and simple determination of chemical oxygen demand (COD) with potassium permanganate as an oxidizing agent by a flow injection technique using a platinum-tube reactor. *Bunseki Kagaku*, 53, 309-314.

[40] Zenki, M., Fujiwara, S. and Yokoyama, T. (2006) Repetitive determination of chemical oxygen demand by cyclic flow injection analysis using on-line regeneration of consumed permanganate. *Analytical Sciences*, 22, 77-80.

[41] Sakai, T., Fujimoto, S., Higuchi, K. and Teshima, N. (2005) Determination of phenols at low levels in water samples using automatic flow injection analysis coupled with on-line solid-phase extraction. *Bunseki Kagaku*, 54, 1183-1188.

[42] Bravo, J. C., Fernandez, P. and Durand, J. (2005) Flow injection fluorimetric determination of β-estradiol using a molecularly imprinted polymer. *Analyst*, 130, 1404-1409.

[43] Xiong, Y., Zhou, H., Zhang, Z., He, D. and He, C. (2006) Molecularly imprinted online solid-phase extraction combined with flow injection chemiluminescence for the determination of tetracycline. *Analyst*, 131, 829-834.

[44] Katarina, Rosi K., Lenghor, Narong. and Motomizu, Shoji. (2007) On-line preconcentration method for the determination of trace metals in water samples using a frilly automated pretreatment system (Auto-Pret AES System) coupled with ICP-AES. *Analytical Sciences*, 23, 343-350.

[45] Seki, T., Hirano, Y. and Oguma, K. (2002) On-line preconcentration and determination of traces of lead in riverwater and seawater by flow injection-flame atomic absorption spectrometry and ICP-mass spectrometry. *Analytical Sciences*, 18, 351-354.

[46] Martin, A. O. , Ruiz da Silva, E. , Laranjeira, M. C. M. and Favere, V. T. (2005) Application of chitosan functionalized with 8-hydroxyquinoline: determination of lead by flow injection flame atomic absorption spectrometry. *Microchimica Acta*, 150, 27-33.

[47] Sung, Y. H. and Huang, S. D. (2003) On-line preconcentration system coupled to electrothermal atomic absorption spectrometry for the simultaneous determination of bismuth, cadmium, and lead in urine. *Analytica Chimica Acta*, 495, 165-176.

[48] Zougagh, M. , Garcia de Torres, A. , Alonso, E. V. and Pavon, J. M. C. (2004) Automatic on line preconcentration and determination of lead in water by ICP-AES using a TS-microcolumn. *Talanta*, 62, 503-510.

[49] Sabarudin, A. , Lenghor, N. , Yu, L. -P. , Furusho, Y. and Motomizu, S. (2006) Automated online preconcentration system for the determination of trace amounts of lead using Pb-selective resin and Inductively coupled plasma-atomic emission spectrometry. *Spectroscopy Letters*, 39, 669-682.

[50] Motomizu, S. , Jitmanee, K. and Oshima, M. (2003) On-line collection/ concentration of trace metals for spectroscopic detection via use of smallsized thin solid phase (STSP) column resin reactors Application to speciation of Cr (Ⅲ) and Cr (Ⅵ) *Analytica Chimica Acta*, 499, 149-155.

[51] Jitmanee, K. , Teshima, N. , Sakai, T. and Grudpan, K. (2007) DRC ICP-MS coupled with automated flow injection system with anion exchange minicolumns for determination of selenium compounds in water samples. *Talanta*, 73, 352-357.

[52] Lunvongsa, S. , Oshima, M. and Motomizu, S. (2006) Determination of total and dissolved amount of iron in water samples using catalytic spectrophotometric flow injection analysis. *Talanta*, 68, 969-973.

[53] Lunvongsa, S. , Tsuboi, T. and Motomizu, S. (2006) Sequential determination of trace amounts of iron and copper in water samples by flow injection analysis with catalytic spectrophotometric detection. *Analytical Sciences*, 22, 169-172.

[54] Lunvongsa, S. , Takayanagi, T. , Oshima, M. and Motomizu, S. (2006) Determination of ultratrace amounts of iron in concentrated acids by flow injection spectrophotometric method based on the catalytic effect of iron ion on the oxidation reaction of N, N-dimethyl-p-phenylenediamine with hydrogen peroxide. *Journal of Flow Injection Analysis*, 23, 25-28.

[55] Bowie, A. R. , Achterberg, E. P. , Fauzi, R. , Mantoura, C. and Worsfold, P. J. (1998) Determination of sub-nanomolar levels of iron in seawater using flow injection with chemiluminescence detection. *Analytica Chimica Acta*, 361, 189-200.

[56] Dasgupta, P. K. (1994) A diffusion scrubber for the collection of atmospheric gases. *Atmospheric Environment*, 18, 1593-1599.

[57] Dasgupta, P. K. , McDowell, W. L. and Rhee, J. (1986) Porous membrane-based diffusion scrubber for the sampling of atmospheric gases. *Analyst*, 111, 87-90.

[58] Philps, D. A. and Dasgupta, P. K. (1987) A diffusion scrubber for the collection of gaseous nitric acid. *Separation Science and Technology*, 22, 1255-1267.

[59] Dasgupta, P. K. , Dong, S. , Hwang, H. , Yang, H. and Genfa, Z. (1988) Continuous liquid-phase fluorometry coupled to a diffusion scrubber for the real-time determination of atmospheric formaldehyde, hydrogen peroxide and sulfur dioxide. *Atmospheric Environment*, 22, 949-963.

[60] Lindgren, P. F. and Dasgupta, P. K. (1989) Measurement of atmospheric sulfur dioxide by diffusion scrubber coupled ion chromatography. *Analytical Chemistry*, 61, 19-24.

[61] Dasgupta, P. K. and Lindgren, P. F. (1989) Inlet pressure effects on the collection efficiency of diffusion scrubbers. *Environmental Science and Technology*, 23, 895-897.

[62] Zhang, G. and Dasgupta, P. K. (1992) Determination of gaseous hydrogen peroxide at parts per trillion levels with a Nafion membrane scrubber and a single-line flow injection system. *Analytica Chimica Acta*, 260, 57-64.

[63] Moskvin, L. N. (1994) Chromatomembrane method for the continuous separation of substances. *Journal of Chromatography A*, 669, 81-87.

[64] Moskvin, L. N. and Simon, J. (1994) Flow injection analysis with the chromatomem-brane-a new device for gaseous/liquid and liquid/liquid extraction. *Talanta*, 41, 1765-1769.

[65] Moskvin, L. N., Simon, J., Loffer, P., Michaolova, N. V. and Nicolaevna, D. N. (1996) Photometric determination of anionic surfactants with a flow injection analyzer that includes a chromatomembrane cell for sample preconcentration by liquid-liquid solvent extraction. *Talanta*, 43, 819-824.

[66] Erxleben, H., Moskvin, L. N., Nikitina, T. G. and Simon, J. (1998) Determination of small quantities of nitrogen oxides in air by ion chromatography using a chromato-membrane cell for preconcentration. *Fresenius' Journal of Analytical Chemistry*, 361, 324-325.

[67] Loffler, P., Simon, J., Katruzov, A. and Moskvin, L. N. (1995) Separation and determination of traces of ammonia in air by means of chromatomembrane cells. *Fresenius' Journal of Analytical Chemistry*, 352, 613-614.

[68] Moskvin, L. N. and Rodinkov, O. V. (1996) Continuous chromatomembrane headspace analysis. *Journal of Chromatography A*, 725, 351-359.

[69] Erxleben, H., Simon, J., Moskvin, L. N., Vladimirovna, L. O. and Nikitina, T. G. (2000) Automized procedures for the determination of ozone and ammonia contents in air by using the chromatomembrane method for gas-liquid extraction. *Fresenius' Journal of Analytical Chemistry*, 366, 322-325.

[70] Wei, Y, Oshima, M., Simon, J. and Motomizu, S. (2002) The application of the chromatomembrane cell for the absorptive sampling of nitrogen dioxide followed by continuous determination of nitrite using a micro-flow injection system. *Talanta*, 57, 355-364.

[71] Wei, Y., Oshima, M., Simon, J. and Motomizu, S. (2001) Application of chromatomembrane cell to flow injection analysis of trace pollutant in ambient air. *Analytical Sciences*, 17 (Suppl), a325-a328.

[72] Wei, Y, Oshima, M., Simon, J., Moskvin, L. N. and Motomizu, S. (2002) Absorption, concentration and determination of trace amounts of air pollutants by flow injection method coupled with a chromatomembrane cell system: application to nitrogen dioxide determination. *Talanta*, 58, 1343-1355.

[73] Sritharathikhun, P., Oshima, M., Wei, Y., Simon, J. and Motomizu, S. (2004) On-line collection/concentration and detection of sulfur dioxide in air by flow injection spectrophotometry coupled with a chromatomembrane cell. *Analytical Sciences*, 20, 113-118.

[74] Sritharathikhun, P., Oshima, M. and Motomizu, S. (2005) On-line collection/ concentration of trace amounts of formaldehyde in air with chromatomembrane cell and its sensitive determination by flow injection technique coupled with spectrophotometric and fluorometric detection. *Talanta*, 67, 1014-1022.

[75] Sritharathikhun, P., Suekane, T., Oshima, M. and Motomizu, S. (2004) On-site analysis of trace amounts of formaldehyde in ambient air using batchwise collection/ concentration method and portable flow injection system. *Journal of Flow Injection Analysis*, 21, 53-58.

9 流动分析方法在医药分析中的应用

J. Martínez Calatayud and J. R. Albert-Carcía

9.1 引言

严格地说,药物分析仅限于药物和兽用制剂,还包括活性成分、辅料和杂质的测定。显然,在这种情况下,确定含量还需要评估它们的稳定性,从而评估任何可降解中间体和最终产品的稳定性。由于在工业上药物的生产过程包括原材料的控制,因此药物分析还涉及确定其他参数,包括含量均匀度、溶解度或溶出度。事实上,药物分析不能仅限于制剂,因为它们通常需要确定各种复杂得多的基质中的特定药物,包括但不限于动物食品、饮料、牛饲料和化妆品,以及许多临床、法医和兽医样品涵盖各种基质类型,如血液、尿液或组织。目前,药物及其代谢物的分析是与环境样品的一个相关课题。分析药物可能是一个比药物分析更准确的术语。

流动技术在药物分析中的使用与流动注射分析(FIA),甚至是与连续分段流动分析一样落后。自该种技术被开发以来,已经有源源不断的测定药物程序和方法被报道,可以预测,由于流动分析是一种非常有价值的实用工具,这些测定程序的数量将继续增长。起初的顺序注射分析(SIA)和它的小型化版本(被称为阀上实验室)以及其他最新的方法,如多路换向分析和多注射器分析的出现,标志着流动分析方法开发和传播进入新阶段。当时出版了一本书,描述了流动注射分析的基础及其围绕药物分析成就的运作模式。正如预期的那样,流动技术已用于所有类型的检测器、流动系统和均质和异质型的物理化学过程。

下面的讨论,主要围绕所涉及的分析化学过程以及所使用的检测器类型展开。对每一种情况下所使用的特定连续方法做进一步的区分是困难的也是没有必要的。

9.2 药物配方分析

9.2.1 分光光度法(紫外-可见光和红外光)均相系统

紫外-可见光吸收检测器是迄今为止最广泛使用的流动分析检测器。事实上,大约三分之一的流动注射分析法采用了该检测器。二极管阵列分光光度计在计算机控制下得到广泛使用,使新型用途的开发成为可能,并极大地促进了其他用途。该分光光度计能够快速记录几个波长甚至全光谱的吸光度,以获得选定波长范围的平均数据,并且仪器中没有移动部件,因此与传统分光光度计相比,再现性大大提高。

基于紫外-可见光的多组分测定很难分析那些以不同浓度存在于样品中,并且具有部分或完全重叠光谱的物质。流动组装-二极管阵列分光光度法提供了一种可靠的途径来分析混合物,例如药物制剂中的混合物。电子衍生化、光谱衍生化和其他化学计量方法为同时测定两种或三种药物提供了非常有用的方法途径。因此,流动注射分析法通过使用0.4s的积分时间,能够同时分离两种、三种甚至四种具有完全重叠光谱的活性成分的混合物(即盐酸乙麻黄碱、盐酸苯肾上腺素、琥珀酸多西拉敏和茶碱)。

9.2.1.1 基于流动的分子吸收分光光度法在药物测定中的应用

分子吸收分光光度法允许通过使用无反应程序(即从它们的天然光谱)来确定药

物。这只需测量给定波长的吸光度,就像测量任何其他物理变量一样。对于固体,只需将样品研磨,用适当 pH 的水溶液或有机溶剂(如甲醇或乙醇)萃取,稀释至所需浓度并读取溶液吸光度即可。该程序已在流动注射分析中用于测定原黄素、3,6-二氨基吖啶和己烯雌酚磷酸盐(磷酸酯)。

该方法的测量精度总是高于涉及一些化学预处理的方法,同时其操作简单。然而,该方法的灵敏度和选择性通常很低,需要进行一些化学衍生反应。药物分析总是需要分析物的衍生化,无论是化学的还是物理的(如光化学,通过用紫外线光源照射样品)。例如,化学衍生化过程涵盖氧化还原、颜色形成或光降解类型的各种反应。

在这方面特别常见的是强无机氧化剂 Ce^{4+}、六氰基铁酸钾(II)、高锰酸钾和重铬酸钾。固相反应器中固定的 MnO_2 和 PbO_2 等不溶性氧化剂(无论是通过其自身溶解度的影响还是在某些载体的帮助下)也经常用于流动技术,也有较温和的氧化剂,例如,Fe^{3+} 和有机反应物,包括氯胺 T 和苯甲酸 2-碘酯。

硫酸中的重铬酸钾已用于通过流动注射分析的氧化法测定吩噻嗪家族的各种成员。Ce^{4+} 已在固相反应器和均相系统中用作氧化剂,包括在单通道流动注射分析组件中,其中将样品注入由金属酸溶液组成的载体-试剂流中。这样铈被用于普鲁卡因胺、酒石酸曲美拉嗪和二苯海拉明等药物的流动注射分析测定。

碱性六氰基铁酸盐(III)的高氧化能力已用于测定异丙肾上腺素。与 4-氨基安替比林结合使用时,这种氧化剂会产生氧化缩合反应,有助于测定酚类化合物,包括一些药物,如硫酸特布他林。对乙酰氨基酚和 N-乙酰基-对氨基苯酚的分光光度法,基于它们与六氰基铁酸盐(III)的氧化,对于后者,N-乙酰基-对苯醌亚胺与苯酚在 80℃ 条件下的氨介质中通过后续反应得到蓝色的 N-(对羟基苯基)-对苯醌亚胺。

在用六氰基高铁酸钾(III)氧化并在氨性介质中与苯酚反应后,通过顺序注射分析测定药物制剂中的对乙酰氨基酚。铁氰化物-氨基安替比林反应已被提议用于药物制剂中非诺特罗的顺序注射分析测定。其他在分光光度法上有用的氧化还原剂包括钒酸盐,在酸性介质中 V^{5+} 还原为 V^{4+} 的过程被建议用于测定某些吩噻嗪。此外,V^{5+} 与异烟肼或异烟酸酰肼形成琥珀色的 1∶1 络合物。介质中的过量试剂使配合物缓慢分解成异烟酸、V^{5+} 和气态氮;虽然反应很慢,但可以通过使用 Os^{8+} 作为催化剂来加速。

氧化剂,如高碘酸盐,偶尔补充少量过氧化氢,已用于测定依米丁、麦角新碱和麦角胺,以及在催化剂存在下测定利血平。最后一种药物在酸性介质中并在作为催化剂的 Mn^{2+} 存在下被高碘酸盐离子氧化;在这些条件下,反应在生成黄色的 3,4-二脱氢利血平处时停止,而不是进行到红褐色的 3,4,5,6-四脱氢利血平。过氧化氢是一种广泛用于传统药物分析的氧化剂,在流动注射分析的早期避免使用,因为它会形成大量气泡,并防止过氧化氢吸收阻挡紫外线区域。通过使用软锰矿反应床破坏过量的过氧化氢,将其分解为水和氧气,并使用除泡器去除氧气泡解决了该问题。

氯胺 T 有助于检测药物制剂中的溴离子,例如地溴铵、格隆溴铵、可马托品、新斯替宁、普鲁士林、吡啶斯的明和东莨菪碱。这种氧化剂已用于间接方法,包括基于对氨基苯甲酸与酸性次氯酸盐的反应以及随后用邻甲苯胺测定残留氯的方法。

碘对 Ce^{4+} 和 As^{3+} 之间氧化还原反应的催化作用已被用于测定药物制剂中的碘。通过监测作为碘浓度函数的 Ce^{4+} 吸光度来进行校准。

顺序注射分析方法已用于检验 Fe^{3+} 对抗坏血酸的氧化动力学,使用 1,10-菲咯啉作为指示剂。该反应被提议用于测定维生素 C。随后,通过分光光度滴定法测定维生素,使用硫酸 Ce^{4+} 作为氧化剂并测量铈离子在 410nm 处吸光度的降低。一种与之非常相似的检测维生素 C 后续方法,包括用高锰酸钾氧化,并监测氧化剂的吸光度下降。

除此以外,固定在顺序注射分析流动池中的三价铁离子还用于测定维生素 C,其中氧化还原指示剂三-(1,10-菲咯啉)-Fe^{2+} 浸渍在全氟磺酸膜中。

形成有色复合物。尽管在基本原理上几乎没有区别,基本的化学过程是相同的,要使用的操作条件取决于金属离子是用作颜色形成试剂,还是包含在相关配方的活性成分中(较少见)。

铁离子已被提议用作生成有色配合物的试剂,Co^{2+} 和 Ni^{2+} 也是如此。注射剂中的 N-乙酰半胱氨酸可以很容易地通过 Ni^{2+} 配合物的形成来确定;将样品注入 0.1mol/L NH_3/NH_4Cl 缓冲液中的 Ni^{2+} 溶液中,并在 415nm 下进行分光光度法测量。使用插入流动注射分析歧管中的分散管,从分析物的峰高和峰宽(即通过假滴定)量化分析物。已使用相同的组件来测定胶囊中的 N-青霉胺,但使用 0.2mol/L 乙酸铵中的 Co^{2+} 作为氧化剂。

药物和 Fe^{3+} 之间螯合物的形成是药物分析中的经典反应。水杨酸、水杨酰胺和水杨酸甲酯离子通过它们的酚基与 Fe^{3+} 反应——在水杨酸甲酯的条件下水解,这已用于测定药物制剂中的土霉素、诺氟沙星和环丙沙星。青霉素 β-内酰胺环的羟基氨解以及与三价铁离子形成的有色络合物也已用于流动注射分析测定青霉素。

水杨酸和乙酰水杨酸已在双进样流系统中测定,方法是将一个样品等分试样注入 NaOH 流体中,该 NaOH 流体被水解并流过一个更长的路径。Fe^{3+}-水杨酸复合物对应的两个流动注射分析信号之一与乙酰水杨酸的含量成正比,另一个(即水解流体的一个)与两种分析物的组合成正比。

铁离子已被用作氧化剂,产生的亚铁离子用于形成用于分光光度监测的有色配合物。一个例子是用 Fe^{3+} 氧化测定扑热息痛,释放的 Fe^{2+} 用于与 2,4,6-三吡啶基-S-三嗪形成复合物。抗坏血酸的测定也类似:样品在注入酸性 Fe^{3+} 流后被氧化,反应产物与 1,10-菲咯啉流合并,由此形成的 Fe^{2+} 有色络合物在 508nm 处监测。使用氯化铁而不是铁离子作为显色试剂测定抗焦虑镇静剂溴西泮。

曲美拉嗪和奋乃静是通过监测它们在盐酸介质中形成的钯螯合物来确定的。为了提高准确度,开发了一种同时测定卡托普利的顺序注射分析方法,有两种变体:一种是在酸性介质中形成 Pd 螯合物,另一种是用 Ag^+ 对分析物进行电位滴定。

金属离子在制剂中作为活性成分。在药物制剂中作为活性成分存在的金属通常通过上述有色螯合物的形成来确定;有一个例外,铁被量化为亚铁离子而不是三价铁离子,这需要事先进行氧化或还原反应。

铁以各种形式存在于药物制剂中。反复用于确定它的程序通常包括使用与顺序注射分析歧管连接的在线流通式透析器,以去除样品中的有色干扰物或浑浊物;透析后,在 667nm 的光度监测下,Fe^{3+} 与 Triton 络合。这需要预先将样品中的 Fe^{2+} 氧化为 Fe^{3+}。

或者，通过插入填充有 Cd 颗粒的色谱柱将铁离子还原为亚铁离子，然后将其与 1,10-菲咯啉反应并在 515nm 处监测反应，从而将总铁测定为 Fe^{2+}。亚铁离子还与 2,2-联吡啶螯合也用于此目的。

最近，药物制剂中的 Fe^{2+} 和 Fe^{3+} 已经成功地通过将后者与 Tiron 螯合产生的配合物形成，并监测这一结果的复合物，接下来是所有铁与过氧化氢及其螯合氧化成 Fe^{3+}。

卡托普利已通过使用流动注射分析和顺序注射分析歧管进行测定，该药物用三价铁离子氧化，所得亚铁离子与 2,2-N-二吡啶基-2-吡啶基腙螯合，得到众所周知的红色配合物。

该领域的其他测定包括在硫代亚硫酸钠存在下，与甲酚酞或者二甲酚橙膳食补充剂中的锌螯合来测定钙。

在流动注射分析分光光度法中，对以形成低溶解度或无溶解度的离子对为基础的痕量测定尤其常见。基本方法已用于各种技术，包括液-液萃取和沉淀。这些实际上是流动注射分析分光光度法分析药物最常用的选择，在比浊法和液-液萃取部分详细说明了它们的用途。

其他衍生化的方法。虽然它们很重要，但在此背景下，它们的使用频率比前几个要低。涉及亚硝酸盐离子的反应就是这样。采用经典的 Bratton-Marshall 反应，游离芳基伯胺在亚硝酸作用下转化为重氮盐，用氨基磺酸除去多余的试剂，再与显色剂偶联，得到偶氮染料。该方法适用于流动注射分析方法，可用于磺胺类药物的定量分析和溶出度测定。

后续的方法都使用了这种化学方法，因为亚硝酸盐试剂是在充满含铜化镉的固相反应器中通过还原硝酸盐溶液而原位制备的。这就避免了将不稳定的亚硝酸盐溶液储存在严格控制的溶液中，并可以经常丢弃它们。

在流动注射分析系统中，偶氮染料的形成被用来与 S-亚硝化结合，作为硫醇的一般检测方法。该方法包括通过与亚硝酸反应形成 S-亚硝基硫醇，并在过量的酸被破坏后与汞离子水解；在后一步骤中释放的亚硝酸形成偶氮染料，用分光光度法测定。在酸性介质中，利用吲哚基团在埃利希试剂中的显色反应，采用顺序注射分析法测定 obopindol，对乙酰氨基酚在酸性介质中与亚硝酸钠发生亚硝化，然后加入氢氧化钠以提高化合物的稳定性。

流动注射分析虚拟滴定法已被用于测定药物，其中化学系统以前用于开发基于单个波长的分光光度读数的方法。用于虚拟滴定的流动组件包括一个呈搅拌室或长而宽的管形式的稀释或分散装置。传统方法更简单、更快捷，并且使用的样品体积更小，而虚拟滴定方法提供更宽的测定范围和准确度略有提高。因为它可以测量更高的浓度，所以虚拟滴定法是分析药物制剂的更好选择，需要最少或不需要稀释。

9.2.1.2　紫外-可见光非均相系统（比浊法，固-液萃取和液-液萃取）

浊度　离子对法可以是一种简单、经济的可替代液-液萃取的药物测定方法。很明显，沉淀物会黏附在管道和流动池壁上，因此无法在常规流动分析中使用比浊法。事实上，最早的流动注射分析浊度测定是在每个样品后插入冲洗流，将通量减半。使用有效的抗絮凝剂可以避免沉淀物的絮凝，确保悬浮液的均匀性。目前，存在快速、成本效益高的控制分析药物配方的程序，该程序基于离子缔合化合物（含有无机或有机平衡离子）的形成及其比浊检测。表面活性剂的存在通常足以确保正确性、可重复操作性，并且几乎不需要冲洗。使用 Triton X-100 作为胶体保护剂，用溴甲酚绿测定氯己定就是这种情况。在某些情况下，连续流（即流动注射分析歧管中的微流）的流体动力学特性足以确保可重复测量，无需任何抗絮凝剂或定期冲洗。一些研究者在不需要任何悬浮稳定剂的情况下成功地测定了各种药物。如用 HgI_4^{2-} 测定驱虫药左旋咪唑；苯乙双胍与钨酸盐离子；硫胺与硅钨酸；苯海拉明和异丙嗪盐酸盐与溴酚蓝；阿米替林与溴甲酚紫。所需的测定程序非常简单，因为样品通常直接注入载体-试剂流中，很少将载体用作反应介质与试剂流合并。

对药物相对映体纯度测试的特别原始的方法是基于对晶体生长的抑制。比浊法基于乙酰氨基酚对 L-谷氨酸和 D-谷氨酸中的晶体生长的抑制作用，通过测量浊度对该影响进行监测。将含有 L-谷氨酸的样品与丙-2-醇流合并，然后将所得溶液注入携带相同醇的流体中，并在 550nm 波长处达到平衡时进行监测。纯水用作空白液和冲洗溶液。程序重复使用 D-谷氨酸而不是异构体 L-谷氨酸。

离子对形成方法有一个变体，包括用有机溶剂提取不溶于水的分析物和平衡离子形成的产物。由于表面活性剂的存在，现在不需要溶剂（甚至不需要固-液萃取），因此程序的受欢迎程度逐渐下降，但这在早期的流动注射分析系统中却是一种常见的选择。在水载体和有机溶剂的汇合点使用分段器，提取产物后需要分离器将有机溶剂驱动到分光光度计和水溶液中。这些早期的歧管随后被更简单、更有效的歧管所取代，通过跨膜透析进行提取；为此，使用了容纳两个平行通道的装置，在其中插入的膜旨在使载体（供体）和溶剂（受体）接触。这些基于膜的系统有助于以氯仿作为萃取剂的萃取液对依诺沙星、西沙必利、司帕沙星和曲唑酮进行测定，以及带有二氯甲烷的氯喹和乙胺嘧啶。

9.2.1.3 红外吸收

通过红外吸收光谱法进行定量的准确性和精确度通常不如紫外-可见光分光光度法。此外，与比尔定律的偏差更为常见，流动池非常狭窄、不切实际且几乎不可复制。除了需要在成型机中使用耐溶剂管之外，流动注射分析红外系统也基本上类似于流动注射分析紫外-可见光系统，无需化学衍生。然而，这在目前的实践中并没有造成特别的问题。最早的流动注射分析-红外分光光度法测定以四氯化碳为溶剂，异氰酸苯酯为载体，不锈钢活塞泵为推进系统。随后以四氯化碳为载体和样品溶剂对布洛芬进行测定，以避免不溶于其中的赋形剂的潜在干扰。流动注射分析-红外分光光度法的其他用途包括用二氯甲烷同时测定乙酰水杨酸和咖啡因。

傅里叶转换红外分光光度法对于获得高灵敏度和高数据采集率以及同时监测特定吸收带的变化特别有用。在流动注射分析阻力分析中首次使用了这种技术并涉及超临界流体，其开发人员提出了特定的流动相作为监测红外光谱区域中某些溶质的最佳选择。流动注射分析-傅里叶红外光谱组合还用于测定水溶液中的胆碱化合物，用圆柱形的内反射池和 ZnSe 晶体作为反射元件。

如果不对样品进行化学衍生，这种方法就并不常用。因此，流动注射分析-傅里叶红外分光光度法方法用于药物制剂中对乙酰氨基酚（扑热息痛）的测定，该方法基于以前的流动注射分析-紫外-可见光分光光度法，需要在测量前将分析物碱性水解为对氨基酚。

此外，还提出了一种复杂的流动体系，该体系集成了多个衍生化过程，可用于通过氢化物生成和傅里叶变换红外光谱法测定药物中的锑。该过程包括在线矿化/氧化样品中有机锑，以及在生产锑之前分别用 $K_2S_2O_8$ 和 KI 将 Sb^{5+} 预还原为 Sb^{3+}。

一个标准流动注射分析系统通过二极管阵列紫外光谱、H-NMR 和红外光谱以及飞行时间质谱的组合，用于表征利多卡因等化合物并以一些药物和相关化合物作为模型进行了测试。该仪器组合可以在线获取 UV、H-NMR、IR 和质谱以及相关成分数据，从而使目标化合物的结构表征接近完整。

9.2.1.4 火焰原子吸收光谱法

由于这项技术在另一章中有详细的论述，本节只讨论它在药物分析中的应用。流动注射分析-火焰原子吸收光谱法的重要性是显而易见的，因为它是在介绍流动注射分析所使用的检测方法的第一本书中的主题。

原子光谱学可以直接测量大量的金属元素，这使得它非常适合临床和法医分析，但很难用于药物分析。然而，在许多其他情况下，人们总是可以使用一种间接的方法来帮助检测各种药物。无论如何，原子吸收光谱技术从来都不是最受欢迎的药物分析检测技术，它在流动注射分析中的大量使用可能是由于几乎所有类型的分析过程都急于自动化，因为在顺序注射分析（一种较新的技术）中从未发生过类似的情况。此外，除了火焰原子吸收光谱外，很少有流动注射分析药物分析应用在这一领域。

在经典的分析化学中，金属离子常被用作沉淀、螯合或氧化还原反应的试剂。事实上，金属离子可以通过适当的反应来确定各种药物和非药物中的有机化合物，最常见的流动注射分析火焰原子吸收光谱法方法可分为两类，即：

（1）采用固-液萃取的方法，包括一些沉淀反应，并对溶解或试剂过量后产生的沉淀进行测量。

（2）使用固-液萃取的固定化试剂（如固相反应器中的金属离子）与目标药物在氧化反应器中或还原反应器中形成可溶性络合物的方法。

9.2.1.5 与分析物形成离子对或中性螯合物的液-液（萃取）体系形成沉淀法

（1）将沉淀溶解 基于在流动歧管中形成不溶沉淀物的方法，因沉淀物是否被溶解而不同。在前一种情况下，保留在过滤器上的沉淀物被清洗并溶解，产生的溶液中的金属离子被监测。氯己定的测定需要使用一个歧管，包括几个阀门，以便在不同的流体之间进行切换，以分配适当的溶液或试剂；将药物插入到含氨铜（Ⅱ）的载体中，以获得沉淀，该沉淀保留在过滤单元中。然后，通过氨水洗涤溶液，洗涤后的沉淀用硝酸流溶解，并驱动到检测器。反相流动注射分析法，左旋咪唑的测定是通过将试剂（HgI_4^{2-}）溶液插入样品载体中，产生的沉淀被保留在过滤器上用于洗涤、溶解和检测。

（2）沉淀物不进行溶解 这些方法可以在更简单的歧管中实现。沉淀物被允许留在过滤器上，其中的金属离子根据其在可产生火焰的溶液中的浓度下降而确定。

利多卡因、普鲁卡因和丁卡因可以通过 Co^{2+} 沉淀来测定。为此,将钴溶液插入蒸馏水载体中以获得正峰,然后插入样品中。以 pH 为 6~7 的 Cu^{2+} 或 Ag^+ 为沉淀剂,采用类似的体系测定五种磺胺类药物。硝基呋喃妥因与硝酸银在氨中发生定量反应,形成沉淀,可通过流动注射在线过滤进行测定。基于碘与 BiI_4^- 在酸性介质中反应形成不溶性离子对,因此,流动注射分析-火焰原子吸收光谱法能够测定乌拉地尔。

在包括固相反应器的流动系统中,样品通过反应床,通过与分析物反应释放的金属离子会产生一个瞬态峰,用作测量的分析信号。反应通常是螯合或氧化还原类型,反应床可用于简单、稳定的流动系统,通常包括用于插入样品的载体通道和位于进样阀和检测器之间的柱。

甘氨酸与碳酸铜以细粉的形式反应确定,该细粉由天然方法固定并构成反应器。天然固定只能用于少数不充分溶解,并且在连续流动条件下保持活性和机械稳定性的试剂。更常见的是,固定试剂需要使用某种类型的载体。

可溶性络合物已被用作测定氨基酸与固定化碳酸铜的基础,正如异烟肼与二氧化锰或安乃近以及昂达森琼与二氧化铅的氧化还原反应物一样。一些基于氧化还原的测定(例如,美沙酮、氯霉素和氯氮䓬的测定)使用还原性金属,例如镉或锌。

用噻肟、亚砜肟和 2,2-N-二氨基二乙胺纤维素微柱,采用流动注射石墨炉原子吸收分光光度法测定沙美特罗辛酸盐和亚叶酸钙中铂的含量。

流动注射分析火焰原子吸收分光光度法通过在线预富集测定复合维生素片中的铜,将分析物沉积到由包覆表面活性剂的氧化铝固定化的 1,5-二苯卡巴酮填充微柱上,然后注入少量氢溴酸进行洗脱。

对有效成分的测定需要使用液-液萃取将这些成分从基质中分离出来。含有萃取分析物的一部分有机相被注入载水器中以获得瞬态输出。

这些方法的实际要求之一是使分离相的流速与检测器雾化器的流速相适应。为了达到这个目的,最常见的选择包括通过与载体试剂合并的通道连续插入样品,以便在相对快速的反应中获得离子对或螯合物,沿着短管段与有机溶剂合并在一个分段。一旦两相分离,其中一相被输送到检测器。有机相的连续流动通过注入阀,在载水器的帮助下进入检测器。该体系已用于通过与二硫化碳在氨水中的反应测定仲胺、安非它明和甲基安非它明,生成的二烷基二硫代氨基甲酸酯衍生物与铜,镍或锌形成螯合物并在分段器和相分离器之间的管段通过甲基异丁基酮(MIBK)萃取。

溴西泮的测定是通过溴西泮铜螯合物与高氯酸盐离子形成离子对,并用甲基异丁基酮萃取。离子对偶法也被用来测定含有各种未鉴定螯合物的可卡因。

与前面的系统不同的是,如上所述,萃取可以跨膜进行。

9.2.2 发光法

近年来,多通道切换组装系统已被证明在一些新的用途上非常有用,例如,在分子连通性研究中对目标化合物行为的理论预测,这依赖于分子拓扑的数学程序。建立了含强氧化剂的有机化合物(主要是药物和杀虫剂)对其在溶液中化学发光行为的预测方程。对药物和农药的预测在 92.7% 的案例中是准确的,该方法随后被应用于特定的药物和农药家族，成功率高达100%（例如多酚）。进一步的应用包括对光诱导荧光和化学发光以及对高锰酸钾影响的预测。

9.2.2.1 流动荧光法在药物分析中的应用

传统上，荧光测定法是确定治疗药物和滥用药物最常用的技术之一。由于其优异的选择性和检测某些微量物质的能力，它的使用无疑得到了促进。此外，荧光技术推荐用于对活性成分纯度的定量。

正确使用荧光测定法需要考虑其与溶剂之间相互作用的潜在影响，这可能会改变发射最大值的强度和位置。极性、黏度和重原子的存在等变量是极其重要的，需要仔细选择和优化所用溶剂或溶剂混合物。因此，奎宁水溶液在加入硫酸后会发出强烈的光。用甲醇代替水作溶剂，大大提高了9-氨基吖啶的测定效率；此外，酒精是乳霜和栓剂的有效溶剂。原黄素也是如此，在二甲基甲酰胺中比在纯水中更容易测定原黄素。事实上，许多流动方法允许对药物进行量化，而不需要衍生化，只需要优化各种因素对发射的影响。因此，流动注射分析将分析物插入0.1mol/L的HCl溶液中，为9-氨基吖啶提供了良好的检测限。此外，荧光测量在两个不同的pH（6和11）中进行，这使得流动注射分析可以测定各种香豆素。一种基于麦角胺的天然荧光的方法，也被用于获得该药物在顺序注射分析体系中的溶出曲线。在少数例子中，由于分析物的水溶性低，需要采用另一种替代介质。

对影响溶剂性能的因素也可以作类似的考虑。因此，有组织的媒介，如表面活性剂和环糊精可以屏蔽荧光团的离子或分子环境的不必要的猝灭效应；此外，它们还减少了对有机溶剂的需求。华法林在β-环糊精中形成包合物，提高了其发光强度，并建立了测定华法林的顺序注射分析方法。

另一种保护发射物不与其环境相互作用并增加其结构刚性的方法是在固体表面固定（吸附）激发态形态；这是固体表面发光方法的基础，显然比在溶液中使用均匀体系更复杂。由于可逆保留的物质（即准备洗脱）的数量相对较少，这种技术具有较高的选择性。药物分析的应用包括四环素的测定，方法是将含有抗生素的缓冲载体与Eu^{3+}溶液合并在同一缓冲液中，并将产生的螯合物保留在流动池内填充的吸附剂床中。

基于激光的激发源虽然应用不广泛，但由于荧光发射强度依赖于激发强度，且高度准直的光束有利于对激发区域的控制，可以扩大荧光药物测定的检测潜力。流动激光荧光法已被用于核黄素、色氨酸和某些维生素等化合物的测定。

与其他技术一样，测定方法通常需要分析物的预先衍生化（如螯合、离子对、氧化还原、酶催化、缩合）。这通常需要使用各种方法，包括光降解、固体或液体萃取或电化学衍生化。在这种情况下，化学衍生化通常涉及使用无机氧化剂。抗坏血酸等强还原剂可与氯化汞氧化成脱氢-L-抗坏血酸，与邻苯二胺螯合所得的喹诺唑啉具有强荧光性。用Cu^{2+}-H_2O_2偶联测定异烟肼，用铜离子溶液测定半胱氨酸和胱氨酸。

大量的方法使用强氧化剂。因此，在固相反应器中固定化的六氰基铁酸盐已用于测定对乙酰氨基酚。此外，Ce^{3+}和Tl^+可以基于它们还原形式（Ce^{4+}和Tl^{3+}）的荧光性离子进行一些间接测定。铈离子在溶液和固相反应器中均被用于间接测定抗高血压的卡托普利和抗组胺苯海拉明以及异丙嗪、曲美拉嗪和三氟拉嗪。铊离子可以通过不同的氧化速率联合测定 L-半胱氨酸和 L-胱氨酸。

采用流动注射分析法测定血清中蛋白质含量，将样品注射到含次氯酸钠的缓冲载体中，并与硫胺和亚硝酸钠混合。过量的次氯酸盐被破坏后，硫胺被氯蛋白氧化成硫色素。

仔细选择所使用的强氧化剂和介质，并阐明浓度和反应时间的影响，使基于分析物之间反应速率差异的多组分测定成为可能。在最近的一项研究中，一种止流系统允许多达五种三环抗抑郁药的检测。氧化剂是 pH 为 4 的高碘酸钾，测定时间为 129~490s。

虽然使用氧化还原反应的荧光流动法可能是药物分析中最常见的方法，但实际上基于所有类型的反应都有有效的替代方法。因此，用流动注射分析和顺序注射分析测定了在 pH 为 10 的并在巯基乙醇存在下的邻苯二醛与利奥西普利的含量。在顺序注射分析体系中，硼酸通过变色酸螯合而被测定。

通过除紫外线照射（如电化学）以外的方法获得的荧光衍生物很少用于流动分析。少数可用的例子之一是在容纳所需电极的流池中，硫胺素对 Ag^+/Ag 电极在+0.4V 电位下氧化为脱氢硫胺素。

适当的辐照可以通过不同的机制破坏有机化合物的化学键，从而促进其降解。这对药物分析来说是一个非常有用的程序，因为它提供了与母体化合物不同性质的光片段（例如，更高或更低的吸收率、荧光、化学发光，更高的电分析活性）。这不仅提高了工艺的灵敏度，而且提高了选择性。在流动法中这种技术的使用提供了巨大的优势，因为它可以使中间和短期光降解产物进行重复性检测。将这种类型的反应纳入一个流动系统总是很容易的。通常将光源放入一个螺旋反应器中就足够了，不需要使用单色灯或单色仪，因为连续光谱灯在预期应用上是相当有效的。大量的荧光法和化学发光法以及少量的紫外-可见光测试法都依赖于分析物首先光降解为发光形态。

用辐射来替代化学试剂有一定的优势，这种替代促进了它在测试中的使用。这些优势包括使用更少的试剂（清洁化学），光源的高稳定性，增加了所需的流动组装的简单性，不需要去除多余的试剂。因此，辐射法已广泛应用于荧光法、化学发光法和电化学测定中。

虽然不完全与荧光测定有关，但固-液非均相系统在这方面应用广泛，通常以固相反应器的形式插在歧管的某一位置。使用固相反应器将不溶的或固定的试剂限制在歧管中的某个地方，并使样品载体溶液循环通过它，这比在溶液中使用试剂更有分析和操作上的优势。

这种目的可以简单地作为样品的前处理，如用 Tl^{3+} 氧化法测定 L-半胱氨酸和 L-胱氨酸。因为 L-胱氨酸的氧化非常缓慢，流动注射分析系统包括一个铜化镉色谱柱，旨将样品在注入载体之前进行氧化。

固相反应器也可以用于原位纯化试剂（离子交换或吸附柱），在使用时产生不稳定的试剂，甚至还需要对试剂溶液进行稀释。典型的例子包括用原位制备的亚硝酸盐试剂测定磺胺嘧啶、磺胺甲基嘧啶、磺胺甲噁唑、磺胺甲氧嘧啶和磺胺甲噻唑，亚硝酸盐试剂是通过在铜化镉微柱上还原硝酸盐溶液得到的。使用反应器来稀释溶液，避免了从母溶液连续稀释时造成结果的不精确。这在涉及使用某些催化剂的测定中特别有用，例如：①通过固定的 Pb^{2+} 反应器来测定安乃近，释放的铅离子随后催化过硫酸钾与邻苯三酚红的反应；②氨基酸的测定，通过释放与分析物成比例的固定 Cu^{2+} 来催化铁离子与硫代硫酸钠的反应。

在样品进样点和检测器之间放置反应器（无论是单个反应器还是多个串联或平行布置的反应器）是衍生分析物的通常选择，目的是测定各种化合物，也用于形态和多组分测定。使用固定化的 Ce^{4+} 并通过监测产生的 Ce^{3+} 的荧光，能够测定一些药物，包括氟非嗪、硫哒嗪和异丙嗪。

最后，衍生化和检测可以通过在流动池中容纳反应器而集成在一起从而可以提高灵敏度和选择性。通过这种方法，将吡哆醛复合物与铍保留在传统的填充有 C_{18} 二氧化硅的流动池中可以对吡哆醛进行检测。随后，该方法用于吡哆酸、吡哆醛和吡哆醛 $5N$-磷酸的联合测定。

9.2.2.2 磷光光度法

1957 年，磷光首次应用于分析。除了包括药物和农药分析在内的几个特别具体的领域外，磷光测定技术几乎没有得到人们的接纳。然而，一旦克服了该技术对极低温度的需要，室温磷光在固体载体和有组织的介质中的普及度就一直在增长。固体支撑物的变体特别适合于流动测试，因为流动池可以很容易地被一个适当的支撑物填充，进而用以保留分析物并进行测量。磷光光度法不受背景光或散射光的干扰，也不受瑞利或拉曼效应的影响。这就提高了信噪比，从而提高了检测范围。

提高选择性的一个例子是对萘西林的测定，虽然很少有 β-内酰胺类抗生素是发荧光的，但萘西林在第 6 位置有一个萘基，具有发光特性。

在停流系统中使用胶束稳定的室温磷光可以对包括萘丁酮在内的各种药物进行动力学测定，制剂、土壤和水果中的萘乙酸，以及配方中的萘夫尼。相应的分析系统针对最有影响的变量进行了优化，包括表面活性剂、pH 调节剂和发射增强剂。测定基于各自的动力学曲线，曲线的最大斜率用作分析物信号。

9.2.2.3 化学发光法

化学发光可以定义为一种化学反应（通常是氧化还原反应）产生电磁辐射的现象，在这种化学反应中，某些产物处于激发态，并在返回基态时以辐射能的形式释放部分或全部激发能。

化学发光法相对于其他基于发射的方法的主要优势可能是，它不需要激发灯，因此没有背景噪声，这增加了简单性、稳健性和成本效益。

早期的化学发光测定使用了一些试剂（鲁米诺、荧光素、罗啡、草酸酯），这些试剂通过强氧化剂的氧化而产生化学发光。分析物不是反应物，而是化学发光反应的抑制剂、增强剂或催化剂。然而，目前使用分析物作为产生化学发光的氧化物质的趋势越来越大。直接方法通常比间接方法具有更高的灵敏度和更低的检出限。这两种方法在药物分析（特别是流动注射分析和顺序注射分析）中得到了广泛应用。

直接化学发光法涉及分析物强烈的反应，通常无机氧化剂包括高锰酸钾（这是最常见且非常有效的）、铁氰化钾、Ce^{4+}、三-（$2,2'$-N-联吡啶）、Ru^{3+}、过氧化氢、氧气或 N-溴代琥珀酰亚胺，或较少见的溴化物或次氯酸钠。发射光强度受实验变量的强烈影响，包括温度、pH 和有组织的介质（表面活性剂、环糊精）的存在。

通过高锰酸钾在强酸性（硫酸或聚磷酸）介质中氧化进行化学发光法测定在数量上超过所有其他方法的总和。这是因为发射物是高锰酸盐的一种还原形式；到目前为止，这种物质被假设为 Mn^{2+}，一些与分析物的中间产物、Mn^{3+}、Mn^{4+}（甚至磷酸盐溶液中的 Mn^{7+} 的中间复合物）。选择硫酸还是聚磷酸也是一个有争议的话题，这方面的研究没有给出明确的结论。聚磷酸似乎在这一过程中起着双重作用，除了提供所需的酸强度外，它还能稳定高锰酸盐还原过程中形成的中间产物，类似于硫酸介质中的多磷离子。

采用固-液一体化萃取的顺序注射分析体系测定沙丁胺醇。分析物被吸附在用羧酸修饰的硅胶中，以促进从基质中分离，随后通过氧化呈现化学发光。

一份研究高锰酸钾在硫酸水溶液中氧化普鲁卡因、苯佐卡因和丁卡因的动力学实验揭示了各自发光时间分布的差异。类似的顺序注射分析系统表明用高锰酸盐测定磺胺类药物时，戊二醛的存在可提高灵敏度。在顺序注射分析中用高锰酸盐确定的其他药物活性成分包括甲氧苄啶和异丙嗪。

基于流动注射分析化学发光法的药物测定方法进行了扫描，对总共 97 种不同分子结构的药物在灯开和关的情况下进行了在线光降解测试。结果显示能够以直接方式（即无辐射）测定多种药物，包括磺胺类药物、噻嗪类药物、烟酰胺、去甲替林、左旋咪唑和苯巴比妥酸等，并且还可以测定其他不具有天然化学发光的药物（例如，氯霉素、右美沙芬、核黄素、麻黄碱、哌嗪酰胺、氯霉唑、茶碱），通过辐射产生化学发光。

已经通过使用称为多路换向的新兴流动方法开发了一种基于化学发光的直接测定方法，用于测定包括药物制剂在内的各种类型的样品中的氢醌。歧管由一组三个通道和三个电磁阀组成，测定在 60℃ 条件下进行。化学过程是用硫酸高锰酸钾氧化氢醌。硫酸奎宁和苯扎氯铵的存在明显增加了光发射。

氧化剂三-（2,2′-N-联吡啶）钌（III）是一种不稳定的物质，需要从三-（2,2′-N-联吡啶）钌（II）制备，经常用于基于化学发光的测定。已经提出了多种通过氧化 $Ru(bipy)_3^{2+}$ 获得活性 $Ru(bipy)_3^{3+}$ 的方法，这些方法包括化学（用酸 Ce^{4+} 或高锰酸盐）、光化学或电化学。$Ru(bipy)_3^{3+}$ 螯合物用于氧化分析物并产生发射物质，激发物质可能是 $[Ru(bipy)_3^{2+}]^*$。

最近，半固体剂型中的吲哚美辛是通过用三-（2,2′-N-联吡啶）钌（III）氧化来测定的，三-（2,2′-N-联吡啶）钌（III）是通过用硫酸铈（IV）氧化在线生成的。

在顺序注射分析系统中，在过二硫酸根离子存在的情况下照射 $Ru(bipy)_3^{2+}$ 会产生氧化物质 $Ru(bipy)_3^{3+}$，后者又可以氧化 L-半胱氨酸和 L-胱氨酸。由于后者的产物是非化学发光的，因此，使用还原柱预先将其转化为 L-半胱氨酸并测量两种分析物的组合信号。

抗生素头孢羟氨苄已通过在顺序注射分析-化学发光系统中与电化学产生的 $Ru(bipy)_3^{3+}$ 反应进行测定。流动池包含一个铂工作电极、一个 AgCl 饱和参比电极和一个钢针作为辅助电极。在铂电极表面由不稳定的 Ru（bipy）$_3^{3+}$ 产生的 Ru（bipy）$_3^{2+}$ 用于氧化分析物。

其他基于化学发光的直接氧化系统包括一种流动进样化学发光法，用于测定色氨酸，该方法在含有过氧化氢、亚硝酸盐和硫酸的介质中使用氨基酸的强化学发光。产生的化学发光归因于过氧亚硝酸对色氨酸的过氧化和环氧化，以及随后产生的二恶烷的分解。

在流动进样装置中研究了柠檬酸戊酯与 NaClO/H_2O_2 系统的化学发光反应。还报道了一种通过与 Ce^{4+}/罗丹明反应测定盐酸普萘洛尔的新流动进样化学发光系统。安乃近钠的测定方法是利用甲醛对溶解的 Mn^{4+} 氧化分析物产生的化学发光的增强作用。还报道了一种基于将过氧亚硝酸激发态的能量转移到分析物来测定吡哌酸的方法，该过氧亚硝酸通过在流动系统中混合酸性过氧化氢和亚硝酸盐在线合成；化学发光是由吡哌啶酸的两个激发态产生的罗丹明 6G 与 Ce^{4+} 反应产生强烈的化学发光，通过这种化学发光建立了一种灵敏、选择性高的药物测定方法。

这里值得特别注意的是基于电化学发光的分析，它利用发生在电极附近的化学发光反应从被动前驱体获得活性氧化剂。这种方法有一些缺陷，包括电极容易结垢、重复性差、涉及的电极面积小、线性范围太短、需要复杂的流动池设计。所获得的电生成氧化剂可用于氧化已知的化学发光试剂，如鲁米诺或过氧化氢，或用于生成分析物的氧化剂。在 KIO_4 中，$MnSO_4$、$AgNO_3$、$CoSO_4$ 和 Cu（NO_3）$_2$ 分别原位生成了 Mn^{3+}、Ag^{2+}、Co^{3+}、[Cu（HIO_6）$_2$]$^{5-}$ 等多种不稳定氧化剂。

最广泛使用的间接化学发光法是基于 5-氨基酞酰肼（鲁米诺）在碱性介质中的氧化。鲁米诺可以被高锰酸钾、高碘酸钾、重铬酸钾、过硫酸盐、氯酸盐、过氧化氢、N-溴代丁二酰亚胺（或 N-氯代丁二酰亚胺）、二氯氰尿酸（或三氯氰尿酸）和电生成的次溴酸盐氧化成 3-氨基酞酸盐（激发态）；然而，到目前为止，最常见的选择是铁氰化钾。这种氧化反应是由大量金属离子和有机化合物催化的。分析物（药物）还可以起到抑制剂、增强剂和催化剂螯合剂等多种作用。例如，Cu^{2+} 螯合物用于甲巯咪唑和咔咪唑的顺序注射分析测定，Co^{2+} 螯合物用于己烯雌酚与钴胺的顺序注射分析测定，它足以酸化介质以释放钴离子（催化剂）。

许多流动注射分析系统已被用于通过活性成分对鲁米诺化学发光的抑制或增强作用来确定活性成分。因此，最近使用反向流动注射分析-化学发光系统来测定多巴胺盐酸盐，该测定方法基于多巴胺盐酸盐在碱性介质中对鲁米诺和六氰基铁酸盐（III）反应的强淬灭作用。还发现安乃近能灵敏地抑制鲁米诺与碱性 $K_3Fe(CN)_6$ 的化学发光反应，化学发光强度的降低与安乃近的浓度成正比，并建立了一种新的药物流动进样方法。用 H_2O_2 催化 Cu^{2+} 氧化鲁米诺释放出的化学发光测定硫酸阿米卡星的方法简便、灵敏；药物与催化剂形成稳定的络合物，从而与分析反应相互作用。

化学发光增强剂也广泛用于流动技术。因此，发现酚妥拉明可以增强在碱性介质中铁氰化钾氧化鲁米诺而产生的发光。此外，基于鲁米诺与 KIO_4-单宁酸体系的化学发光反应，存在一种用于测定盐酸多巴胺的流动进样-化学发光方法。根据药物对鲁米诺与过氧化氢化学发光反应的显著增强作用，开发了一种基于化学发光的测定乙酰螺旋霉素的方法。

一种灵敏的基于流动进样化学发光的抗精神病药利培酮的检测方法是利用其对鲁米诺与过氧化氢反应的催化作用，药物对发射光强度的提高与浓度成正比。

FI-CL 系统已用于通过偏最小二乘校准同时测定抗坏血酸和 L-半胱氨酸。随后的方法依赖于两种化合物以不同速率将 Fe^{3+} 定量还原为 Fe^{2+} 的能力。产生的亚铁离子用鲁米诺/O_2 系统检测；因此，使用偏最小二乘法可以测量和处理铁离子与抗坏血酸和半胱氨酸反应的不同时间发出的光量。

鲁米诺可被电化学氧化，例如，用于测定异烟肼和诺氟沙星，溶解氧形成超氧自由基，可氧化鲁米诺（例如，在测定 β-葡萄糖中），氧化程度主要取决于反应条件。

液-液多相系统和固-液多相系统的使用已被证明对提高鲁米诺反应的选择性是有效的。一个例子是在有机介质（十六烷基甲基溴化铵胶束）中与四氯金酸盐（III）离子配对后，用二氯甲烷萃取氯丙嗪。固-液萃取已用于将鲁米诺及其氧化剂（铁氰化物、重铬酸盐、高碘酸盐）分别保留在两种不同的树脂中，这些树脂共同装填在一个在线放置在检测器流动池前面的微型柱中。

迄今为止，SIA-CL 尤其是 FIA-CL 方法的数量远远多于其他基于化学发光的流动方法。用于化学发光抑制的多泵法是后者中的案例。降血糖药物二甲双胍通过其对 Cu^{2+} 离子的抑制（清除）作用确定，其中 Cu^{2+} 用于催化鲁米诺和过氧化氢之间的化学发光反应。

这里还特别值得注意的是基于亚硫酸根离子与强氧化剂氧化的分析系统。该反应产生低强度的光，但可以通过有机化合物的氧化来增强；被还原的物质可能是二氧化硫，它作为中间体，将其激发能转移到相邻的发色团。喹诺酮格帕沙星产生低强度的化学发光，镧系元素的存在显著增强发光；二氧化硫的激发能被形成的喹诺酮-镧系元素络合物吸收。

与荧光测定一样，在线光源的使用已被证明在基于化学发光的测定中非常有效。用于此目的的组件非常简单，并且能够进行多种测定，例如磺胺类的测定（由于在线源的存在，其与高锰酸钾的化学发光显著增强）、氯霉素与氧化剂的作用、头孢菌素与高锰酸盐在乙二醛存在下加速反应。之后的方法也已用于获得市售的口服固体磺胺甲噁唑制剂的溶出曲线。乳酸盐也以这种方式测定，维生素 K_3 也是如此，两者都使用鲁米诺反应。

最近研究了分子印迹过程（MIP）在连续流动分析方法中的潜力，以提高基于化学发光法测定的选择性和稳健性。该过程包括在目标分析物存在下功能性和交联单体的共聚，随后去除后者，留下与所得分子印迹聚合物中的目标分析物互补的结合位点。在流动进样系统中，由此获得的 MIP 颗粒用于填充到流动池中。该技术被认为是特定分子识别的简单而可靠的选择，并已被用于测定肾上腺素等化合物。

9.2.3 电化学法

9.2.3.1 电导测定

流动电导测定法（不常见，在药物分析中更不常见）依赖于各种类型的流动池。一些例外包括测定抗坏血酸的方法，其中将一部分样品注入 5mmol/L NH_3 载流中，并将混合物通过流动池,以测量由此产生的电导率的增加。在另一种方法中,将部分乙酰胆碱标准溶液注射到 pH7.4 的磷酸盐缓冲液载体中,并通过含有固定在玻璃珠上的乙酰胆碱酯酶的聚乙烯管。然后将溶液与 H_2SO_4 合并并通过 PTFE 膜扩散池。通过膜扩散的乙酸被输送到水载体中的电导池内。而后的方法用于测定药物制剂中的乙酰胆碱。

9.2.3.2 电位计

电位技术具有多种优势,不仅仪器简单,而且还具有选择性高、灵敏度高、响应性强以及在分析检测之前几乎不存在化学反应的特点。

与间歇电位检测不同,连续流动电位检测中的分析物-电极接触时间是恒定的,因此精度总是更好。此外,流动测量的动力学性质使得其能够测量分析物的响应而不受其他活性成分或赋形剂的干扰,因此提高了选择性。而且,分析物与电极表面之间的短接触时间将中毒风险降至最低并延长电极的使用寿命。

对于药物制剂和复杂基质中的药物分析,离子选择性电极是其他昂贵耗材的有效替代品。这些电极不需要复杂的样品处理-样品颜色或浊度没有不利影响。通常,它们所需要的只是调整介质的 pH 和离子强度。有关电位检测器的更多详细信息,参见第 3 章。

离子选择性电极可用于两种不同的模式,具体取决于它们是对分析物还是对某些其他物质敏感(即目标化合物是否可以以直接或间接方式确定)。

(1) 直接测定 提出了一种由环氧树脂和石墨组成的管状电极,用于片剂和溶液中巴比妥和苯巴比妥的测定。采用聚氯乙烯(PVC)固定化邻硝基苯基醚中季铵盐的保留方法,对其进行了测定。一个类似的体系被用来测定五种不同的药物配方中的苯甲酸盐。其他基于膜支撑的苯甲酸盐离子对的电极也用于此目的(例如,基于三烷基甲基铵盐的电极测定血清中的胆碱酯酶,以及基于四庚基铵酰胺的电极)。

维生素 B_1 和维生素 B_6 已通过使用管状选择性电极在没有参比溶液的情况下进行测定。将维生素-四氯苯硼酸盐溶于邻硝基苯基辛基醚中制备电极膜,并将其固定在聚氯乙烯上。在多路换向流动系统中使用离子选择电极测定药物配方中的维生素 B_6。该歧管包括一个三通电磁阀,其连接到开关阀,并使用电位检测器促进各种程序的自动化。

药剂中的克拉维酸钾是在顺序注射分析系统中测定的,该系统包括两个电位检测器,一个用于检测克拉维酸,另一个用于检测钾。这种配置保证了检测结果的准确性,并对克拉维酸盐降解情况进行了评估,并且还可以检测流动系统中的故障。该基本原理随后应用于测定氯离子,使用基于同一离子载体的电极和光极。

药物制剂中的二环胺盐酸盐是通过使用流动注射分析结合各种塑料膜电极来测定的,并对使用寿命、可用的 pH 范围、工作浓度范围和温度进行了表征。这些含有邻苯二甲酸二丁酯增塑剂的塑料电极由基质中的屈他维林-硅钨酸盐、硅钼酸盐、磷钨酸盐、磷钼酸盐或四苯基硼酸盐离子结合物组成,采用类似的方法测定盐酸邻苯二甲酸丁酯。用分散在同一增塑剂中的聚硅酸硅钨酸盐、硅钼酸盐、磷钨酸盐或磷钼酸盐离子缔合物测定盐酸甲贝弗林。

选择性电极与多路换向相结合用于测定各种配方中的对乙酰氨基酚。歧管包括酶反应器以将乙酰氨基酚水解为4-氨基苯酚。与现有选择相比，后续的方法提供了多种优势，可用于测定复杂基质（如人血清）中的这种分析物-不会对选择性电极造成干扰。

（2）间接测定　目前有大量用于确定药物的选择性电极。然而，药物也可以通过衍生化分析物和监测某些物质（无论是反应物还是反应产物）来确定。这是经典药物分析中广泛使用的方法。

1-氟-2,4-二硝基苯（FDBN）是含有多种官能团的选择性试剂，包括胺、氨基酸、酚、硫醇、肼、酰肼和叠氮化物中的官能团，这有助于使用动力学电位法测定多种药物。检测基于氟化物选择性电极，该电极用于测量1-氟-2,4-二硝基苯与相关药物反应中释放的氟离子量。

在流动注射分析系统中也通过使用药物与1-氟-2,4-二硝基苯的反应并通过氟化物选择性电极监测释放的氟离子来测定异舒普林。该反应由十六烷基甲基溴化铵催化。

在流动注射分析电位测定药物制剂和生物流体中的屈他维林离子之前，已构建了用于盐酸屈他维林的碳糊电极，并对其成分、寿命、可用pH范围、响应时间和温度进行了全面表征。该电极是基于两种离子交换剂的混合物（即溶解在磷酸三甲苯酯中作为糊状液体的屈他维林-硅钨酸盐和屈他维林-四苯基硼酸盐）。这种改良的碳糊电极用于各种滴定剂对屈他维林铵离子进行电位滴定中的终点指示。

在电位分析中值得特别注意的是那些涉及用玻璃电极监测pH变化的分析。在用含有固定化青霉素酶的固相反应器测定青霉素V时，对内酰胺环水解时产生一种酸，该酸用玻璃电极检测；对酸的定量基于到达检测器的样本和先前通过反应器的样本之间的信号差异。

通过固定化青霉素酶，玻璃电极可修饰为酶电极。为此，这种酶与戊二醛薄膜交联，该薄膜被包在两个不同的玻璃电极表面。

采用三种不同的方法联合测定了总青霉素V和青霉酸，方法如下：①在氯化汞存在下，青霉酸将钼砷酸还原为钼蓝，通过分光光度法检测所形成的染料；②在碘被还原为碘离子时，碘-淀粉化合物的吸光度下降的情况下，流动注射分析对测定青霉酸的经典程序的调整适应；③测定青霉素水解为青霉酸时pH的变化。在这三个步骤中，青霉素在含β-内酰胺酶的固相反应器中被定量转化为青霉酸。

最后，金属电极也被用于电位测量。因此，预氧化镍电极有助于酒精、氨基酸和碳水化合物等的测定。

9.2.3.3　极谱法

历史上，人们认为伏安法可以推导极谱法，这是海洛夫斯基在1920年开发的一种电分析技术。目前，极谱法被简单地认为是伏安法的一个分支。与其他电分析技术的最大区别是它使用了一个汞滴电极作为工作电极。此外，它的特别之处还在于没有对流传质，因此，事实上，极限电流是完全通过扩散控制的。

极谱测量在很大程度上依赖于温度和溶解氧的存在。因此，用于连续流动工作的最简单的极谱系统需要一个高度精确的恒温器。通过使用适当的表面活性剂，氧的氧化可以转移到更负的电位，或者使用差分技术来抑制不必要的信号。

汞滴电极难以集成到流动注射分析系统中，因此，到目前为止发展了少量的流动注射分析极谱测定方法，包括药物配方中的青霉酸、硝酸异山梨酯和别嘌呤醇。

在极谱法测定别嘌呤醇时，金属汞的氧化伴随并形成通式为 HgL_x^{2-nx} 的高不溶性螯合物，其中 x 可为 1 或 2。在磷酸盐缓冲液中，别嘌呤醇的极谱包含两个信号，其中一个是由于形成了一层不溶物质的薄膜，抑制了电极反应。所需要的样品处理也非常简单：它足以制成粉末片剂，将粉末溶解在碱性介质中，过滤并稀释到硼酸盐缓冲载体中用于注射。苄青霉素酸是青霉素的降解产物，因此，它可以用来确定青霉素纯度（青霉素是极谱惰性，青霉酸会给出阳极波）。根据氨苄西林、邻氯西林、卡苄西林和苯氧甲基青霉素水解得到的各种青霉素类的极谱图，它们都给出了相似的半波电位但略有不同的阳极波。因为这些青霉素都没有极谱活性，从它们的水解溶液中获得的反应只能归因于它们各自的青霉素酸的存在。在此基础上，使用连续流组件将汞滴正面输送到载体和样品进口。固相反应器和极谱检测的联合使用是流动注射分析测定抗坏血酸的基础。该歧管包括一个反应器，该反应器包含抗坏血酸氧化酶并被固定化在琼脂糖上，用于测量酸的酶氧化产生的氧气量。

已建立了一种基于汞滴电极吸附测定片剂中维生素 B_2 的顺序注射分析系统，并用于监测维生素在水溶液中的光降解。

9.2.3.4 电流滴定法

伏安法包括一系列电分析技术，通过电流-电位曲线中提供的分析信息，对这些曲线的详细分析使大量的电分析测定得以实现，这些电分析测定可根据是否依赖电位扫描（伏安法）或使用恒定电位（安培法）分为两大类。在连续流动系统中，大多数电分析工作都集中在电池设计、微型化和配置上，特别是开发新型的、具有创新形状的微型化电极，其表面由特殊材料构成。大多数流动注射分析和顺序注射分析电分析系统非常简单，在检测前不需要样品的衍生化。然而，有些确实需要化学衍生化，详情请参阅第 4 章。碳基电极，包括碳糊电极和玻碳电极，是最受欢迎的安培探测器之一，这可能是因为它们可以提高高度可重现性的操作。

氰化物和羟胺是含有哌啶肟盐的制剂的两种分解产物，用于抵抗胆碱酯酶中毒。这两种产物都是用带有银电极，Ag/AgCl 作为参比电极，玻碳作为辅助电极的电解池测定的。在通过扩散分离 HCN 之后，酸通过单独的通道被输送到安培检测器。羟胺用三碘氧化法测定，所得亚硝酸盐用比色法测定。

金属电极在这里不如碳电极常见。可以使用铜电极通过两种方式氧化测定氨基酸和肽。在相对于 Ag/AgCl 约 0.0V 的中性或微碱性溶液中，该反应涉及形成氨基酸-Cu^{2+} 与电极表面上铜阳极溶液中的离子形成螯合物。

金属电极包括基于其他种类（氧化物）和几种金属组合的电极。构建了一种用于电化学检测三环抗抑郁药（丙咪嗪和去丙咪嗪）的锑掺杂氧化锡电极，并采用循环伏安法和差分脉冲伏安法在流动注射分析体系中进行了测试。其他研究者使用氧化镍电极测定咖啡因和水杨酸。最初的电极是一根镍丝，在上面电解形成一层氧化镍。

所谓的化学修饰电极有助于解决电流分析中的一些经典问题，并因此扩大了电化学检测器在流动系统中的使用。这些电极具有特意结合到其表面的特定化学功能。许多多羟基化合物在通过将亚铜盐沉积在玻璃碳表面上制备的电极上进行电催化氧化。这种类型的电极已经过多种物质的测试，包括碳水化合物（单糖、二糖和寡糖）、氨基糖、糖醇、抗生素（硫酸链霉素、硫酸卡那霉素、洋地黄毒苷和红霉素）以及醛糖酸、糖醛酸和醛糖二酸。化学修饰电极在流动检测中的意义源于电催化现象。事实上，这些电极可以催化溶质的氧化或还原反应，这些溶质表现出比传统电极高的过电位。一些表面结合的氧化还原介质加速电子转移和降低工作电位的能力促进了这一点。改变传统电极活性的最常见方法之一是将生物催化剂加入碳糊基质中。已发现通过添加钴酞菁修饰的碳糊电极可将各种物质的电氧化所需的电位降低数百毫伏，并证明可用于测定巯基化合物，例如半胱氨酸、N-乙酰半胱氨酸和高半胱氨酸。一种固定方法是将酶加入硅脂中，然后用硅脂填充石墨表面的微孔，这提供了一层非常靠近石墨中敏感部位的硅树脂。以这种方式固定的酪氨酸酶已用于流动注射分析测定各种酚类化合物。

将酶与电极结合的一种常见方法是在电极表面放置一层固定膜，将茶碱氧化酶包埋在铂盘电极上，对临床样品中的茶碱进行电流测定。

通过附加一个变量——渗透选择性，来扩展先前的方法，可以进一步提高电极选择性和灵敏度。为此，目的电极被涂上一层适当的膜，该膜能够选择到达电极的特定成分或阻碍分析物快速扩散的成分。除了不易中毒外，涂层电极通常比未涂层电极表现出更广泛的应用范围，对含有对乙酰氨基酚、多巴胺和奋乃静等分析物混合物的生物液体的比较研究表明了这一点。在玻碳电极上涂上一层邻菲咯啉铁（II）膜，可以通过流动注射分析安培法测定生理样品（尿液）中的对乙酰氨基酚，而不受抗坏血酸或尿酸的干扰。在渗透选择性方面，全氟磺酸有助于排除通常存在于尿液中的阴离子（如抗坏血酸盐、尿酸盐）。全氟磺酸涂层还被用于研究以异丙嗪为模型化合物的阳离子药物的电极选择性和稳定性。

9.2.3.5 连续流动伏安法

自1980年以来，人们一直在研究应用于流动注射分析中的连续流动伏安法。例如，在抗坏血酸存在的情况下，通过将样品插入配备多孔网状玻碳工作电极的库仑流通检测器中来测定多巴胺，在此处停止流动为了便于电极捕获初始电解的产物和多巴胺的再生物以获得分析信号。该过程的效率取决于抗坏血酸的不可逆氧化和多巴胺的准可逆行为。

基于吸附溶出伏安测量的顺序注射分析系统允许使用样品抽吸装置通过在简单歧管中的汞电极处吸收来测定各种物质，例如核黄素。

伏安法可以测定来自10个不同家族的19种药物。对分析物进行循环伏安法，并将结果与通过电流测量法获得的结果进行比较，证明其在测定痕量分析物方面更加灵敏和可靠。

因为背景电流（基线）是通过氢气释放、氧溶解的减少或其他溶剂氧化等作用而形成，为了避免它们受到上述因素的影响需要在分析前对样品进行脱气。在测定包括氯丙嗪、对乙酰氨基酚和去甲肾上腺素在内的各种药物时，已根据重现性、线性和检测限对这种方法进行了评估。

复合维生素制剂中的金属离子Cu和Zn通过将片剂溶解在硝酸中来测定。在HNO_3或KNO_3作为支撑电解质在0~0.8V下电沉积后进行伏安扫描。

9.2.3.6 连续流量安培法

安培法是在适当的电压作用下，测量被分析物氧化或还原所产生的电流强度。因此，安培法与库仑法特别相似，两者之间唯一区别是被氧化或还原的分析物的比例。安培法测量要求使用一个流动池来容纳工作电极、参比电极和辅助电极，反应发生在第一个电极。选择的电位是从伏安图中选择的；为了使灵敏度和选择性最大化，所选择的电流通常与分析物的极限电流一致。辅助电极避免电流通过参比电极，从而避免参比电极在外加电压下发生不必要的变化。为确保检测器不会对其他样品成分作出反应需要仔细选择操作条件。因此，如果被分析物是样品（例如在药物分析中）中唯一的电活性（电氧化或电还原）物质，那么实际上测量是特定的。然而，如果存在除分析物以外的任何电活性物质，并且其氧化或还原电位落在施加的电位窗口内，则目标物质不能在没有干扰的情况下被测定。

在连续流动系统中，安培法是一种非常常见的检测方法，因为只要使用合适的工作电位，它除了选择性强外，还具有较低的检测限。电流法在流动注射分析中最早的应用之一是用网状玻碳流通电极测定抗坏血酸、肾上腺素和左旋多巴，该电极在电流法和库仑法两种模式下工作。在 pH 为 5.5 的醋酸缓冲液中，在+0.19V 下以饱和甘汞为参比电极，进行了抗坏血酸的安培测定。在另一个应用中，抗坏血酸在 293nm 处通过分光光度法检测，抗坏血酸和尿酸在+0.6V 处通过安培法检测。

顺序注射分析系统能够同时测定 L-甲状腺素、D-甲状腺素和 L-三碘甲状腺原氨酸，前者使用电流生物传感器，后两者使用化学修饰碳糊构建的免疫传感器。

青霉素是通过金电极上的脉冲电流法测定的，使用 Ag/AgCl 作为参比电极，铂作为辅助电极。该测试在强酸介质中灵敏度最高，但化学稳定性不足。醋酸-醋酸盐缓冲液提供了更高的稳定性，但灵敏度略低。

双电流法是一种电流法模式，使用两个极化电极，适用于测量被施加低电压（10~500mV）时所通过的电流强度。在流动系统中，这可以通过使用两个相同的浸泡在流动液中的铂电极来实现。如果溶液在施加电压的情况下，含氧化和还原形式的可逆（或准可逆）氧化还原电对，则还原物质可以被氧化，氧化物质可以被还原。这种行为已在氧化还原电对中被观察到，包括 "Br_2/Br^-、I_2/I^-、Ce^{4+}/Ce^{3+}、Ti^{4+}/Ti^{3+}、VO_3^-/VO_2^+、Fe^{3+}/Fe^{2+} 和 $Fe(CN)_6^{3-}/Fe(CN)_6^{4-}$"。因此，$Fe(CN)_6^{3-}/Fe(CN)_6^{4-}$ 电对被用在碱性介质中在高温下以六氰基铁酸盐（III）氧化分析物的形式来测定糖浆中的还原糖。使用配有两个极化为 200mV 的铂电极的电池进行检测。蔗糖需要预先通过注入 1mol HCl 水解为碳水化合物；一旦糖被水解，酸就会在强碱性介质中与氧化剂合并。

Fe^{3+}/Fe^{2+} 电对在盐酸介质中的可逆性被用于含有铂电极的流动注射分析双安培法中测定复合维生素配方中的铁含量。

间接双安培方法基于分析物、过量试剂和反应产物的氧化还原反应，分别提供可逆所需的氧化还原对的氧化和还原形式。异丙嗪和硫利达嗪的间接双安培测定表明，Fe^{3+} 是比 $Fe(CN)_6^{3-}$、Ce^{5+}、VO_3^-、Ti^{4+}、I_2 和 Br_2 更有效的氧化剂。在 20℃ 下进行检测，在两个工作电极之间使用 150mV 的电压。

碘-叠氮化物反应很容易被元素硫和含硫化合物诱导。化合物之间的诱导时间差异很大；因此，虽然一些化合物的诱导时间短至 30s，甚至更短，但在其他化合物中，例如二硫化物（胱氨酸、维生素 B_1、磺胺噻唑），诱导反应是在三碘化物存在下，叠氮化物缓慢成功裂解 C-S 或 S-S 键后才开始的，需要几个小时才能完成。在流动注射分析系统中，可以通过将样品溶液注入含有碘-叠氮化物混合物的载体中并通过双安培法监测碘消耗量来测定诱导时间短的化合物。

9.3 制药行业的流动程序分析仪

化学和生化反应可以通过许多变量进行监测。流动方法为监控工业生产和质量控制过程提供了理想的方法。制药行业投入大量精力开发控制程序（主要是流动注射分析和顺序注射分析类型）以实时获取有用信息，提高生产过程的一致性、质量和产量。流动方法尽量避免或至少最小化样品制备的需要，这有助于随时快速获取评估生产过程状态所需的数据。此外，同时测定几个变量的能力可以大大提高官方认可的方法所能获得的信息内容。自从设计了连续流动方法以来，人们对于开发稳健、高效的系统且在反应介质中直接处理样品并同时测定多种分析物的兴趣从未停止增长。

9.3.1 药品生产过程分析

如表 9.1 所示，流动系统也被用于监控某些药物的生产过程。因此，顺序注射分析被证明是有效的监测青霉素工业生产发酵过程的手段。所使用的特定系统被设计用来监测丝状真菌黄青霉（Penicillium chrysogenum）的培养过程中的三个主要参数（即葡萄糖、青霉素和乳酸）。作为碳源和能源的葡萄糖必须在生产阶段保持受控水平。乳酸是青霉菌培养的主要控制因素，因为目标产物是抗生素。通过使用葡萄糖氧化酶（GOD）和乳酸氧化酶（LOD）来测定葡萄糖和乳酸，这两种氧化酶被固定在固相酶促反应器中以便于催化聚葡萄糖和乳酸的氧化反应，被分别氧化成聚葡萄糖酸-5-内酯和丙酮酸酯。两种反应都会产生过氧化氢，通过向反应介质中加入鲁米诺和 $K_3FE(CN)_6$ 后产生的化学发光，被用于测定产生的过氧化氢。青霉素是通过在含有固定化青霉素酶作为催化剂的反应器中水解青霉素酸来测定的。所得青霉素酸与碘反应，这降低了碘与鲁米诺反应中产生的化学发光。除了青霉素酸与碘的反应以及后续测量抗生素的可降低的吸光度的碘-淀粉复合物之外，初始系统随后被改进为使用先前的间接方法在线测定葡萄糖和青霉素。该系统足够可靠，在发酵超过 400h 后仍可准确测定青霉素。

表 9.1　　　　　　　　　生产过程监控评估

被分析物	样品	检测方法	线性范围 R.S.D./%	吞吐量/h^{-1}/ 检测时间间隔/h	参考文献
青霉素、葡萄糖、乳酸	黄青霉 (P. chryogenum)	化学发光法	0.01~1.200g/L 0.01~7.000g/L 0.005~5.000g/L 2.06、2.56、2	30/160 20/160 20/160	[1]

9 流动分析方法在医药分析中的应用

续表

被分析物	样品	检测方法	线性范围 R.S.D./%	吞吐量/h^{-1}/检测时间间隔/h	参考文献
青霉素、葡萄糖	黄青霉 (*P. chryogenum*)	化学发光法、紫外-可见分光光度法	0.1~1.8g/L 0.01~7.000g/L	—/425 —/350	[2]
吗啡	罂粟花 (*P. somniferum*) 的水提物	化学发光法 (250~600nm)	2.5×10^{-6} ~ 3×10^{-6} mol/L 1.4	100/144	[3]
吗啡	罂粟花 (*P. somniferum*) 的无水提取物	化学发光法 (250~600nm)	0.01~1g/L b	120/144	[4]
西咪替丁、溴西泮、双氯芬酸和残留物	残留水-样本	质谱法	—	—	[5]

两种顺序注射分析系统被提出用于监测和控制吗啡的工业生产过程，其中涉及一系列植物罂粟花（*Papaver somniferum*）的水和非水溶剂的提取。在含六偏磷酸钠的酸性介质中，药物与高锰酸钾发生化学发光反应，发出的光通过光纤从流池转移到光电倍增管，从而测定药物浓度。

流动法也被用于控制制药工业废水。因此，采用流动系统测定了某制药厂废水中的几种物质及其代谢物。该测定方法包括前处理步骤，固相的液-液萃取，然后插入到流动歧管中，并通过质谱（FIA MS）或串联质谱（FIA MS-MS）进行检测。

9.3.2 药物自动溶出和释放测试

溶出度检测的重要性体现在许多方面，其中之一就是有利于药品质量的控制。获得有关溶出过程动力学的全面信息有助于新剂型的开发。药典官方正式认可的溶出试验，在预设时间从溶出介质中提取样品。由此获得的分析信息不足以描述溶出动力学，并妨碍了新剂型的开发和评估。此外，官方方法的样品采集效率很低，而且多个多溶出容器（通常是六个）同时操作时，是不切实际的、困难的，甚至是不可能实现的。因此，持续监测溶解介质中的活性成分需要使用自动化系统。

在理想的情况下，自动化系统在这方面提供了一些优势，包括：①能够进行快速测定和样品预处理，这有利于实时监测；②表现出高通量（因此，也具有高分辨率）；③使用适量的样品，在溶出介质中引起最小的体积变化，特别是在高采样率时；④可同时控制单一检测器进行的若干溶出试验，并可同时测定同一配方中的若干成分；⑤可使样品以稳定的方式进行预处理，并可长期连续检测；⑥可连续监测基线和在线重新校准检测系统；⑦试剂消耗量低。

流动系统具有前面提到的特点。最近，在流动系统中评估了通过在线过滤直接从溶解介质中收集样品、在多溶解容器中同时进行测试以及使用化学计量方法进行多组分测定的能力（表9.2）。

最早实现上述理想目标的尝试始于流动注射分析建立之初。因此，一些有效的新分析方法不仅用于测定目标分析物，而且还用于获得其溶解谱。结果是大量使用各种类型检测器的分散程序阻碍了溶出度测试的自动化。此类程序是研究的主题。然而，通过使用推荐的标准程序同时测定总体曲线和使用导数分光光度法测定最多两个单独的曲线，随后又尝试改进获得溶解曲线的程序。

表9.2　药物制剂的溶出度测定和药物释放测试，1998—2007 年

药物制剂的溶出曲线			
药物	流动系统检测	线性范围 R.S.D/%	参考文献
布洛芬	顺序注射分析，紫外-可见光分光光度法（222nm）	0~0.12g/L 0.5	[6]
乙酰水杨酸 咖啡因-非那西丁	顺序注射分析，紫外-可见光分光光度法（220~310nm）	44~220mg/L 30~1500mg/L 10~50mg/L 0.4~0.5	[7]
利血平	流动注射分析，光致荧光法 $\lambda_{exc}=386nm$, $\lambda_{em}=490nm$	0.01~0.75mg/L 1.4	[8]
异烟肼	流动注射分析，化学发光法	$5\times10^{-7}\sim1\times10^{-4}$mg/L 3.0	[9]
布美他尼	顺序注射分析，荧光测定法 $\lambda_{exc}=314nm$, $\lambda_{em}=370nm$	0.05~10mg/L 0.46	[10]
乙酰水杨酸	顺序注射分析，Selec. electr	0.05~10mmol/L 0.20	[11]
酒石酸麦角胺	顺序注射分析，荧光测定法 $\lambda_{exc}=236nm$, $\lambda_{em}=390nm$	0.03~0.61mg/L <0.86	[12]
磺胺嘧啶-甲氧苄啶 阿米替林-奋乃静	流动注射分析，无反应紫外-可见光分光光度法 两个同步曲线，顺序注射分析，紫外-可见光分光光度	—	[13]
芦丁三水合物- 抗坏血酸	顺序注射分析，紫外-可见光分光光度法（262nm）	2~20mg/L 0~100mg/L 0.4~0.7	[14]
磺胺甲噁唑	流动注射分析光致荧光法	0~8mg/L 1.4	[15]
磺胺甲噁唑-甲氧苄啶	流动注射分析，无反应紫外-可见光分光光度法 两个同步曲线，顺序注射分析，紫外-可见光分光光度法	—	[16]

续表

	药物制剂的溶出曲线		
药物	流动系统检测	线性范围 R.S.D/%	参考文献
溴己新-阿莫西林	流动注射分析无反应紫外-可见光分光光度法 三个同步曲线	—	[17]
氢氯噻嗪-卡托普利	流动注射分析紫外-可见光分光光度法（273nm 和250nm） 三个同步曲线	10~50mg/L 5~25mg/L	[18]
磺胺甲噁唑-甲氧苄啶	流动注射分析多变无反应。三个同时配置的紫外-可见光分光光度计（257nm 和247nm）	0~200mg/L 0~90mg/L	[18]
盐酸哌唑嗪	顺序注射分析荧光测定法 $\lambda_{exc} = 244$nm, $\lambda_{em} = 389$nm	0.02~2.43mg/L 1.9	[19]
氯喹	流动注射分析 Selec electr	0.01~100mmol/L	[20]
法莫替丁	流动注射分析紫外-可见光分光光度法（265nm）	20~60mg/L	[21]
卡马西平	流动注射分析紫外-可见光分光光度法（288nm）	$1.08 \times 10^5 \sim 6.48 \times 10^{-5}$mol/L 1.92	[22]
布美他尼	顺序注射分析荧光测定法 $\lambda_{exc} = 314$nm, $\lambda_{em} = 370$nm	—	[23]
盐酸普萘洛尔	顺序注射分析溶解和渗透组装	—	[24]
乙酰水杨酸	流动注射分析		[25]
吲哚美辛	顺序注射分析		[26]
水杨酸	—		[27]
水杨酸	顺序注射分析荧光法同时释放测试		[28]
利多卡因和丙胺卡因	顺序注射分析紫外-可见光分光光度法		[29]
曲安奈德和水杨酸	顺序注射分析液相色谱		[30]

第一个用于溶出度测试的顺序注射分析系统是为评估布洛芬片剂和胶囊的溶出度曲线而开发的。在线过滤后收集样品，并通过监测吸光度确定分析物。随后，该系统的高度灵活性用于偏最小二乘校准同时监测阿司匹林片剂中乙酰水杨酸、非那西丁和咖啡因的溶解。紫外-可见光吸收、选择性电极、荧光、双安培滴定法、化学发光等，已被提议作为同时处理一种或多种药物溶出曲线的流动系统的检测器。流动注射分析、顺序注射分析和多换向组件也已被用于此类应用。

通过使用分别包含一个和两个电磁阀的歧管，结合导数分光光度法，多路换向方法还能够获得两种不同配方中磺胺甲恶唑-甲氧苄啶和氢氯噻嗪-卡托普利混合物中单个药物的溶解曲线。多路换向系统的使用大大减少了试剂消耗（相对于传统流动程序，最多可减少25倍）。

9.3.3 膜扩散

研究新的药物释放机制依赖于它们穿透组织和通过血管扩散的能力。这需要开发有效的方法来理解、优化和预测药物在普遍使用之前的释放和吸附相关特性。此类方法倾向于在体外而不是体内进行测试，以促进严格、可重复的控制。然而，体外结果只有在它们与相应体内过程的各个方面准确相关时才有效。这促进了旨在开发能够尽可能模仿分子在体内穿透人体的方式的实验设计的研究。目前最广泛使用的体外方法用于吸附测试的膜的模拟渗透。实验通常需要使用模拟上皮组织的各种膜、适当调整的环境、各种协议和扩散细胞来模拟身体内外的条件。

扩散或膜渗透细胞通常由包含供体和受体的两个隔室组成，这些隔室由模拟皮肤或某些其他组织的膜隔开。供体隔室可以是开放式单室隔室或封闭式两隔室。后者的优点是可以不限速率使用供体，这有助于监测具有高渗透系数的分析物的渗透曲线。

根据流体输送到受体隔室的方式，扩散池可以是间歇或流动类型。在间歇池中，受体介质在其隔室中不断搅拌，然而，培养基不会被替换，而只是被补充以抵消因样品提取造成的体积损失。受体隔室具有相对较大的容积以促进分析物的均化和稀释。然而，在流动池中，受体介质连续循环通过受体隔室，因此介质不断更新。出于这个原因，流动池更擅长模拟皮肤上的血流。然而，与间歇池不同的是，流动池要求受体流的体积尽可能小。因此，腔室容积和流速在流动系统中至关重要；事实上，受体室和收集的馏分的体积都应该尽可能小，而流速正好相反。在任何情况下，如果混合发生在适当的条件下，这些变量对渗透曲线没有影响。

表 9.3 列出了药物通过各种类型膜扩散研究的精选参考资料。

表 9.3 　药物配方领域的膜扩散研究

被分析物	检测方法	线性间隔	R.S.D./%	样本数/h^{-1}/检测间隔时间/h	参考文献
咖啡因 氨茶碱	紫外-可见光分光光度法（272nm）	0.4~20mg/L	—	170/2	[31]
利多卡因	紫外-可见光分光光度法（212nm）	2.5~200mg/L	1.89 2.44	5/4	[32]
吲哚美辛	荧光测定法（λ_{exc} = 330nm，λ_{em} = 385nm）	0.05~10mg/L	2.3	120/6	[33]
水杨酸	紫外-可见光分光光度法（385nm）	0.05~10mg/L	0.52	120/6	[34]

在一项关于咖啡因和氨茶碱通过人工亲脂性膜扩散的研究中，使用多路换向方法插入受体溶液。该膜由浸渍有 N 和 S_2 脂质的硝酸纤维素载体组成。研究了表面活性剂对扩散过程的潜在影响。通过使用插入受体溶液路径中的流动池监测药物。在每次测试中获得的大量测量结果（在 340nm 范围内）可能有助于促进人工膜的动力学研究并阐明赋形剂对扩散过程的影响。

与顺序注射分析系统耦合的 Franz 流动池用于确定半固体制剂中活性成分的释放程

度，使用各种材料（聚碳酸酯、纤维素混合酯、聚四氟乙烯、有机硅、聚偏二氟乙烯）和可变孔径的膜进行比较含有 1%吲哚美辛、2.5%利多卡因和丙胺卡因的凝胶的释放曲线。选择孔径为 0.4μm 的聚碳酸酯膜，以比较先前药物在凝胶和发油中使用各种程序的释放速率。在 6h 内成功监测了三种半固体制剂中吲哚美辛的释放曲线，在没有人为干预的情况下成功监测了利多卡因和丙胺卡因 4h 的释放曲线。最近，使用相同的方法来评估含有 3% Franz 活性成分的三种半固体制剂中水杨酸的释放。测试期间不需要操作员控制。最多可使用六个 Franz 池的能力使该系统对于作为质量控制分析一部分的放行测试，以及对于监控产品特性的前后变化、批次监控和产品开发也非常有用。

9.3.4 用于筛选潜在药物的功能性细胞检测

配体或反应物与细胞受体的结合不一定引起生理反应。识别和表征新药需要通过使用功能的测试来确认候选化合物的效率，以便根据它们是引发还是抑制生物反应将物质分类为刺激剂或拮抗剂。如果药物与受体的结合确实引起了反应，那么可以通过监测一些参数来评估反应，例如，细胞溶质钙的释放、细胞内 pH 的变化、酸的释放、葡萄糖或氧的摄取，甚至膜电位的变化。这种类型的手动测试经常涉及将生物材料持续暴露于可变药物浓度下，而不是在每个浓度下都使用新样品。这会降低受体的敏感性甚至破坏生物材料从而影响反应。最近，通过使用顺序注射分析结合适当的试剂进行连续进样-微珠进样，生物测试实现了自动化；该系统使用细胞培养物并测量各种细胞反应变量（表9.4）。细胞被吸附到微滴表面，少量所得悬浮颗粒用作被分析的代表性样品。因为在受控暴露期间插入的每部分刺激剂都与新鲜的细胞反应，所以反应是高度可重现的，并且以传统方式培养细胞时遇到的典型生物变异性会受到抑制。此外，微珠进样方法可以获得用刺激剂直接手动添加到培养基中时无法提取的动力学信息。根据与受体相互作用的动力学行为，以及反应的持续时间和最大强度，人们可以区分特定的拮抗剂并评估它们的效率。

表 9.4　　　　　　　　　　　细胞流动测定

被分析物	样品	检测方法	线性范围，测定间隔/s	参考文献
细胞内钙	用乙酰胆碱、毛果芸香碱和阿托品处理的 CHO Ml 细胞	Fura-2-Am 荧光测定法	-500	[35]
氧	用阿莫巴比妥处理的 CHO Ml 细胞	荧光法-铂-卟啉复合物	180	[36]
细胞内的 pH	用氨基甲酸乙酯处理 CHO Ml 细胞	BCECF-AM 荧光法	700	[37]
葡萄糖	叠氮化的肝细胞细胞	紫外-可见光分光光度法（340nm）	0.1~5.6mmol/L 120	[38]
葡萄糖乳酸	叠氮化的肝细胞细胞	紫外-可见光分光光度法（340nm）	0.1~5.6mmol/L 10 0.05~1.00mmol/L 30	[39]

用含有大鼠毒蕈碱受体的中国仓鼠卵巢细胞进行拮抗剂的功能测试。细胞在装载有荧光敏感的细胞内钙（Fura-2-Am）的微滴上培养并悬浮在缓冲溶液中。顺序注射分析用于自动将少量悬浮液吸入喷射环室，然后，微柱上捕获的细胞与目标药物相互作用。通过监测细胞溶质钙的瞬时释放（受体被药物刺激的结果）来评估药物引起的细胞反应。该系统还用于通过分析由恒定浓度的乙酰胆碱和可变浓度的拮抗剂的混合物组成的一系列样品来检查毒蕈碱拮抗剂对细胞内钙的反应的影响。

另一种类似的系统已被用于通过摄氧量来测量细胞刺激。氧气在细胞的有氧代谢中起着核心作用，在有氧氧化途径中充当最终电子受体，因此，测量摄氧量有助于研究涉及有氧机制的生化反应。后续改进的方法用于通过监测摄氧量来检查戊巴比妥和乙酰胆碱的影响。该传感器测量了固定在微滴上的铂-卟啉复合物的磷光中氧的存在而引起的降低，微滴也用作细胞培养的底物。刺激细胞的毒蕈碱受体被发现增加了氧的摄取量，这反映在某些药物引起的代谢变化上。

微珠进样系统通过荧光强度监测细胞外 pH。两种测定都不需要物理吸引力，两者之间的唯一区别是指示剂在细胞内用于测量细胞内 pH 或黏附到外膜以测量细胞外 pH 的方式。该研究是使用卡巴胆碱作为刺激剂进行的。暴露于刺激剂中导致酸的释放和胞质溶胶的酸化。这两种变化都与卡巴胆碱浓度成正比，从而可以构建浓度-反应曲线。

阀上实验室（LOV）概念可与微珠进样方法一起使用，以评估原位活细胞对葡萄糖的摄取。葡萄糖是细胞碳和能量的主要来源，它的摄取可用于评估化学刺激后的代谢状态或研究细胞缺氧之前、期间和之后的代谢影响。在最早将阀上实验室用于生物测试时，肝细胞（TABX2S）被培养在微滴上，随后在控制温度下填充并嵌入含有微生物反应器的微柱中。葡萄糖的使用量是通过嵌入微滴的液体中测量的，该液体是通过酶促过程嵌入微滴的。通过两步酶促反应测定葡萄糖，其反应产物为 NADH。还在叠氮化物存在下进行了测试，叠氮化物是细胞色素 C 氧化酶（一种参与葡萄糖有氧光降解的酶）的抑制剂。在叠氮化物的存在下，葡萄糖只能厌氧降解，因此细胞增加了对这种来源能量的消耗，以抵消这种降解途径造成的能量不足。

阀上实验室（LOV）-微珠注射组合系统被设计用于测定培养细胞的葡萄糖摄取和乳酸挤出。将乳酸挤出速率与细胞葡萄糖摄取相结合，提供了一种鉴定细胞培养中普遍存在的代谢方式的方法。通过比较这两个参数，可以估计有氧或无氧产生的能量比例。类似地，代谢机制可用于区分初级和次级培养物——初级培养物往往是有氧的，甚至可以阐明某些化学物质对活细胞的影响。化学物质对细胞代谢的影响可以通过比较它们存在时的葡萄糖摄取量和乳酸挤出来评估。这种方法还可用于控制可能引起代谢改变的环境影响。例如，可以通过对载体脱气来模拟缺血事件（例如，撞击、心搏骤停）。Schulz 等提出的系统，用于比较两种肝细胞的代谢机制，即：TABXS2，它产生过量的 Bcl-xL（一种调节主要代谢物通透性的线粒体膜蛋白）和 TABX1A，它不产生 Bcl-xL。与之前的研究一样，细胞被培养、包装并嵌入含有微生物反应器的微柱中。通过监测受体液体中的葡萄糖摄取和乳酸挤出，利用基于相同酶测定（即 NADH 生产）的定量检测，建立细胞的代谢机制。测试也在叠氮化物存在的情况下进行，这导致葡萄糖摄取和乳酸挤出增加。

参考文献

[1] Min, R. W., Nielsen, J. and Villadsen, J. (1995) *Analytica Chimica Acta*, 312, 149.

[2] Min, R. W., Nielsen, J. and Villadsen, J. (1996) *Analytica Chimica Acta*, 320, 199.

[3] Barnett, N. W., Lewis, S. W. and Tucker, D. J. (1996) *Analytical Chemistry*, 355, 591.

[4] Barnett, N. W., Lenehan, C. E., Lewis, S. W., Tucker, D. J. and Essery, K. M. (1998) *Analyst*, 4, 601.

[5] Schröder, H. F. (1999) *Waste Management*, 19, 111.

[6] Liu, X. Z. and Fang, Z. L. (1998) *Analytica Chimica Acta*, 358, 103.

[7] Liu, X. Z., Liu, S. S., Wu, J. F. and Fang, Z. L. (1999) *Analytica Chimica Acta*, 392, 273.

[8] Chen, H. and He, Q. (2000) *Talanta*, 53, 463.

[9] Li, B., Zhang, Z. and Liu, W. (2001) *Talanta*, 54, 697.

[10] Solich, P., Polydorou, C. K., Koupparis, M. A. and Efstathiou, C. E. (2001) *Analytica Chimica Acta*, 438, 1331.

[11] Paseková, H., Sales, M. G., Montenegro, M. C., Araujo, A. N. and Polasek, M. (2001) *Journal of Pharmaceutical and Biomedical Analysis*, 24, 1027.

[12] Legnerová, Z., Sklenárová, H. and Solich, P. (2002) *Talanta*, 58, 1151.

[13] Moreno Galvez, A., García Mateo, J. V. and Martínez Calatayud, J. (2002) *Journal of Pharmaceutical and Biomedical Analysis*, 27, 1027.

[14] Legnerová, Z., Satinský, D. and Solich, P. (2003) *Analytica Chimica Acta*, 497, 165.

[15] Catala Icardo, M., Garcia Mateo, J. V., Fernandez Lozano, M. and Martínez Calatayud, J. (2003) *Analytica Chimica Acta*, 499, 57.

[16] Moreno Galvez, A., Gomez Benito, C. and Martínez Calatayud, J. (2003) *Journal of Flow Injection Analysis*, 20, 8.

[17] Vranic, E., Catalá Icardo, M. and Martínez Calatayud, J. (2003) *Journal of Pharmaceutical and Biomedical Analysis*, 33, 1039.

[18] Tomšů, D., Catala Icardo, M. and Martínez Calatayud, J. (2004) *Journal of Pharmaceutical and Biomedical Analysis*, 36, 549.

[19] Legnerová, Z., Huclová, J., Thun, R. and Solich, P. (2004) *Journal of Pharmaceutical and Biomedical Analysis*, 34, 115.

[20] Saad, B., Zihn, Z. M., Jab, M. S., Rahman, I. A., Saleh, M. I. and Mahsufi, S. (2005) *Analytical Sciences*, 21, 521.

[21] Tzanavaras, P. D., Verdoukas, A. and Balloma, T. (2006) *Journal of Pharmaceutical and Biomedical Analysis*, 41, 437.

[22] Çomoglu, T., Gönül, N., Sener, E., Dal, A. G. and Tunel, M. (2006) *Journal of Liquid Chromatography and Related Techniques*, 29, 2677.

[23] Tzanavaras, P. D. and Themelis, D. G. (2007) *Analytica Chimica Acta*, 588, 1.

[24] Motz, S. A., Klimundová, J., Schaefer, U. F., Balbach, S., Eichinger, T., Solich, P. and

Lehr, C. -M. (2007) *Analytica Chimica Acta*, 581, 174.

[25] Solich, P. , Ogrocká, E. and Schaefer, E. (2001) *Pharmazie*, 56, 787.

[26] Solich, P. , Sklenárová, H. , Huclová, J. , Stínsky, D. and Schaefer, U. F. (2003) *Analytica Chimica Acta*, 499, 9.

[27] Klimundová, J. , Sklenárová, H. , Schaefer, U. F. and Solich, P. (2005) *Journal of Pharmaceutical and Biomedical Analysis*, 37, 893.

[28] Klimundová, J. , Mervartová, K. , Sklenářová, H. , Solich, P. and Polášek, M. (2006) *Analytica Chimica Acta*, 366, 573.

[29] Klimundová, J. , Šatinský, D. , Sklenářová, H. and Solich, P. (2006) *Talanta*, 69, 730.

[30] Chocolouš, P. , Holik, P. , Šatinský, D. and Solich, P. (2007) *Talanta*, 72, 854.

[31] Sales, M. G. F. , Reis, B. F. and Montenegro, M. C. B. S. M. (2001) *Journal of Pharmaceutical and Biomedical Analysis*, 26, 103.

[32] Klimundová, J. , Satinský, D. , Sklenárová, H. and Solich, P. (2005) *Talanta*, 69, 730.

[33] Solich, P. , Sklenárová, H. , Huclová, J. , Satinský, D. and Schaefer, U. F. (2003) *Analytica Chimica Acta*, 499, 9.

[34] Klimundová, J. , Sklenárová, H. , Schaefer, U. F. and Solich, P. (2004) *Journal of Pharmaceutical and Biomedical Analysis*, 37, 893.

[35] Hodder, P. S. and Ruzicka, J. (1999) *Analytical Chemistry*, 71, 1160.

[36] Lahdesmaki, I. , Scampavia, L. , Beeson, C. and Ruzicka, J. (1999) *Analytical Chemistry*, 71, 5248.

[37] Lahdesmaki, I. , Beeson, C. , Christian, G. D. and Ruzicka, J. (2000) *Talanta*, 51, 497.

[38] Schulz, C. M. and Ruzicka, J. (2002) *Analyst*, 127, 1293.

[39] Schulz, C. M. , Scampavia, L. S. and Ruzicka, J. (2002) *Analyst*, 127, 1583.

拓展阅读

[1] Abdulrahman, L. K. , Al – Abachi, A. M. and Al – Qaissy, M. H. (2005) Flow injection - spectrophotometeric determination of some catecholamine drugs in pharmaceutical preparations via oxidative coupling reaction with p-toluidine and sodium periodate. *Analytica Chimica Acta*, 538, 331-335.

[2] Catalá Icardo, M. , Fernandez Lozano, M. and Martinez Calatayud, J. (2003) Enhanced flow injection-chemilumino-metric determination of sulphonamides by on-line photochemical reaction. *Analytica Chimica Acta*, 499, 57-69.

[3] Catalá Icardo, M. , Lahuerta Zamora, L. , Antón Fos, G. M. , Martínez Calatayud, J. and Duart, M. J. (2005) Molecular connectivity as a relevant new tool for predicting analytical behavior: A survey of chemiluminescence and chromatography. *Trends in Analytical Chemistry*, 24, 782-791.

[4] Gallignani, M. , Ayala, C. , Brunetto, M. R. , Burguera, M. and Burguera, J. L. (2003) Flow analysis – hydride generation – Fourier transform infrared spectrometric determination of antimony in pharmaceuticals. *Talanta*, 59, 923-934.

[5] Gerardi, R. D. , Barnett, N. W. and Lewis, S. W. (1999) Analytical applications of tris (2, 2′- bipyridyl) ruthenium (III) as a chemiluminescent reagent. *Analytica Chimica Acta*, 378, 1-41.

[6] Gilbert López, B. , Llorent-Martínez, E. J. , Ortega-Barrales, P. and Molina Díaz, A. (2007) Development of a multicommuted flow-through optosensor for the determination of a ternary pharmaceutical

mixture. *Journal of Pharmaceutical and Biomedical Analysis*, 43, 515-521.

[7] Gómez-Taylor, B., Palomeque, M., García Mateo, J. V. and Martínez Calatayud, J. (2006) Photoinduced chemiluminescence of Pharmaceuticals. *Journal of Pharmaceutical and Biomedical Analysis*, 41, 347-357.

[8] Grudpan, K., Kamfoo, K. and Jakmunee, J. (1999) Flow injection spectrophotometric or conductometric determination of ascorbic acid in a vitamin C tablet using permanganate or ammonia. *Talanta*, 49, 1023-1026.

[9] Martinez Calatayud, J. (1996) *Flow Injection Analysis of Pharmaceuticals. Automation in the Laboratory*, Taylor and Francis, London.

[10] Martínez Calatayud, J. and Catalá Icardo, M. (2005) Flow injection analysis. Clinical and pharmaceutical applications. *Encyclopedia of Analytical Sciences*, 2nd edn, Academic Press, 0127641009.

[11] Martínez Calatayud, J. (2005) *Spectrophotometry. Pharmaceutical Applications*. Encyclopedia of Analytical Sciences, 2nd edn, Academic Press, 0127641009.

[12] Murillo Pulgarin, J. A., Alañon Molina, A. and Perez-Olivares Nieto, G. (2004) Determination of hydrochlorothiazide in pharmaceutical preparations by time resolved chemiluminescence. *Analytica Chimica Acta*, 518, 37-43.

[13] Pimenta, A. M., Montenegro, M. C. B. M. S., Araujo, A. N. and Martinez Calatayud, J. (2006) Application of Sequential Injection Techniques to flow analysis. *Journal of Pharmaceutical and Biomedical Analysis*, 40, 16-34.

[14] Ruedas Rama, M. J., Ruiz Medina, A. and Molina Diaz, A. (2004) Bead injection spectroscopy-flow injection analysis (BIS-FIA): an interesting tool applicable to pharmaceutical analysis: Determination of promethazine and trifluoperazine. *Journal of Pharmaceutical and Biomedical Analysis*, 35, 1027-1034.

[15] Sales, M. G. F., Tomás, J. F. C. and Lavandeira, S. R. (2006) Flow injection potentiometric determination of chlorpromazine. *Journal of Pharmaceutical and Biomedical Analysis*, 41, 1280-1286.

10 连续流动分析的工业和环境应用

Kees Hollaar 和 Bram Neele

10.1 引言

连续式流动分析仪（CFA）或分段式流动分析仪（SFA）的优势在于所有的样品、所采用的标准以及系统质量检查的处理方法和分析条件是相同的，样品可以进行预处理，这使得它非常适合工业和环境领域中的许多不同方面的应用。虽然该系统在技术上看起来很复杂，但它们的稳健性已经得到了充分证明，并在许多样本流量很大的实验室中使用，例如农业实验室和环境实验室。该技术已被证明具有对复杂样品的操作能力，也具有复杂化学自动化中无可争议的优势。此外，对更多的样品进行分析的需求不断增长，这种需求正被转化为将复杂的手动化学实现高度自动化的努力。在线紫外消解和在线蒸馏法分析氰化物就是一个很好的例子：基于在线紫外消解，对总氮进行分析，用镉还原法测定硝酸盐。此外，有机溶剂的萃取可以用最小体积的萃取液在微尺度上进行。对于这一提取过程，一个典型例子就是测定啤酒中的苦味。除了可以对样品进行完整处理，而且还可以同时分析同一样品的一些参数，这对用户来说是一个非常有吸引力的前景，因为它节省了分析的周转时间。

通过使用这些自动化方法，最大限度地减少实验室的人力资源，实现了人力成本的显著降低。有趣的是，如今在某些领域的实验室人员是短缺的，这却成了自动化发展的一个契机。但同样重要的是，大多数的除去人类参与的研究表明，分析自动化不一定会导致人员冗余，因为在外部环境中遇到的问题越来越多，这就需要在现有实验室中增加更多新的部门[1]。由于自动化程序的工作量要小得多，因此与化学品的使用和处置相关的成本也降低了。通过在整个样品处理过程中对样品的自动条形码进行读取，可以避免人工处理造成的转录错误。这种自动化是可以实现的，因为在数字时代，其数字潜力是巨大的。随着现代数字化技术的发展，检测精度至少提高了 2 倍。全自动化系统控制扩大了分析的数量，分析因子也由一个扩大到 2 个以上。白天的办公时间可用于样品的制备，而自动分析是在实验室操作员工作以外的时间进行，没有操作员的干扰。

由于分析方法的普遍商业化，分析方法的规范化趋势正在增长。当不同的商业化实验室进行分析时，更需要结果的可比性，任何组织都不应违背该趋势。因此，对于使用的方法和方法的分析标准都必须是明确的，而这些标准是由环保局、国际标准化组织、水和废水检验标准方法、中欧酿造分析委员会、美国酿造化学家协会、欧洲酿酒公约等许多全国性的相关组织制定的。自动化和标准化实现了分析结果的准确性和可比性。

10.2 环境和工业领域概述

流动分析法的应用很广泛，因此不可能给出所有应用领域和特定应用领域的详细完整概述。此处仅列出了主要应用领域及其特定测试指标和测试参数，以及已知的优缺点。列表中列出的应用程序的是一个有限的概述，让人们初步对特定应用程序中的自动化应用程序有一定的印象。这里有多种检测范围和方法可供使用。

10.2.1 在环境领域的应用

环境实验室有大量的自动化应用程序（表10.1）。在世界各地的许多实验室中，对硝酸盐+亚硝酸盐、亚硝酸盐、氨和磷酸盐进行同时分析。经常使用氯化物或硫酸盐来建立一个5道或6道同时系统。对不同的供应商而言，这些系统的每个应用程序每小时能够执行高达60或80个样品，在这个配置下，每小时可产生300~500个分析结果。在环境实验室中有了越来越多的复杂分析应用，如总氰化物、酚指数，亚甲基蓝活性物（MBAS）。一种更现代的方法被用来分析水样本中总氮和总磷，但不需要通过离线消解来处理样品。在分析过程中，通过紫外辐射和高温加热进行了消解。与之前采用的人工或（半）自动检测的（半）离线消解方法相比，这种方法快速且成本低。这些全自动应用系统的一大优点就是没有使用用于消解的高浓度强酸和有毒催化剂。这些自动化的流动分析系统的缺点是不能处理大颗粒样品，在将样品进入自动化系统之前，都要进行彻底地均质和混合，这将在一定程度上克服这一缺点。但是，以干物质来表示颗粒量，其值不能超过100mg/L。世界上一些国家的法规要求使用总氮而不是凯氏氮，政府当局发现不仅凯氏氮部分测试对环境有害，而且以硝酸盐的形式存在的氮也是对环境有害的。分析海水样本的实验室是目前使用全自动流动分析技术的用户群体。全自动流动分析技术可检测到极低浓度的营养物质，达到亚 μg/L 水平；Kerouel 和 Aminot 用荧光法对氨进行检测，达到 1.5nmol/L 的水平[2]。

表 10.1　　　　　　　　　　环境应用概述

酸度（总）	电导率	硝酸+亚硝酸盐
碱度（总）	铜	亚硝酸盐
铝（水解）	氰化物（总、游离、WAD）	氮（总）
氨基酸（总）	溶解有机碳	非离子物
氨	甲醇	酚（总）
硼	氟化物	磷酸盐（正磷酸盐、完全水解）
溴化物	甲醛	高锰酸钾
钙	硬度（总）	COD 值
碳酸（总）	肼	硅酸盐
化学需氧量（COD）	Fe^{2+}/Fe^{3+}	钠/钾
氯化物	镁	硫酸盐
氯	锰	硫化物
胆碱酯酶抑制剂	甲基蓝活性物质	亚硫酸盐
Cr^{6+}	钼	总有机碳
颜色	镍	浊度
		尿素

10.2.2 在植物和土壤中的应用

从土壤或植物样品中提取的物质通常是自然带色的：在土壤中总是有中等浓度到高浓度的有色有机物以腐殖酸的形式存在，特别是在分析那些被建议施肥的土壤。需要注意的是，如果样品萃取物的 pH 在分析过程中发生变化，腐殖酸开始在溶液中凝结和沉淀，这可能会导致流动分析出现问题。这些有机物还可以干扰反应过程，如活化镉还原硝酸盐的过程。因此，使用流动分析仪的一个有力论据就是其本身很可能有一个透析器，透析器会阻断腐殖酸和背景颜色进入反应过程。其采用的可同时分析的应用程序组是不同于环境测试的应用程序组：这里含氮化合物包含氨和硝酸盐，与之一起的磷酸盐和钾（NPK）都是非常重要的。然而，不同的样品提取方法存在很大差异，而且土壤或植物样品的重量与提取液的比值在分析范围内也存在很大差异。土壤与水按照 1∶60 比例进行混合，获取的提取物用于测定"水中磷酸盐"（PW），土壤与氯化钙（0.01mol/L）按照 1∶10 比例混合，获取的提取物用于测定磷酸盐，比较这两种测试结果完全不同。比较以农业为目的的分析或和环境为目的进行分析会发现一个主要的区别。后者总是要求"总"测定，要求在分析之前进行其他提取程序甚至消解程序，而农业领域要求的是"对植物可用"的分析物。在关于土壤样品的环境调查领域，流动分析法的重要应用是对总氰化物、酚指数、总氮和总磷等指标的检测。而对土壤和植物分析的延伸，其实是对肥料和肥料原料的分析，该延伸的分析必须获得精确和准确的结果，因为化肥中分析物最终浓度的不准确性将涉及数千吨产品，极大地影响成本。流动分析法在肥料工业中非常重要的应用是对氨、硝酸盐、磷和钾的分析，精确性对肥料而言是最重要的。所测定的分析物清单见表 10.2。

表 10.2　植物和土壤应用概述

醋酸	氰化物（总，游离）	果胶
铝	溶解有机碳	苯酚（总）
氨基酸（总）	氟化物	磷酸盐（邻位、总、可水解）
氨基氮（游离）	半乳糖（D+）	蛋白（溶）
氨	葡糖酸（D）	还原糖（总）
硼	葡萄糖/果糖/蔗糖	硅酸盐
溴化物	Fe^{2+}/Fe^{3+}	钠/钾
钙	乳酸（L/D）	淀粉
碳水化合物	赖氨酸（L）	糖（总）
总碳	镁	硫酸盐
总氯	苹果酸（L）	亚硫酸盐
碳酸	锰	总有机碳
柠檬酸	硝酸+亚硝酸盐	尿素
电导率	亚硝酸盐	
铜	氮（总）	

10.2.3 在制药领域的应用

在药物分析中，使用连续流动分析设备的需求不断增长。可追溯性、数据完整性以及可确保操作一致性的特定的标准操作程序（SOP）是制药实验室的必备条件。分析的全面自动化和数字数据归档的准确性是必不可少的。当使用流动分析技术时，这些需求恰好得到了满足。完整的运行分析数据可以在运行期间和之后得到有效验证。在相同条件下分析样品，通过存档，使得验证结果和所有测试数据仍然可用。无论是测定发酵液中的葡萄糖还是药丸和粉末中的葡萄糖，自动化以及对所有分析过程中步骤的控制性对于制药实验室都很重要。但是通常，在其他应用领域，它们甚至无法被识别名称。流动系统在应用方面有些分散，因为制药公司并不都生产相同的产品。与所有实验室一样，实验中有很多耗时和且工作量大的测定，因此其自动化通常被广泛接受。根据样品的所需要分析的量，需要在半自动方法、手动方法或连续流分析法之间做出选择。目前在制药领域应用的最新主题是分析维生素或前体维生素。连续流分析的一个优点就是将分析这些参数所需的不同步骤（例如提取）完全自动化。测定的分析物清单如表 10.3 所示。

表 10.3　制药应用概述

乙酸	甲醛	亚硝酸盐
氨基酸（总）	甲酸	氮（总）
氨	半乳糖（D+）	苔黑素（糖值）
硼	葡糖酸（D）	青霉素
钙	葡萄糖/果糖/蔗糖	pH
泛酸钙	谷氨酸（L）	苯丙氨酸
碳水化合物	甘油	磷酸（邻位、总、可水解）
碳	己糖醛酸	蛋白质
碳酸盐（总）	乳酸（L/D）	维生素 B_6
氯化物	乳糖	还原糖（总）
头孢菌素	赖氨酸（L）	钠/钾
柠檬酸	苹果酸（L）	淀粉
DNA	吗啡	糖（总）
乙醇	硝酸+亚硝酸盐	甲氧苄氯嘧啶
氟化物		

10.2.4 在啤酒和葡萄酒领域的应用

在啤酒分析中，生产控制和质量稳定是非常重要的。世界各地的酿酒商销售的其实是酒的味道，这意味着要分析与味道有关的几个参数。流动分析法在对啤酒中的苦味、游离氨基氮、多酚、双乙酰、麦芽中 α-淀粉酶、β-葡聚糖的测定中得到了广泛的应用。这些方法并不总是来自传统的光度检测类型，还包括紫外范围内的检测方法和荧光检测方法。例如，与人工检测相比，流动分析法在苦味分析中使用少量溶剂，这是一个重要的优点。这种自动化的应用大大提高了样品的检测量，这本身就非常方便，而且又减少了提取溶剂的体积和节省了大量的成本。一个与欧洲法规有关的参数引起了人们的兴趣，那就是啤酒中总二氧化硫的分析。欧洲的法规对二氧化硫的浓度做了要求，这是必须遵守的。几乎所有的自动化应用程序（表 10.4）都符合 EBC（欧洲酿酒公约）、ASBC（美国酿酒化学家协会）标准或 MEBAK（欧洲酿酒委员会）。

表 10.4　啤酒和葡萄酒分析中的应用概述

醋酸	乙醇	可溶性蛋白质
酸度（总）	葡聚糖（β）	还原糖（总）
氨基氮（游离）	葡聚糖酶（β）	山梨酸
花色苷	葡萄糖/果糖/蔗糖	淀粉
苯甲酸	甘油	糖（总计）
苦味	羟基脯氨酸	二氧化硫（总、游离）
钙	铁（总游离）	酒石酸
二氧化碳	乳酸（L/D）	硫代巴比妥酸
柠檬酸	苹果酸（L）	浊度
颜色	硝酸盐＋亚硝酸盐	黏度
氰化物（总）	pH	挥发性酸度
密度	磷酸盐（邻位，总）	
二乙酰	多酚	

在葡萄酒应用中，一个非常重要的优点是可以同时使用多参数系统，同时分析八个参数是很常见的。单个样本输出的结果包括葡萄糖、果糖、蔗糖、苹果酸、柠檬酸、总还原糖、总二氧化硫、游离二氧化硫等，和一些其他指标如挥发性酸度。这些参数也可方便地用作对果汁和其他饮料的分析。pH 和密度也可以同时测量，不是因为这些测量的复杂性，而是因为其他参数额外的自动化优势。如果自动化执行所有其他应用程序而手动执行这两个应用程序，这显然是不明智的。

10.2.5 在烟草领域的应用

就其性质而言，对烟草的应用可以被视为对植物测定，这些特定的烟草植物只有经过一些香料和其他添加剂的处理才能在吸烟时变得"美味"。尽管如此，这些烟叶仍然是植物。提取程序和分析程序主要是在生产香烟、雪茄或其他烟之前对原材料进行。通常，流动分析仪上的常规分析是针对氨、硝酸盐＋亚硝酸盐、还原糖和总生物碱（尼古丁）进行的。使用"一体式"方式提取提取物，由于提取物样品通量高且颜色深，因此每个应用程序都配备透析器。应用方法通常根据 Coresta（与烟草相关的科学研究合作中心 Cooperation Center for Scientific Research Relative to Tobacco，Coresta）和国际标准化组织（ISO）而定。可以自动化处理不同的提取程序，测定氰化物、淀粉和苹果酸，也可用于特定的分析目的。除了对烟草本身进行分析外，还对纸（用于生产卷烟）和烟草烟雾（表 10.5）进行分析。

表 10.5　　　　　　　　　　烟草分析中的应用概述

醋酸	葡萄糖/果糖/蔗糖	蛋白质
生物碱（总）	乳酸（L/D）	还原糖（总）
氨基酸（总）	苹果酸（L）	山梨酸
氨基氮（游离）	薄荷醇	淀粉
氨	烟碱	糖（总计）
碳水化合物（总）	硝酸盐＋亚硝酸盐	钾
氯化物	氮（总）	挥发性酸度
柠檬酸	果胶	挥发性碱
氰化物（总）	磷酸盐（邻位、总、可水解）	

10.2.6　在食品领域的应用

首先，在乳粉中，硝酸盐和亚硝酸盐的浓度受到了良好的监测，以确保人体尤其是婴儿不会摄入过高剂量的这些化合物。牛乳样品中蛋白质和脂肪的分离是通过长时间的透析来完成的，因为亚硝酸盐的检测浓度小于 $2\mu mol/L$。而蛋白质分析，即经凯氏定氮消解后测量的总氮，这是食品部门流动分析中最常见的。虽然像 DUMAS 这样的技术在蛋白质测定方面是具有竞争力的，但 DUMAS 方法几乎只用于固体样品，因为燃烧分析仪可以省去消解步骤。但是采用离线湿消解法既可以测定固体食品中的氮和也可测定饮料中的氮，然后采用光度法测定。可检测分析物的完整列表见表 10.6。

表 10.6　　　　　　　　　　食品分析中的应用概述

醋酸	甘油	蛋白
氨基酸（总、游离）	组胺	吡哆醇
氨	羟基脯氨酸	丙酮酸
抗坏血酸	碘	还原糖（总）

续表

苯甲酸	$Fe^{2+} + Fe^{3+}$	核黄素
碳酸氢盐	乳酸（L/D）	钠/钾
咖啡因	乳糖	山梨酸
钙	赖氨酸（L）	淀粉
碳水化合物（总）	苹果酸（L）	糖（总）
二氧化碳	烟酸	二氧化硫
氯化物	烟酰胺	硫胺素
柠檬酸	烟酸	有机碳总量
肌酸/肌酸酐	硝酸盐+亚硝酸盐	维生素 B_1
氰化物（总）	亚硝酸盐	维生素 B_2
密度	氮（总）	
乙醇	苦黑酚（糖值）	泛酸
氟化物	果胶	维生素 B_6
半乳糖（D+）	苯丙氨酸	维生素 C
葡萄糖/果糖/蔗糖	磷酸盐（正磷酸盐、总、水解）	维生素 PP
谷氨酸（L）	多酚	挥发性酸

10.3 流动分析方法的应用及其范围

本章讨论了几种应用：用于环境分析的总氮、总磷、总氰化物和酚指数的检测，还将讨论烟草行业的总还原糖和啤酒行业的苦味的检测。

对新的自动化应用程序的需求是无穷无尽的，尤其是对耗时且费力的应用程序进行自动化的压力更大。需要自动化的主要是一些复杂的应用程序，例如消解、蒸馏和提取。其他个别应用程序的检测范围也有待改进。这些应用程序主要需要更改硬件和设计新组件。在许多实验室，工作人员会根据个人和当地需求去开发系统，如果发布开发的系统，其他人会根据他们的需求复制这些开发的系统。然而，经常发生即使这样也还要进行更改的情况，因此需要对更改的硬件或新开发的硬件进行标准化。本章进一步讨论的应用程序的标准化。要了解连续式流动分析仪系统的基本工作原理，可以阅读本书前面章节的扩展说明。

自动化分析系统以模块化的方式构建，以提供最大的灵活性来配置它，这样可以满足本地用户的需求（图10.1）。在有限的硬件条件下，用户可以很容易地切换到不同的应用程序。系统的核心是内置的应用程序。

自动化应用程序是分析系统中的一个单独的模块（图10.2）。通过这种配置，分析仪可以很容易地在一个同步分析系统中作为单一通道或多通道进行处理。

样品的选取　　　　　化学区　　　　　数据处理

图 10.1　自动化连续流动系统的典型配置

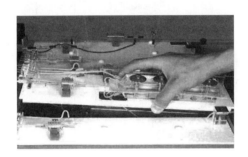

图 10.2　自动化应用程序的典型设计

10.3.1　总氮

总氮测定的自动化程序是基于以下反应[4-6]：为了消解复杂的氮化合物，首先将样品与过二硫酸钾/氢氧化钠溶液混合。然后加入硼砂缓冲液，使液流通过紫外在线消解器。在离开紫外消解器时，流体被加热到 107℃。紫外消解器和加热过程中花费的时间约为 20min，以确保完全消解氮化合物。为了催化消解过程，可以将钛离子添加到消解试剂中。由于所有氮都被氧化成硝酸盐，因此下一步就是借助经典方法——改良的 Griess 反应来测定这种硝酸盐。该方法首先通过添加铜活化的镉将硝酸盐还原为亚硝酸盐，然后亚硝酸盐将与磺胺和 α-萘基乙二胺二酸盐反应。在 540nm 处测量显色的红紫色复合物，并且与样品中最初存在的氮成正比。

该过程需要在化学模块中添加透析器（图 10.3），以避免过量的过硫酸盐破坏镉还原柱的活性。

该分析程序适用于各种水样，如地表水、地下水、废水、饮用水、海水，也适用于土壤提取物。最低检测限低于 1μg/L，这对于所有样品类型来说都足够低了。该应用程序十分稳健，不会受到酸性和海水样品基质的影响。图 10.3 中的流程图显示了自动应用模块确定总氮所需的硬件组件。

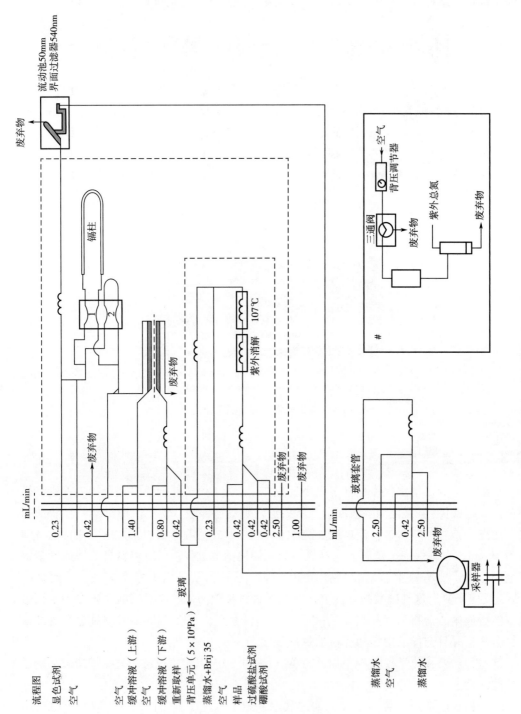

图10.3 自动化分析测试氨的设计流程图

在紫外氧化过程之前添加稀释回路装置，这提供了额外的灵活性。通过这种方式，可以确定更高的应用范围，例如，即使在 1~500μg/L 的基本应用设计中，仍然可以确定高范围（>200mg/L）。这种自动化方法灵敏度高，结合数字检测器，更加适用于较宽的检测范围。

该化学模块的另一个特点是可以将其用于对硝酸盐和亚硝酸盐的分析检测。通过断开消解部分的硬件，样品跟随模块中改良的 Griess 反应部分。这种硬件的"双重"使用带来了成本的节约。而且从低浓度氮（μg/L）到高浓度氮（mg/L），曲线的线性响应都非常好。由于所使用的试剂长期稳定，因此在分析程序开始运行时，只需进行一次校准。对于像这样的复杂应用，稳定性和信噪比也非常出色。最初，校准曲线是用标准系列建立的，然后样品按批次进行，并按照用户定义的顺序进行质量控制。许多样品会进行超过 10h 的长时间测试运行，这是常规操作。

10.3.2 总磷酸盐

测定总磷酸盐的自动化程序基于以下反应[7-10]：将样品与酸性过氧化物二硫酸钾溶液混合，水解复合磷酸盐化合物，生成正磷酸盐。该流动样品被引导通过紫外消解器。有机磷酸盐复合物在紫外线辐射和氧气的作用下分解成正磷酸盐。然后将硫酸加入流动样品中，加热到 107℃。无机磷酸盐复合物在该部分应用程序中被水解为正磷酸盐。在紫外和高温下，总水解过程的时间大约是 15min。水解后，正磷酸盐用成熟和可靠的方法分析检测，使用七钼酸铵，由于该方法对 pH 敏感，在加入七钼酸铵溶液之前，先用氢氧化钠溶液中和高温加热器流出的酸性样品。正磷酸盐在流动中会发生反应，在酒石酸锑（Ⅲ）氧化钾的催化下形成磷钼酸络合物。这种络合物被 L（+）抗坏血酸还原为深蓝色的复合物。将蓝色复合物在 880nm 处测量，测量该光谱下磷酸盐的特异性。当显色反应的 pH 保持在 pH<1 以下时，将不会测量到硅酸盐的干扰。通过添加透析器，避免了废水样品自身极高背景色的干扰。

该分析方法适用于各种水样，如地表水、地下水、废水、饮用水、海水，也适用于土壤提取的样品。由于这种自动化应用非常稳健，其最低检测限低于 1μg/L，因此对海水和土壤样品的分析也非常有吸引力。正如总氮测量所提到的那样，由于信号的质量很好，这使得分析检测的结果达到亚微克水平成为可能。这种应用程序可以与总氮同时运行。在实践中，大多数样品必须对两种分析物进行分析，如果将这些应用程序结合在一个系统中，将在没有操作人员参与的情况下使样品检测量增加一倍。对于这两种应用，是按照国际标准[12]，样品在 pH 为 2 的条件下保存。

自动化分析测试磷酸盐的典型流程图如图 10.4 所示。

10.3.3 氰化物

总氰化物是游离氰化物离子的总和，包括所有有机结合的氰化物、配合物和简单金属氰化物，以及金和铂的氰化物，但钴配合物和硫氰酸酯结合的氰化物除外。由于氰化物在酸性介质中会高温催化分解，钴配合物不能被完全确定，且钴离子是这种分解反应的催化剂。

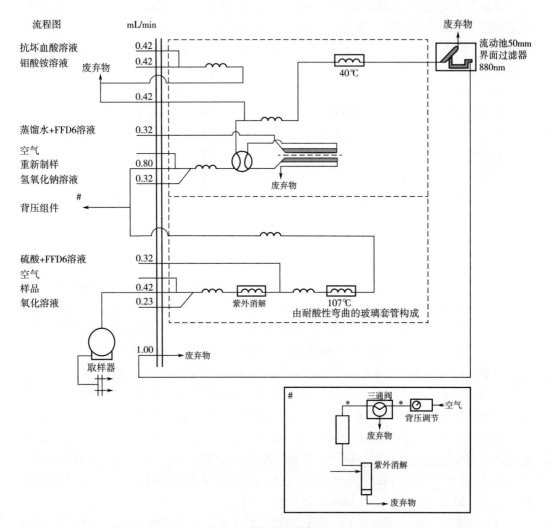

图 10.4 自动化分析测试磷酸盐的典型流程图

游离氰化物被确定为易释放氰化物。它是由游离氰化物离子和简单金属氰化物的总和。有机氰化物尚未确定。

弱酸解离的氰化物基本上是游离氰化物，CN、HCN 和弱金属形成氰配合物如 $[Cd(CN)_4]^{2-}$、$[Mn(CN)_5]^{3-}$、$[Ni(CN)_4]^{2-}$，以及汞氰化物 $Hg(CN)_2$。铁配合物不包括在内。用这种方法测定的酸分离氰化物，就等于易氯化氰化物（CATC）和有效氰化物。

10.3.3.1 总氰化物

测定总氰化物的自动化程序基于以下反应[11,12]：络合氰化物在 pH 为 3.8 的连续流动液中被紫外辐射分解。UV-B 灯和硼硅玻璃分解线圈（仅允许约 310nm 及以上的光透过）用于阻挡波长小于 290nm 的紫外线，从而防止硫氰酸盐转化为氰化物，在该波长及其以下波长能量下氰化物还会被分解。在分解之后，在 125℃ 下运行在线蒸馏装置，将释放的氰化氢与消解试剂分离。以恒定流量进行蒸馏，在蒸馏盘管中会产生一个小真空。然后在馏出液中用分光光度法测定馏出液中的氰化氢。为了与氰化物发生特定的颜色反应，按以下顺序添加一些试剂溶液：先加入氯胺-T，与它形成氯化氰。然后加入吡啶-4-碳酸和1,3-二甲基-巴比妥酸，它们与氯化氰形成红色络合物，在600nm处进行颜色检测。

10.3.3.2 游离氰化物

测定游离氰化物的自动化程序基于以下反应，同时使用相同的硬件配置，但关闭紫外线灯。将硫酸锌溶液加入样品流中以沉淀氰化铁酸锌络合物中存在的任何铁氰化物。样品中的游离氰化氢由在线蒸馏装置在125℃、pH3.8和真空条件下分离。然后用分光光度法测定馏出物中的氰化氢，如总氰化物测定程序中所述。

自动化分析测试总氰化物的典型流程图如图10.5所示。

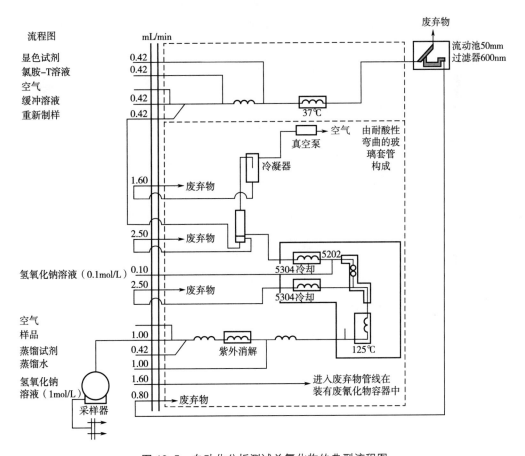

图10.5 自动化分析测试总氰化物的典型流程图

10.3.4 酚指数

酚指数代表一组酚类化合物，如具有羧基、卤素、甲氧基或磺酸基团的酚，邻位、间位取代的酚类和对位取代的酚类。这些化合物是可蒸馏的，但是这些化合物的挥发性是不同的。如果酚类化合物的挥发性低，则可蒸馏量小于100%，未完全计入总结果。测量波长通常设置为505nm。测定邻位和间位取代酚的自动化程序基于以下反应[13-16]：为确定酚指数，样品通过在线蒸馏装置在酸性介质中蒸馏。将馏出物重新取样到缓冲溶液中，将pH控制在10.0。然后加入碱性氰化铁和4-氨基安替比林溶液。存在的酚类化合物将与这些化学物质反应形成红棕色复合物，在505nm处测量。

为了建立稳定的检测信号，分析系统所处的环境起着重要的作用。化学品的纯度和制备的试剂溶液必须为分析纯。在做准备工作时，必须小心地清洁玻璃器皿和移液器，特别是六氰基铁酸盐溶液必须避光。针对有色络合物的测量，已经讨论了很长时间。可在460~520nm测量其吸光度。通常，505nm被广泛接受。可以预料的是在正常值的检测范围内吸光度会较低。100μg/L 酚将产生大约 0.035AU 的吸光度。要想在低 μg/L 水平进行检测，检测器的分辨率和信号稳定性必须很高。在这种情况下，可以达到 <100μg/L 酚的检测限，它还导致±0.3%的变异系数。

图10.6所示为自动化分析测试酚指数的典型流程图。

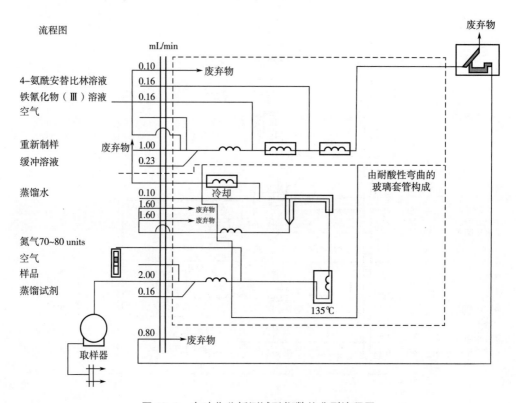

图10.6　自动化分析测试酚指数的典型流程图

10.3.5　总还原糖

烟草中存在的还原糖或还原性碳水化合物可以在提取后测定。顾名思义，身并不是直接确定特定的糖的含量，而是根据其还原能力来测定的。蔗糖属于还原糖类，在测定前需要水解。

检测总糖的自动化程序基于以下反应[17,18]：在单独提取后，样品与醋酸溶液混合并通过透析膜。如果存在蔗糖，首先在95℃下使用盐酸进行在线水解，将蔗糖转化为葡萄糖和果糖，这两种糖都是还原糖。透析后，样品在碱性介质中与对羟基苯甲酸肼（PAHBAH）混合。为了形成黄色的腙复合物，流动样品被加热到85℃。在410nm处测量由此形成的有色复合物。

在流程图（图10.7）中，自动化应用程序中包含蔗糖的水解步骤。水解可以通过加入酸或酶，例如转化酶或β-葡萄苷酶，水解必须在透析前进行。这是因为不同分子大小的透析速度不同。如果不需要蔗糖测定，一个捷径可以跳过水解步骤——将进样器的进样管直接连接到图中的再进样管。

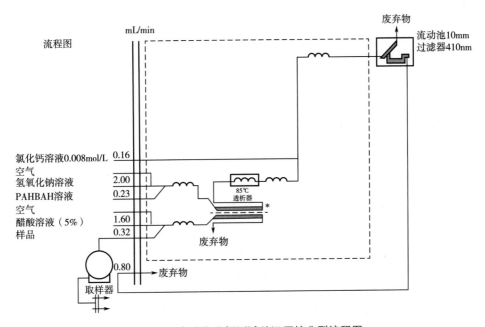

图10.7 自动化分析测试总还原糖典型流程图

很明显，各种还原糖（如葡萄糖和果糖）与显色剂的反应必须产生相同的颜色强度。在校准中，标准品可以由葡萄糖或果糖制成。添加纯蔗糖作为标准品对蔗糖水解步骤进行质量检查。对照纯葡萄糖校准标准进行检查，形成的葡萄糖和果糖的理论浓度。这种检查在使用酶的情况下特别重要。

10.3.6 苦味

苦味是啤酒品尝中的描述性术语。苦味是由一组异α-酸和其他影响味道的成分形成的，这些成分在酿造过程中从啤酒花进入啤酒。

苦味测定的自动化程序基于提取方法[19-21]：酸化后，样品在异辛烷中提取。在275nm处测量异α-酸的光度吸收。

有机溶剂异辛烷的相分离程序被内置于自动化应用程序中。它是一种内部经过涂层处理以获得平滑连续分离的小型装置。抽气取决于温度，抽气盘管是恒温的。如果有15~20℃恒定温度的自来水可用，则可使用该自来水；如果没有，则可能需要内部冷却的水循环浴。为了输送溶剂确实需要特殊的泵管，普通的管往往不耐溶剂腐蚀。被称作 Acidflex（氟橡胶类型）泵管可用于输送溶剂。

图 10.8 显示了自动化分析测试苦味的典型流程图。

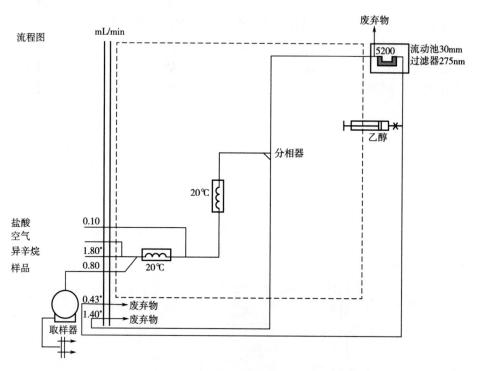

图 10.8 自动化分析测试苦味的典型流程图

苦味检测很难找到人工校准标准。实际上，这意味着实验室需使用已知样品。这些样品经过手动分析后，在 4℃ 下至少可稳定 3 个月。也可以将样品分成小份并在 -20℃ 下冷冻。需要时，只需解冻部分样品即可使用。

由于异 α-酸来自啤酒花，因此也可以使用这些纯化的酸作为标准。与啤酒中测量的苦味相比，这些酸的强度是未知的。因此，这些异 α-酸需要通过添加手动程序进行单独量化。

如今，人们观察到越来越多的啤酒苦味变化。测定范围 5~40 个苦味单位（BU），主要在 20~35BU。啤酒中仍会出现异常浓度的苦味，可高达 80BU。在这些情况下，将围绕该浓度建立标准化，以提供可靠的结果。化学应用不需要改变，可以处理从低到高的浓度。

10.4 流动分析应用程序的开发

自动流动分析应用程序正在取代手动应用程序。其不仅增加了样品通量，而且减少劳动力、实验室空间、化学品和废物的使用，所以自动流动分析是非常高效、准确和可靠的。通常，对于蒸馏应用，每天 10 个样品的数量已经足以实现自动化（例如，氰化物）。

自动化应用程序是在操作员干扰最小的情况下测定样品的分析结果，它主要基于湿化学，其次是比色测量。自动化应用不包括样品保存，也不包括从固体样品中提取液体。必须知道和理解自动化应用程序的所有步骤。在每个步骤中任何原始样本的变化都可以被识别：稀释、试剂添加、反应时间、温度、萃取、蒸馏等。每个步骤都包括一个条件，例如，对于稀释，条件为稀释因子；对于添加试剂，条件为体积。

列出所有这些步骤，将可能自动化的内容和不可能自动化的内容（自动化程序过程中的挑战）分别列出。当此步骤无法自动化时，之前的所有步骤都不能用于该自动化，因为它会破坏"流动"分析的过程。步骤中自动化的可能性是给定好的，它从分析过程中的某个点开始，一直持续到检测器中的测量。此列表是手动程序自动化的基础。要将手动步骤转换为自动步骤，必须非常清楚地了解流动分析仪的功能和特性[3]。下面给出一个简单的例子。

自动化土壤样品中氨的手动程序步骤：

(1) 称取 10g 干土；
(2) 加入 50mL 蒸馏水；
(3) 混合 1h；
(4) 过滤纸过滤器；
(5) 吸取 5mL 样品滤液到 50mL 量瓶中；
(6) 加入 10mL 缓冲溶液；
(7) 混合 5min；
(8) 加入 5mL 试剂 R1；
(9) 加入 10mL 试剂 R2；
(10) 加入 5mL 试剂 R3；
(11) 混合；
(12) 将烧瓶置于 40℃ 水浴中 15min；
(13) 冷却并加蒸馏水至 50mL；
(14) 在 10mm 比色皿中测量 660nm 处的吸光度；
(15) 用蒸馏水代替 R3 测量空白。

自动化必须从步骤 5 开始。泥浆的过滤不能在流动分析仪中进行，必须跳过。直到最后，所有步骤都是有自动化的可能性。由于我们从自动化中排除了前四个步骤，因此很明显，报告的最终结果已针对提取中的样品摄入量进行了校正。值得注意的是，在步骤 15 中，空白测量，手动操作需要相当多的额外工作（这是一个挑战），但在自动化应用程序中使用透析器的功能，因此空白测量不需要任何额外的工作。透析器避免了样品的空白颜色对测量的影响。通常，与流动分析中使用的体积相比，手动过程中使用的体积更大。总端流量为 1.6~2.0mL/min。为了方便体积转换，尽可能地保持比率恒定是十分重要的，如果无法保持比率恒定，可能需要修改试剂浓度。反应时间转换为混合盘管的长度（=体积）和流经的流量。由于混合条件非常充分，因此反应得到优化，并不总是需要像手动过程那样花费全部时间。此外，所有样品、校准标准和QC均在相同的环境下处理。对于温度的要求则需要通过加热（或冷却）反应盘管来解决。这在自动化应用程序的任何地方都是如此。最后，了解用于测定分析物的化学品是非常重要的。缓冲容量、反应的pH、对温度和光的敏感性等，在手动程序和自动化程序中都起着重要作用。

图10.9显示了自动化分析土壤中氨的典型流程图。

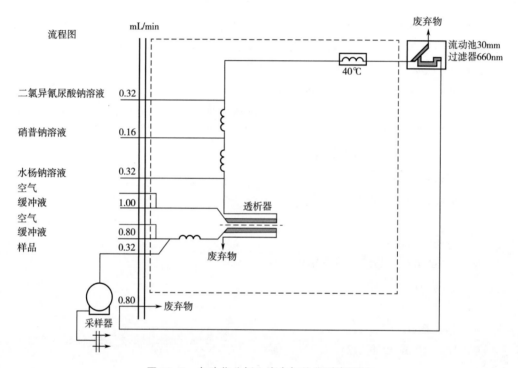

图10.9 自动化分析土壤中氨的典型流程图

10.5 连续流动分析的趋势

连续流分析仪的分析优势在于始终如一的质量和样品通量。尽管市场上有替代的自动化分析仪，例如，可以执行一些简单直接应用的离散分析仪，但它们一次只能确定一个参数。可同时分析五六个或更多的参数是连续流分析仪的优势，可以提高分析的检测量。流动注射分析法的一些缺点正在减缓流动系统在所有应用领域的实施，这些就是小内径管堵塞和具有各种基质、颗粒样品的处理和样品背景颜色等技术问题。离散分析仪也必须部分考虑这一点，这些分析仪还受到低检测限的影响。

连续式流动分析仪几乎不需要考虑所有这些问题。对于具有不同分析步骤的冗长而复杂的应用程序，如提取、消解和蒸馏，连续式流动分析仪将始终支持实施这些技术，并配置强大而可靠的分析仪。数字检测器的使用带来了更广泛的检测范围和高水平的准确读数。现在的数字检测器使用20位及以上的模数转换器，使得检测器分辨率低于 0.0003AU。与早期的 8 位 AD 转换器相比，该系统在几十年的检测中得到了显著的改进。自动样品稀释和超大规模稀释的系统的智能化将减少投入工作中的劳动量。将数据采集和分析仪控制程序集成到实验室网络中，实现了远程数据访问和分析仪控制的功能，即使是在家中的操作员也可以使用。

参考文献

[1] Valcárcel, M. and Luque de Castro, M. D. (1988) *Automatic Methods of Analysis*, Elsevier, Amsterdam, pp. 143-149.

[2] Kérouel, R. and Aminot, A. (1997) Fluorimeric determination of ammonia in sea and estuarine waters by direct segmented flow analysis. *Marine Chemistry*, 57, 265-275.

[3] Furman, W. B. (1976) *Continuous Flow Analysis, Theory and Practice*, Marcel Dekker, New York.

[4] Houba, H. J. G., Novozamsky, I., Uittenbogaard, J. and van der Lee, J. J. (1987) Automatic determination of total soluble nitrogen in soil extracts. *Landwirtschaji Forschung*, 40, 295-302.

[5] Kroon, H. (1993) Determination of nitrogen in water; comparison of a continuous flow method with on-line UV digestion with the original Kjeldahl method. *Analytica Chimica Acta*, 276, 287-293.

[6] Skalar Automated N&P Analyser, Skalar Publication No. 0711293.

[7] Standard Methods for Examination of Water and Waste Water. 15th edition 1980 APHA-AWWA-WPCF, pp. 410-425.

[8] Boltz, D. F. and Mellon, M. G. (1948) Spectrophotometric determination of phosphate as molybdiphosphoric acid. *Analytical Chemistry*, 20, 749-751.

[9] Walinga, I., van Vark, W., Houba, V. J. G. and van der Lee, L. L. (1989) Plant analysis Procedures, Part 7, Department of Soil Science and Plant Nutrition, Wageningen Agricultural University, Syllabus, pp. 138-141.

[10] ISO 15681-2. Determination of orthophosphate and total phosphorus contents by flow analysis, Part 2: Method by continuous flow analysis (CFA).

[11] ISO 14403: 2002. 1st edition 2002-03-01. Water quality-Determination of total cyanide and free cyanide by continuous flow analysis.

[12] ISO 5667-3: 2003. Water quality-Sampling-part 3: Guidance on the preservation and handling on water samples.

[13] Standard method for the examination of water and wastewater. 21st edition, Chapter 5, Method 5530 Phenols, 2005.

[14] ISO 14402. Water quality-Determination of phenol index by flow analysis.

[15] EPA, method 420.2.

[16] ASTM, Standard Test Methods for the Phenolic Compounds in Water, D1783-01.

[17] Davis, R. E. (1976) A combined automated procedure for the determination of reducing sugars and nicotine alkaloids in tobacco products using a new reducing sugar method. *Tobacco Science*, 32, 39-44.

[18] ISO 15154. Tobacco, Determination of the content of reducing carbohydrates; Continuous flow analysis method.

[19] Cooper, A. H. and Hudson, J. R. (1961) *Automated Analysis Applied to Brewing*, Brewing

Industry Research Foundation, pp. 436-438.

[20] European Brewery Convention, Analytica EBC, 5th edition, 2006 method 9.8.

[21] The American Society of Brewing Chemists, Methods of Analysis of the ASBC, 2004, Beer Bitterness method 23D.